OXFORD GRADUATE TEXTS IN MATHEMATICS

OXFORD GRADUATE TEXTS IN MATHEMATICS

Books in the series

1. Keith Hannabuss: *An Introduction to Quantum Theory*
2. Reinhold Meise and Dietmar Vogt: *Introduction to Functional Analysis*
3. James G. Oxley: *Matroid Theory*
4. N.J. Hitchin, G.B. Segal, and R.S. Ward: *Integrable Systems: Twistors, Loop Groups, and Riemann Surfaces*
5. Wulf Rossmann: *Lie Groups: An Introduction through Linear Groups*
6. Qing Liu: *Algebraic Geometry and Arithmetic Curves*
7. Martin R. Bridson and Simon M. Salamon (eds): *Invitations to Geometry and Topology*
8. Shmuel Kantorovitz: *Introduction to Modern Analysis*
9. Terry Lawson: *Topology: A Geometric Approach*
10. Meinolf Geck: *An Introduction to Algebraic Geometry and Algebraic Groups*
11. Alastair Fletcher and Vladimir Markovic: *Quasiconformal Maps and Teichmüller Theory*
12. Dominic Joyce: *Riemannian Holonomy Groups and Calibrated Geometry*
13. Fernando Villegas: *Experimental Number Theory*
14. Péter Medvegyev: *Stochastic Integration Theory*
15. Martin A. Guest: *From Quantum Cohomology to Integrable Systems*
16. Alan D. Rendall: *Partial Differential Equations in General Relativity*
17. Yves Félix, John Oprea, and Daniel Tanré: *Algebraic Models in Geometry*
18. Jie Xiong: *Introduction to Stochastic Filtering Theory*
19. Maciej Dunajski: *Solitons, Instantons, and Twistors*
20. Graham R. Allan: *Introduction to Banach Spaces and Algebras*
21. James Oxley: *Matroid Theory, Second Edition*
22. Simon Donaldson: *Riemann Surfaces*
23. Clifford Henry Taubes: *Differential Geometry: Bundles, Connections, Metrics and Curvature*
24. Gopinath Kallianpur and P. Sundar: *Stochastic Analysis and Diffusion Processes*
25. Selman Akbulut: *4-Manifolds*

4-Manifolds

SELMAN AKBULUT

OXFORD
UNIVERSITY PRESS

OXFORD
UNIVERSITY PRESS

Great Clarendon Street, Oxford, OX2 6DP,
United Kingdom

Oxford University Press is a department of the University of Oxford.
It furthers the University's objective of excellence in research, scholarship,
and education by publishing worldwide. Oxford is a registered trade mark of
Oxford University Press in the UK and in certain other countries

First Edition published in 2016

Published in the United States of America by Oxford University Press
198 Madison Avenue, New York, NY 10016, United States of America

British Library Cataloguing in Publication Data
Data available

Library of Congress Control Number: 2016932194

ISBN 978–0–19–878486–9

To the memory of my father Fahrettin Akbulut

Preface

This book is an attempt to present the topology of smooth 4-manifolds in an intuitive, self-contained way, as it has developed over the years. I have tried to refrain from long-winded pedagogic explanations, instead favoring a concise, direct approach to key constructions and theorems, and then supplementing them with exercises to help the reader fill in the details which are not covered in the proofs. In short, I have tried to present the material directly in the way that I have learnt and thought to myself over the years. When there was a chance to explain some idea with pictures I always went with pictures staying true to the saying "A picture is worth a thousand words". Clearly the material covered in this book is biased towards my interest. I have tried to arrange the material in historical order when possible, while avoiding being encyclopedic with complete referencing when discussing various results. For the sake of simplicity, I have made no attempt to cover all related results around the materials discussed, and later I have chosen various examples, showing how to analyze them using techniques developed in the earlier chapters.

I want to thank many of my friends and collaborators, Robion Kirby, Cliff Taubes, Rostislav Matveyev, Yakov Eliashberg, Ron Fintushel, Burak Ozbagci, Anar Akhmedov, Selahi Durusoy, Cagri Karakurt, Firat Arikan, Weimin Chen, and Kouichi Yasui who have helped me to understand this rich field of 4-manifolds as well as patiently tutoring me in the related fields of symplectic topology and gauge theory over the years. And special thanks to Cliff Taubes, John Etnyre, Burak Ozbagci, Kouichi Yasui, Cagri Karakurt and Rafael Torres for reading the manuscript and making many valuable suggestions to improve the exposition. I thank NSF and MPIM for supporting this project, and Yasar Gencel for providing a pleasant environment in Akbuk, where a large portion of this book was written. Finally, I want to thank Selahi Durusoy for helping me to format the entire book into its present form.

Contents

Chapter 1

4-manifold handlebodies

Let M^m and N^m be smooth manifolds. We say M^m is obtained from N^m by *attaching* a k-handle and denote $M = N \smile_{\varphi} h^k$, if there is an imbedding $\varphi : S^{k-1} \times B^{m-k} \hookrightarrow \partial N$, such that M is obtained from the disjoint union N and $B^k \times B^{m-k}$ by identification

$$M^m = [N \sqcup B^k \times B^{m-k}]/\varphi(x) \sim x \qquad \text{(and the corners smoothed)}$$

Here $\varphi(S^{k-1} \times 0)$ is called the *attaching sphere*, and $0 \times S^{m-k-1}$ is called the *belt sphere* of this handle. We say ∂M is obtained from ∂N by *surgering* the $k-1$ sphere $\varphi(S^{k-1} \times 0)$. There is always a Morse function, $f : M \to \mathbb{R}$ on M, inducing a handle decomposition $M = \cup_{k=0}^{m} M_k$, with $\phi = M_{-1} \subset M_0 \subset .. \subset M_m = M$, and M_k is obtained from M_{k-1} by attaching k-handles [M1], [Mat]. This is called a handlebody structure of M . If M^m is closed and connected we can assume that it has one 0-handle and one m-handle. So in particular in the case of dimension four, M^4 is obtained from B^4 (the zero handle) by attaching 1-, 2- and 3-handles and then capping it off with B^4 at the top (the 4-handle).

In [S] Smale, and also in [W] Wallace, defined and studied handlebody structures on smooth manifolds. Smale went further by turning the operations on handlebody structures into a great technical machine to solve difficult problems about smooth manifolds, such as the smooth h-cobordism theorem, which implied the proof of the topological Poincare conjecture in dimensions ≥ 5. The two basic techniques which he utilized were the *handle sliding* operation and *handle canceling* operation [M4], [Po], [H].

Given a smooth connected 4-manifold M^4, by using its handlebody we can basically see the whole manifold as follows. As shown in Figure 1.1, we place ourselves on the boundary S^3 of its 0-handle B^4 and watch the feet (the attaching regions) of the 1- and 2-handles. The feet of the 1-handles will look like a pair of balls of the same color, and the feet of 2-handles will look like imbedded framed circles (framed knots) which might go over the 1-handles. This is because the 2-handles are attached after the 1-handles.

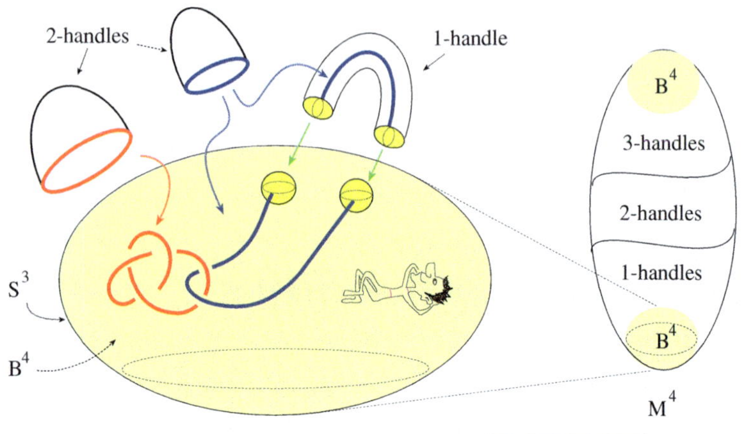

Figure 1.1: Handlebody of a 4-manifold

Fortunately, for closed 4-manifolds, we don't need to visualize the attaching S^2s of the 3-handles, because of an amazing theorem of Laudenbach and Poenaru [LP] which says that they are determined by their 1- and 2-handles, i.e. if there are 3-handles they are attached uniquely! (in [Tra] this was extended to the case of simply connected manifolds with nonempty connected boundary). Here both B^n and D^n will denote n-balls.

The feet of a 1-handle in S^3 will appear as a pair of balls B^3 (Figure 1.2), where on the boundary they are identified by the map $(x, y, z) \mapsto (x, -y, z)$ (with respect to the standard coordinate axis placed at their origins).

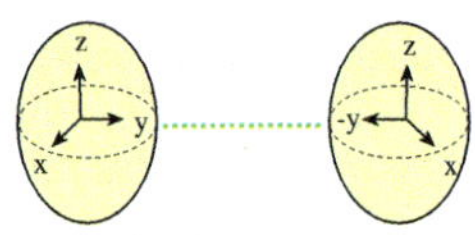

Figure 1.2

Any 2-handle which does not go over the 1-handles is attached by an imbedding $\varphi : S^1 \times B^2 \to \mathbb{R}^3 \subset S^3$, which we call a *framing* . This imbedding is determined by the knot $K = \varphi(S^1 \times (0,0))$, together with an orthonormal framing $\mathbf{e} = \{u, v\}$ of its normal bundle in $\mathbb{R}^3$. This framing $\mathbf{e}$ determines the imbedding φ by

$$\varphi(x, \lambda, \nu) = (x, \lambda u + \nu v)$$

Also note that by orienting K and using the right hand rule, one normal vector field u determines the other normal vector v. Hence any normal vector field of K determines a framing, which is parametrized by $\pi_1(SO(2)) = \mathbb{Z}$. We make the convention that the normal vector field induced from the collar of any Seifert surface of K is the zero framing u_0 (check this is well defined, because linking number is well defined). Once this is done, it is clear that any $k \in \mathbb{Z}$ corresponds to the vector field u_k which deviates from u_0 by k-full twist. Put another way, u_k is the vector field when we push K along it, we get a copy of K which has linking number k with K (e.g. Figure 1.3). So we denote framings by integers. The following exercise gives a useful tool for deciding the k-framing.

Exercise 1.1. *Orient a diagram of the knot K, and let $C(K)$ be the set of its crossing points, define writhe $w(K)$ of a knot K to be the integer*

$$w(K) = \sum_{p \in C(K)} \epsilon(p)$$

where $\epsilon(p)$ is $+1$ or -1 according to right or left handed crossing at p. Show that the blackboard framing (parallel on the surface of the paper) of a knot is equal to its writhe.

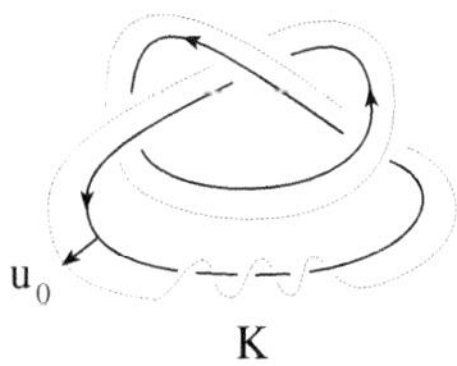

Figure 1.3: $w(K) = 3$ and u_0 is the zero framing

When the attaching framed circles of 2-handles go over 1-handles their framing can not be a well-defined integer. For example, by the isotopy of Figure 1.4 the framing can be changed by adding or subtracting 2. One way to prevent this framing changing isotopy is to fix an arc, as shown in Figure 1.2, connecting the attaching balls of the 1-handle, and make the rule that no isotopy of framed knots may cross this arc. Only after this unnatural rule, can we talk about well-defined framed circles. Most of the work of [AK1] is based on this convention. Readers can compare these constructions with [K3], [GS] and [Sc], where 4-manifold handlebodies are also discussed.

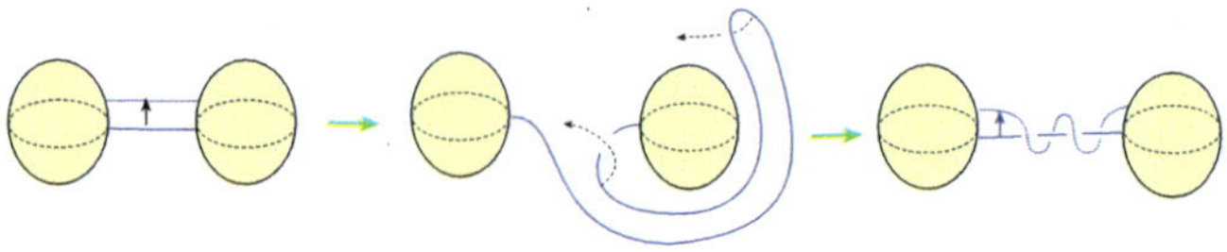

Figure 1.4

1.1 Carving (invariant notation of 1-handle)

Carving is based on the following simple observation: If the attaching sphere $\varphi(S^{k-1}\times 0)$ of the k-handle of $M = N^m \smile_{\varphi} h^k$, bounds a disk in ∂N, then M can be obtained from N by excising (drilling out) an open tubular neighborhood of a properly imbedded disk $D^{m-k-1} \subset N$. This is because we can cancel the k-handle by the obvious $(k+1)$-handle and get back N (Section 1.3); we can then undo the $(k+1)$-handle by puncturing (carving) it, as indicated in Figure 1.5. In particular if M^4 is connected, attaching a 1-handle to M^4 is equivalent to carving out a properly imbedded 2-disk $D^2 \subset B^4$ from the 0-handle $B^4 \subset M$. This simple observation has many useful consequences (e.g. [A1]).

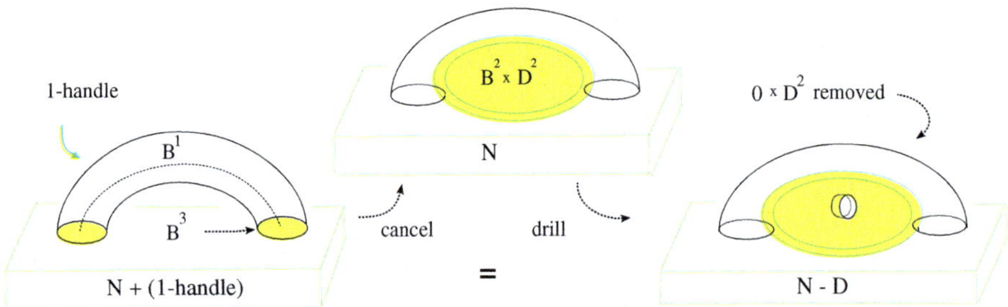

Figure 1.5: Attaching a 1-handle is the same as drilling a 2-disk

To distinguish the boundary of the carved disk $\partial D \subset S^3$ from the attaching circles of the 2-handles, we will put a "dot" on that circle . In particular, this means a path going through this dotted circle is going over the associated 1-handle. So the observer in Figure 1.1 standing on the boundary of B^4 will see the 4-manifold as in Figure 1.6. As discussed before, the framing of the attaching circle of each 2-handle is specified by an integer, so the knots come with integers (in this example the framings are 2 and 3).

This notation of 1-handle has the advantage of not creating ambiguity on the framings of framed knots going through it, as it happens in Figure 1.4. Also this notation can be helpful in constructing hard to see diffeomorphisms between the boundaries of 4-manifolds (e.g. Exercise 1.4, Section 2.5).

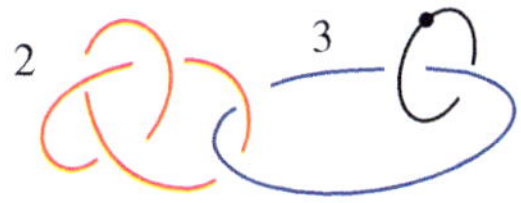

Figure 1.6

We will denote an r-framed knot K by K^r, which will also denote the corresponding 4-manifold (B^4 plus 2-handle). Since any smooth 4-manifold M^4 is determined by its 0, 1 and 2-handles, M can be denoted by a link of framed knots $K_1^{r_1}, K_2^{r_2}, \ldots$ (2-handles), along with "circle with dots" $C_1, C_2, \ldots$ (1-handles). Note that we have the option of denoting the 1-handles by either dotted circles, or pairs of balls (each notation has its own advantages). Also for convenience, we will orient the components of this link:

$$\Lambda = \{K_1^{r_1}, \ldots, K_n^{r_n}, C_1, \ldots, C_s\}$$

We notate the corresponding 4-manifold by $M = M_\Lambda$. For example, Figure 1.6 is the 4-manifold M_Λ with $\Lambda = \{K_1^2, K_2^3, C\}$, where K_1 is the trefoil knot, and K_2 is the circle which links K_1 , and C is the 1-handle. Notice that a 0-framed unknot denotes $S^2 \times B^2$; it also denotes S^4 if we state that there is a 3-handle is present. Also a ± 1-framed unknot denotes a punctured (that is a 4-ball removed) $\pm\mathbb{CP}^2$.

Exercise 1.2. *Show that $S^1 \times B^3$ and $B^2 \times S^2$ are represented by a circle with dot, and an unknotted circle with zero framing, and they are related to each other by surgeries of their core spheres (i.e. the zero and dot are exchanged in Figure 1.7).*

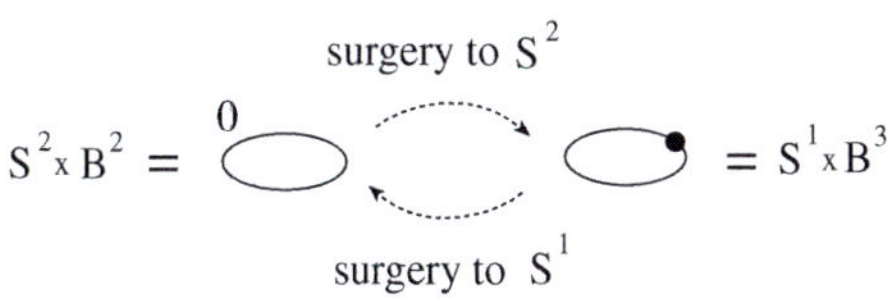

Figure 1.7

Exercise 1.3. *Show that $M_{\Lambda \sqcup \Lambda'}$ corresponds to the boundary connected sum $M_\Lambda \natural M_{\Lambda'}$.*

Exercise 1.4. *Show that the so-called Mazur manifold W in Figure 1.8 is a contractible manifold, and surgeries in its interior (corresponding to zero and dot exchanges on the symmetric link) gives an involution on its boundary $f : \partial W \to \partial W$, where the indicated loops a, b are mapped to each other by the involution f.*

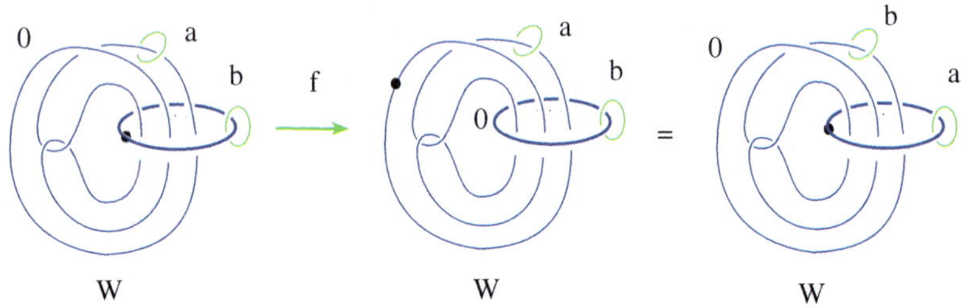

Figure 1.8

Definition 1.1. *A knot $K \subset S^3$ is called a slice knot if it bounds a properly imbedded smooth 2-disk $D^2 \subset B^4$. If this disk is a push-off an immersed disk $f : D^2 \looparrowright S^3$ with ribbon intersections then it is called a ribbon knot (i.e. double points of f are pairs of arcs, one is an interior arc the other is a properly imbedded arc). Here putting a "dot" on a slice knot we will mean removing an open tubular neighborhood of D^2 from B^4.*

1.2 Sliding handles

As explained in [S] and [H] we can change a given handlebody of M to another handlebody of M, by sliding any k-handle over another r-handles with $r \le k$ (Figure 1.9).

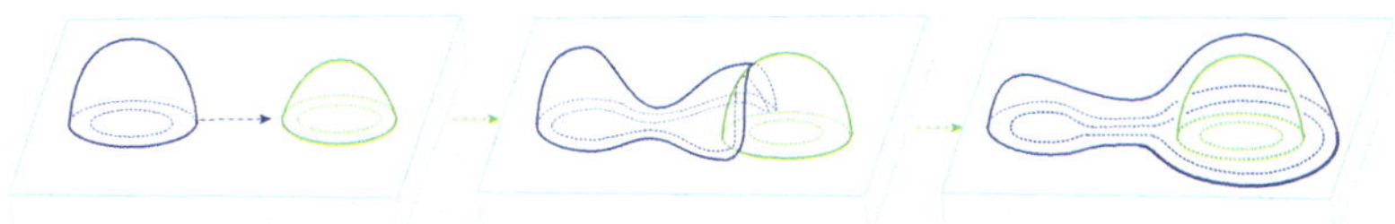

Figure 1.9: Sliding a 2-handle over another 2-handle

In particular in a 4-manifold we can perform the following handle sliding operations: We can (a) slide a 1-handle over another 1-handle, or (b) slide a 2-handle over a 1-handle or (c) slide a 2-handle over another 2-handle. This process alters the framings of the attaching circles of the 2-handles in a well-defined way as follows.

Exercise 1.5. *Let $\Lambda = \{K_1^{r_1}, \dots, K_n^{r_n}, C_1, \dots, C_s\}$ be an oriented framed link with circle with dots, let μ_{ij} be the linking number of K_i and C_j, and λ_{ij} be the linking number of K_i and K_j. Also let K_i' denote a parallel copy of K_i (K_i pushed off by the framing r_i), and C_i' be a parallel copy of the circle C_i. Show that the handle slides (a), (b), (c) above corresponds to changing one of the elements of Λ as indicated below ($j \neq s$).*

(a) $C_i \mapsto C_i + C_j' :=$ *The circle obtained, by connected summing C_i to C_j' along any arch which does not go through any of the C_k''s.*

(b) $K_i^{r_i} \mapsto K_i + C_j' := \bar{K}_i^{\bar{r}_i}$ *The framed knot obtained, by connected summing K_i to C_s' along any arc, with framing $\bar{r}_i = r_i + 2\mu_{ij}$*

(c) $K_i^{r_i} \mapsto K_i + K_j' := \bar{K}_i^{\bar{r}_i}$ *The framed knot obtained, by connected summing K_i to K_j' along any arc, with framing $\bar{r}_i = r_i + r_j + 2\lambda_{ij}$*

These rules are best explained by the following examples (Figure 1.10):

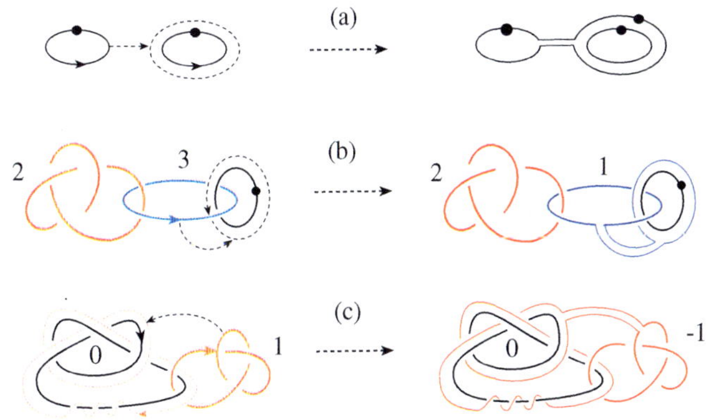

Figure 1.10: Examples of handle slides (a), (b), (c)

Exercise 1.6. *Show that a 1-framed unknot represents $\mathbb{CP}^2$, and a pair of trivially linked 0-framed unknots (as in the first picture of Figure 1.14) represent $S^2 \times S^2$ (see Chapter 3).*

Remark 1.2. *It is hard to make sense of 0-framing for a knot in a general 3-manifold $K \subset Y^3$, unless Y is specified as $Y = \partial M_\Lambda$, where $\Lambda = \{K_1^{r_1}, \dots, K_n^{r_n}\}$, in which case 0-framing means the 0-framing measured from the boundary of the 0-handle B^4 of M_Λ.*

1.3 Canceling handles

We can cancel a k-handle h^k with a $(k+1)$-handle h^{k+1} provided that the attaching sphere of h^{k+1} meets the belt sphere of h^k transversally at a single point (as explained in [S] and [H]). This means that we have a diffeomorphism:

$$N \smile_{\varphi} h^k \smile_{\psi} h^{k+1} \approx N$$

For example, in Figure 1.5 canceling 1- and 2-handle pairs, of a 3-manifold, were drawn. Any 1-handle, and a 2-handle whose attaching circle (framed knot) goes through the 1-handle geometrically once, forms a canceling pair. If no other framed knot goes through the 1-handle of the canceling pair, simply erasing the pair from the picture corresponds to canceling operation. Figure 1.11 gives three equivalent descriptions of a canceling handle pair, so if you want to cancel just erase them from the picture.

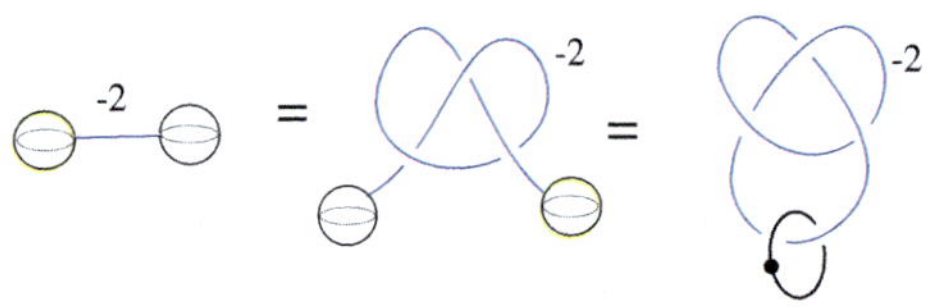

Figure 1.11

It follows from the handle sliding description that, if there are other framed knots going through the 1-handle of a canceling handle pair, the extra framed knots must be slid over the 2-handle of the canceling pair, before the canceling operation is performed (erasing the pair from the picture as in Figure 1.12).

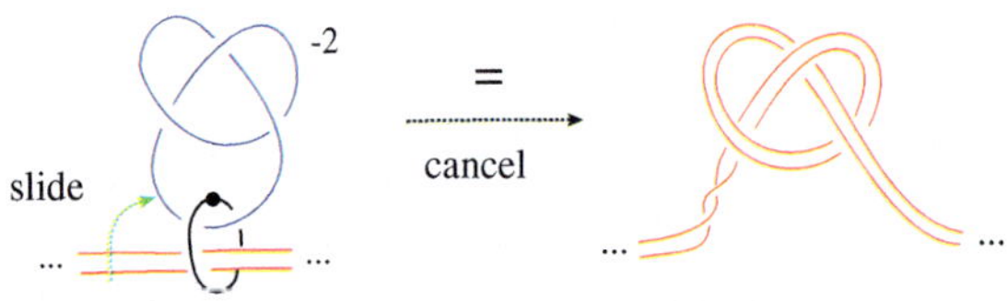

Figure 1.12

Notice that since 3-handles are attached uniquely, introducing a canceling 2- and 3-handle pair is much simpler operation. We just draw the 2-handle as a 0-framed unknot, which is $S^2 \times B^2$, and then state that there is a canceling 3-handle on top of it. In a handle picture of a 4-manifold, no other handles should go through this 0-framed unknot to be able to cancel it with a 3-handle.

Exercise 1.7. *Verify the diffeomorphisms of Figure 1.13 by handle slides (the integers* ± 1 *across the stands indicate* ± 1 *full twist). Also show that this operation changes the framing of any framed knot, which links the* ± 1 *framed unknot* r *times, by* r^2.

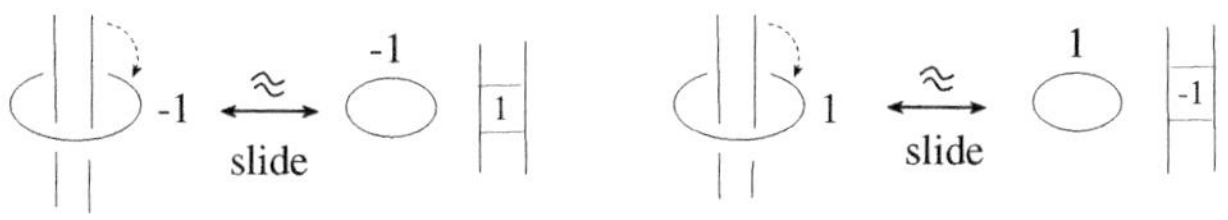

Figure 1.13

Exercise 1.8. *Show that* $S^2 \times S^2 \# \mathbb{CP}^2$ *and* $\mathbb{CP}^2 \# \mathbb{CP}^2 \# \bar{\mathbb{CP}}^2$ *are diffeomorphic to each other by sliding operations, as shown in Figure 1.14 (the identification of these two 4-manifolds is originally due to Hirzebruch).*

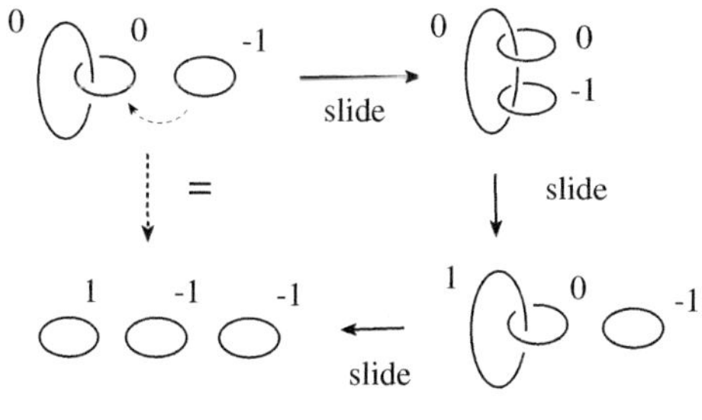

Figure 1.14

The operation $M \mapsto M \# \pm \mathbb{CP}^2$ is called a *blowing up* operation, and its converse is called a *blowing down* operation. If $C^{\pm 1}$ is an unknot with ± 1 framing in a framed link Λ, then by the handle sliding operations of Exercise 1.7, we obtain a new framed link Λ' containing $C^{\pm 1}$ such that $C^{\pm 1}$ does not link any other components of Λ'. So we can write $\Lambda' = \Lambda'' \sqcup \{C^{\pm 1}\}$. Therefore we have the identification $M_\Lambda = M_{\Lambda''} \# (\pm \mathbb{CP}^2)$, and hence a diffeomorphism of 3-manifolds $\partial M_\Lambda \approx \partial M_{\Lambda''}$. Sometimes the operations $\Lambda \leftrightarrow \Lambda''$ on framed links are called *blowing up/down* operations (of the framed links -Figure 1.15).

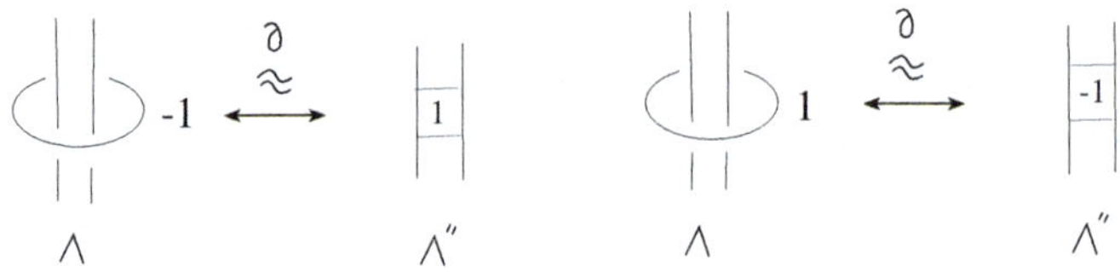

Figure 1.15: Blowing up/down operation

Recall that the set Λ of framed links encodes handle information on the corresponding 4-manifold $M = M_\Lambda$, which comes from a Morse function $f : M \to \mathbb{R}$. Cerf theory studies how two different Morse functions on a manifold are related to each other [C]. In [K1], by using *Cerf theory*, Kirby studied the map Φ defined by $\Lambda \mapsto \partial M_\Lambda$, from framed links to 3-manifolds. It is known that the map Φ is onto [Li] (see Chapter 5).

$$\Phi : \{\text{Framed links}\} \mapsto \{\text{closed oriented } 3-\text{manifolds}\}$$

In [K1] it was shown that under Φ any two framed links are mapped to the same 3-manifold if and only if they are related to each other by the handle sliding operation of Exercise 1.5 (c), and blowing up or down operations. For this reason, in the literature manipulating framed links by these two operations is usually called *Kirby calculus.*

1.4 Carving ribbons

In some cases carving operations allow us to slide a 1-handle over a 2-handle, which is in general prohibited (the attaching circle of the 2-handle might go through the 1-handle). But as in the configuration of Figure 1.16, we can imagine the carved disk (colored yellow inside) as a trough (or a groove) just below the surface; then sliding it over the 2-handle (imagine as a surface of a helmet) by the indicated isotopy makes sense.

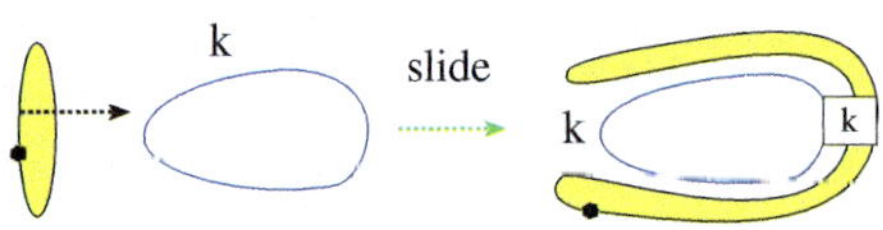

Figure 1.16

Exercise 1.9. *Justify the isotopy of Figure 1.17, by first breaking the 1-handle into a 1-handle and a canceling 1- and 2-handle pair, as in the first picture of Figure 1.17, followed by sliding the new 2-handle over the 2-handle of Figure 1.16, then at the end by canceling back the 1-and 2-handle pair. Also verify the second picture of Figure 1.17.*

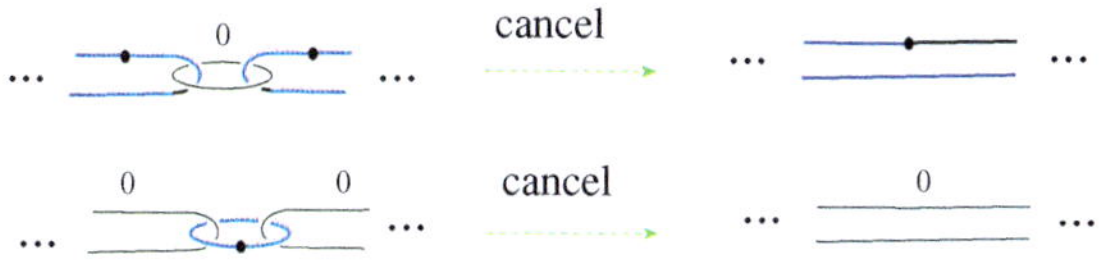

Figure 1.17

This move can turn a 1-handle, which is a carved disk in B^4 bounding an unknot (in a nonstandard handlebody of B^4), into a carved ribbon disk bounding a ribbon knot (in the standard handlebody of B^4), as in the example of Figure 1.18.

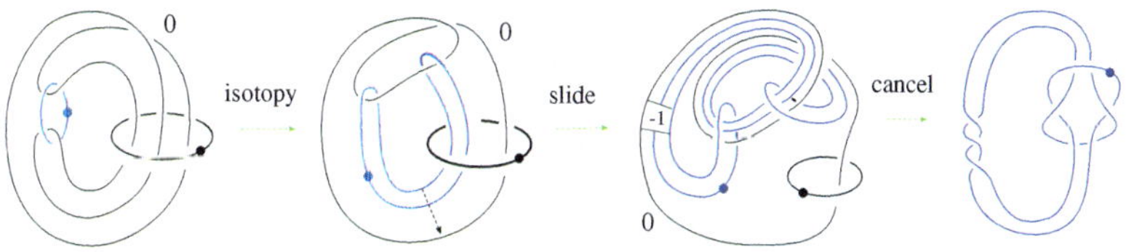

Figure 1.18

We can also consider this operation in reverse, namely, given a ribbon knot, how can we describe the complement of the slice disk which the ribbon knot bounds in B^4? i.e. what is the ribbon complement obtained by carving this ribbon disk from B^4? Clearly by ribbon moves (cutting and regluing along the bands), we can turn any ribbon knot into a disjoint union of unknotted circles, which we can use to carve B^4 along the disks they bound, getting some number of connected sum $\#_k(S^1 \times B^3)$. Now we can do the reverse of the ribbon moves, which describes a cobordism from the boundary $\#_k(S^1 \times S^2)$ of the carved B^4 to the ribbon complement in S^3. During this cobordism, every time two circles coalesce the complement gains a 2-handle, as in Figure 1.19. To sum up, to construct the complement of the ribbon which K bounds in B^4, we perform k ribbon moves (for some k) turning K into k disjoint unknots $\{C_j\}_{j=1}^k$, and after every ribbon move we add a 2-handle, as indicated in Figure 1.19, and put dots on the circles $\{C_j\}_{j=1}^k$.

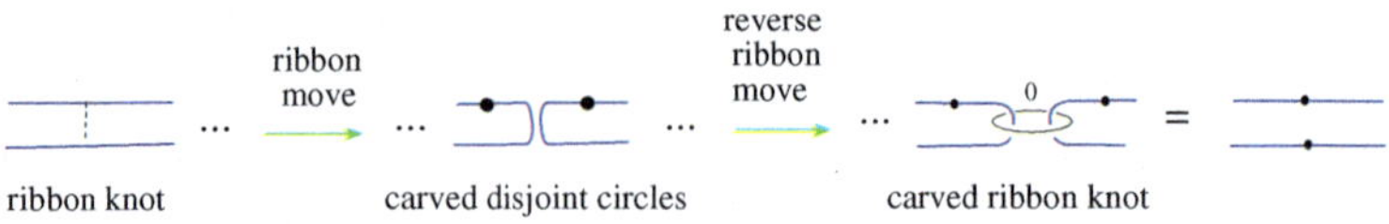

Figure 1.19

For example, applying this process to the ribbon knot K of Figure 1.20 gives the handlebody shown at the bottom right of the Figure (this knot appears in [A2] playing important role in exotic smooth structures of 4-manifolds). Note that this process turns carving a ribbon from the standard handlebody picture of B^4 into carving the trivial disks from a non-standard handlebody picture of B^4. Clearly the top left picture is the handlebody of the complement of an imbedding $S^2 \hookrightarrow S^4$, obtained by doubling this ribbon complement in B^4, as in Figure 1.21 (we build this handlebody from the bottom $\#_2(S^1 \times B^3)$ to top by attaching 2-handles corresponding to the reverse ribbon moves).

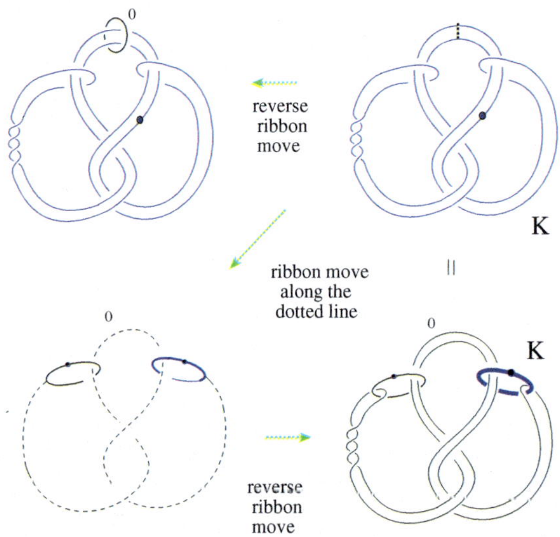

Figure 1.20

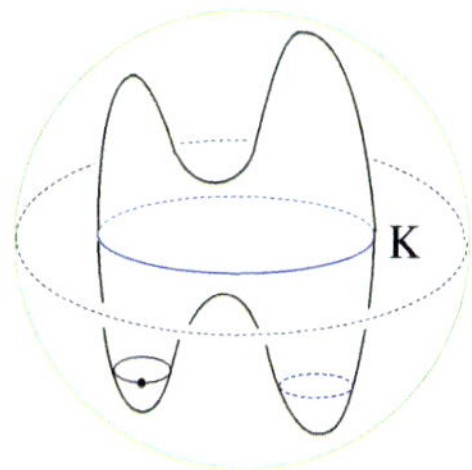

Figure 1.21

Exercise 1.10. *Show that the indicated ribbon move to the ribbon knot of Figure 1.22 gives the ribbon complement in B^4 (the second picture), which consists of two 1-handles and one 2-handle; and that the third picture (with an extra 2-handle, and an invisible 3-handle) is the complement of the double of this ribbon disk (a copy of S^2) in S^4. The last diffeomorphism is obtained by sliding the small 0-framed 2-handle over the 1-handle.*

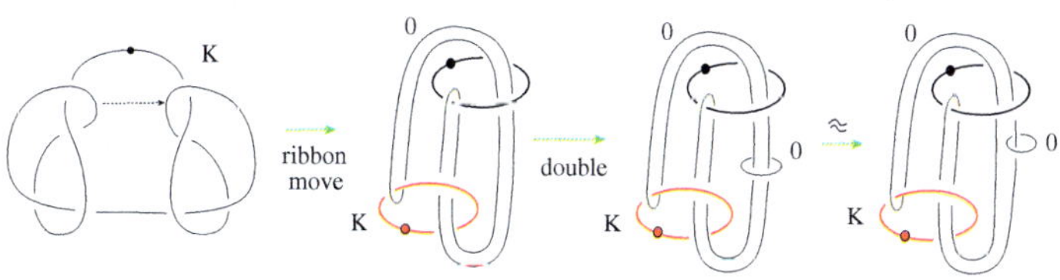

Figure 1.22

We can define an imbedding $S^2 \subset S^4$ by doubling a ribbon disk $D \subset B^4$ (consistent with the doubling process of Section 3.1). Figure 1.23 ⇒ Figure 1.24 is the process of carving the ribbon disk D from B^4, then doubling what is left; ending up with $S^4 - S^2$.

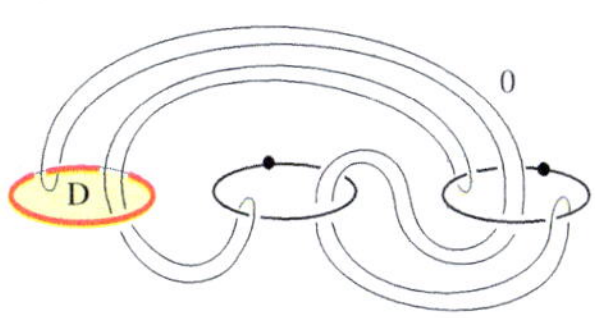

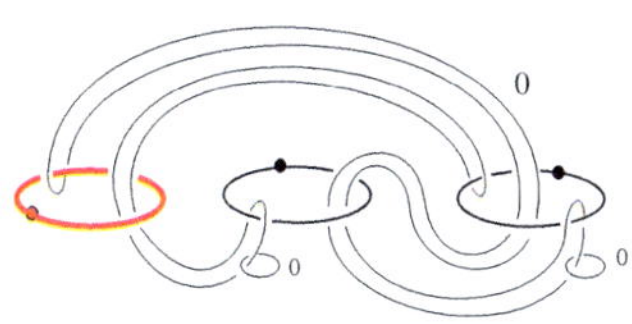

Figure 1.23: Ribbon disk $D \subset B^4$ Figure 1.24: Doubling $B^4 - D$

1.5 Non-orientable handles

If we attach a 1-handle $B^1 \times B^3$ to B^4 along a pair of balls $\{B^3_-, B^3_+\}$ with the same orientation of ∂B^4 we get an orientation reversing 1-handle, which is the non-oriented B^3-bundle over S^1. So to an observer, located at the boundary of B^4, this will be seen as a pair of balls with their interiors identified by the map $(x, y, z) \mapsto (x, -y, -z)$. To demonstrate this in pictures, we use the notation adapted in [A3], by drawing a pair of balls with small arcs passing through their centers, which means that the usual oriented 1-handle identification is augmented by reflection across the plane which is orthogonal to these arcs (Figure 1.25). When drawing the frame knots going through these handles,

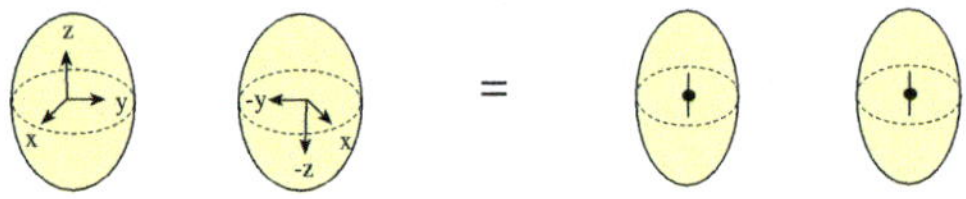

Figure 1.25

one should not forget which points of the spheres ∂B^3_- and ∂B^3_+ correspond to each other via the 1-handle. Also framings of the framed knots going through these handles are not well defined, not only because of the isotopies of Figure 1.4, but also going through an orientation reversing handle, the framing changes sign. For that reason when these handles are present, we will conservatively mark the framings by circled integers (in the knowledge that when pushing them through the handle, the integer changes its sign). For the same reason, any small knot tied to a strand going through this handle will appear as its mirror image when it is pushed though this handle, as indicated in Figure 1.26.

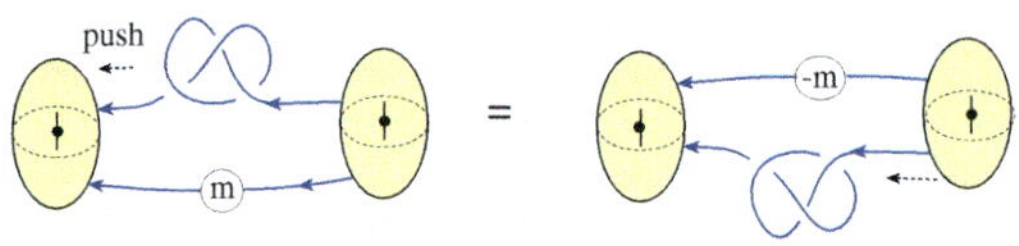

Figure 1.26

Sometimes to simplify the handlebody picture, we place one foot of an orientation reversing 1-handle at the point of infinity ∞ (continue traveling west to ∞ you will appear to be coming back from the east with your orientation reversed), as in the example of Figure 1.27. Notice that when we attach B^4, an orientation reversing 1-handle, we get the orientation reversing D^3 bundle over S^1, denoted by $D^3 \tilde{\times} S^1$.

Exercise 1.11. *Show that the diffeomorphism type of the manifold $M(p,q)$ in Figure 1.27 depends only on parity of $p+q$, it is $D^2 \times \mathbb{RP}^2$ if the parity is even, or else it is $D^2 \tilde{\times} \mathbb{RP}^2$ (a twisted D^2-bundle over $\mathbb{RP}^2$). Also prove that there is an identification $\partial(D^2 \tilde{\times} \mathbb{RP}^2) \approx \partial(D^3 \tilde{\times} S^1)$.*

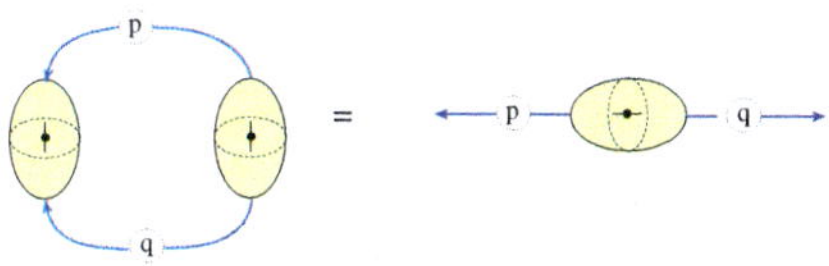

Figure 1.27: M(p,q)

Now if $D^2 \tilde{\times} \mathbb{RP}^2 \subset M^4$, we define the *blowing down* $\mathbb{RP}^2$ operation as:

$$M \mapsto (M - D^2 \tilde{\times} \mathbb{RP}^2) \smile_\partial (D^3 \tilde{\times} S^1)$$

Figure 1.28 shows how this operation is performed in a handlebody.

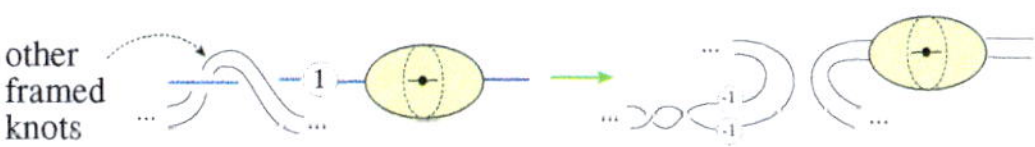

Figure 1.28: Blowing down $\mathbb{RP}^2$ operation

By this, in [A3], an exotic copy of $S^3 \tilde{\times} S^1 \# S^2 \times S^2$ was constructed (the second picture of Figure 1.29). A version of this manifold is discussed in Section 9.5. Here, an exotic copy of M means, a manifold homeomorphic but not diffeomorphic to M.

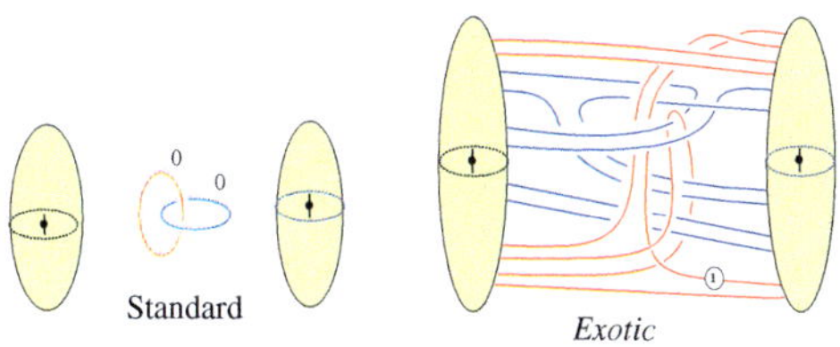

Figure 1.29: Standard and exotic $S^3 \tilde{\times} S^1 \# S^2 \times S^2$

1.6 Algebraic topology

Consider the Eilenberg–McLane space $K(\mathbb{Z},2) = \mathbb{CP}^\infty$, and let $[M^4, K(\mathbb{Z},2)]$ be the homotopy classes of maps from M to $K(\mathbb{Z},2)$. Then any $\alpha \in H_2(M^4;\mathbb{Z})$ can be represented by a closed oriented surface $F = f^{-1}(\mathbb{CP}^2) \subset M$, where $f : M \to \mathbb{CP}^3$ is a map transversal to $\mathbb{CP}^2$ corresponding to α under the following identifications:

$$\alpha \in H_2(M^4;\mathbb{Z}) \approx H^2(M;\mathbb{Z}) = [M^4, K(\mathbb{Z},2)]$$

Now let M^4 be a 4-manifold given as $M = M_\Lambda$, where $\Lambda = \{K_1^{r_1},\dots,K_n^{r_n},C_1,\dots,C_s\}$. Notice, changing the orientation of the manifold M_Λ corresponds to changing Λ to $-\Lambda$

$$-\Lambda = \{-K_1^{-r_1},\dots,-K_k^{-r_k},C_1,\dots,C_s\}$$

where $-K$ denotes the mirror image of the knot K. So in short, $-M_\lambda = M_{-\Lambda}$. Also notice that, in Λ, replacing the dotted circles $C_1,..,C_s$ with 0-framed unknots does not change the boundary ∂M_Λ (Exercise 1.2). Algebraic topology of M^4 can easily be read off from its handlebody description $\Lambda = \{K_1^{r_1},\dots,K_n^{r_n},C_1,\dots,C_s\}$. For example to compute $\pi_1(M)$ for each circle-with-dot C_i, we introduce a generator x_i, then each loop K_j gives a relation $r_j(x_1,\dots,x_s)$ recording how it goes through the 1-handles. To compute the homology groups, to any $\alpha = \sum_j c_j K_j$ with $c_j \in \mathbb{Z}$, we associate $[\alpha] \in H_1(M)$ (by thinking K_j as loops). When $[\alpha] = 0$, we can view α as a CW 2-cycle and think of $\alpha \in H_2(M)$.

The intersection form $(\alpha,\beta) \mapsto \alpha.\beta$ on $H_2(M)$, induced by the cup product

$$q_M : H_2(M;\mathbb{Z}) \otimes H_2(M;\mathbb{Z}) \longrightarrow \mathbb{Z}$$

can be computed from these 2-cycles. For example, if $\alpha = \sum_i c_i K_i^{r_i}$ and $\beta = \sum_j c'_j K_j^{r_j}$, then $\alpha.\beta = \sum_{i,j} c_i c'_j L(K_i,K_j)$, where $L(K_i,K_j)$ is the linking number of K_i and K_j. In particular, $L(K_j,K_j)$ is given by the framing r_j of K_j. We say q_M is even if $\alpha.\alpha$ is even for each $\alpha \in H_2(M;\mathbb{Z})$, otherwise we call q_M odd. The signature $\sigma(M)$ of M is defined by the signature of the bilinear form q_M , i.e. if $b_2^+(M)$ and $b_2^-(M)$ are the dimensions of maximum positive and negative subspaces for the form q_M on $H_2(M;\mathbb{Z})$, then

$$\sigma(M) = b_2^+(M) - b_2^-(M)$$

Intersection form is also defined for compact oriented manifolds with boundary:

$$q_M : H_2(M,\partial M;\mathbb{Z}) \otimes H_2(M,\partial M;\mathbb{Z}) \longrightarrow \mathbb{Z}$$

Remark 1.3. *(Additivity of the signature) Let $M^4 = M_1 \cup_Y M_2$ be a compact oriented smooth manifold consisting of a union of two compact smooth manifolds glued along their common boundary Y^3; then $\sigma(M) = \sigma(M_1) + \sigma(M_2)$. This is because the classes of $H_2(M)$ come in three types: (1) surfaces contained entirely in M_1 or M_2, (2) surfaces intersecting Y transversally along homologically nontrivial loops $C \subset Y$, (3) surfaces contained in Y^3. The Poincare duals of type (2) classes (the dual surfaces to the loops C in Y) are type (3) classes. type (3) classes have self intersection zero, hence they and their duals don't contribute to the signature because q_M on them is the hyperbolic pair,*

$$\begin{pmatrix} 0 & 1 \\ 1 & * \end{pmatrix}$$

Hence the only contribution to the signature calculation comes from the type (1) classes.

Exercise 1.12. *Show that if a closed 4-manifold M bounds, then $\sigma(M) = 0$.*

Let M be a closed 4-manifold, then its second Betti number can be expressed as $b_2(M) = b_2^+(M) + b_2^-(M)$, and Poincare duality implies that the intersection form q_M is unimodular, that is, the natural induced map $H_2(M;\mathbb{Z}) \to H_2(M;\mathbb{Z})^*$ is an isomorphism. If M is closed and simply connected, the intersection form q_M determines the homotopy type of M ([Wh], [M2], [MS]). Furthermore, homotopy equivalent, simply connected, closed smooth 4-manifolds $M \simeq M'$ are stably diffeomorphic ([Wal]), i.e. for some k:

$$M \# k(S^2 \times S^2) \approx M' \# k(S^2 \times S^2)$$

In the topological category much more is true. By the celebrated theorem of Freedman [F], homotopy equivalent, simply connected, closed smooth 4-manifolds are homeomorphic to each other (there is also a relative version of this theorem, where the closed manifolds are replaced by compact manifolds with homeomorphic boundaries). On the other hand homeomorphic 4-manifolds don't have to be diffeomorphic to each other (e.g. Chapter 9). Also by [F], every even unimodular bilinear form can be realized as the intersection form of a closed, simply connected, topological 4-manifold, and in the odd case there are two such manifolds realizing that intersection form. By contrast, in the smooth case, there are many restrictions on unimodular bilinear forms to be represented as the intersection form of closed smooth 4-manifolds [D3] (see Section 14.1). So not every topological manifold has a smooth structure unless it is noncompact:

Theorem 1.4. *([FQ]) Connected, noncompact 4-manifolds admit smooth structure.*

Any closed, oriented 3-manifold Y^3 bounds a compact, smooth, oriented 4-manifold M^4, consisting of 0- and 2-handles (Chapter 5). If $\Lambda = \{K^r\}$, the following alternative notations will be used:

$$M_\Lambda = B^4(K, r) = K^r \tag{1.1}$$

So $\partial(K^r)$ will be the 3-manifold obtained by surgering S^3 along K with framing r. If $K \subset Y$, we will denote the 3-manifold obtained by surgering Y along K with framing r by the notations $Y(K, r)$ or $Y_r(K)$ (provided the framing r in Y is well defined). For the sake of not cluttering notations we will also abbreviate $Y_0(K) = Y_K$ (e.g. Section 13.11) (not to be confused with knot surgery notation X_K, when X is a 4-manifold (Section 6.5).

Chapter 2

Building low-dimensional manifolds

Just as we visualized 4-manifolds by placing ourselves on the boundary of their 0-handles (Figure 1.1) and observing the feet of 1- and 2-handles, we can visualize 2 and 3-manifolds by placing ourselves on the boundary of their 0-handles. In this way we get analogous handlebody pictures such as in Figures 2.1 and 2.2; except in this case we don't need to specify the framings of the attaching circles of the 2-handles. The 3-manifold handlebody pictures obtained this way are called *Heegaard diagrams*. Clearly we can thicken these handlebody pictures by crossing them with balls to get higher-dimensional handlebodies, as indicated in these figures. For example, the middle picture of Figure 2.1 is a Heegaard diagram of $T^2 \times I$ and the last picture is $T^2 \times B^2$; and Figure 2.2 is a Heegaard diagram of S^3, and a handlebody of $S^3 \times B^1$.

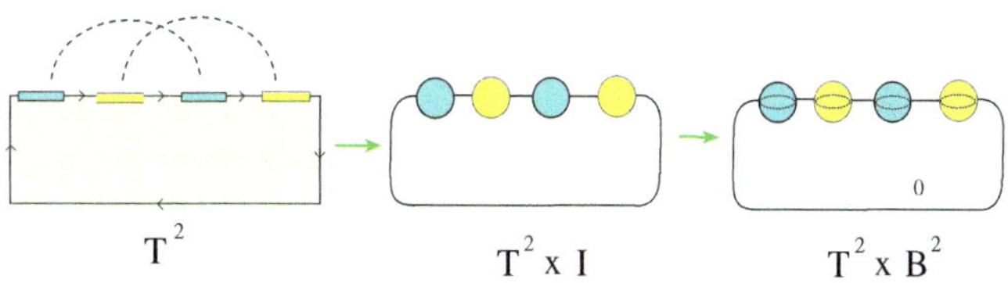

Figure 2.1

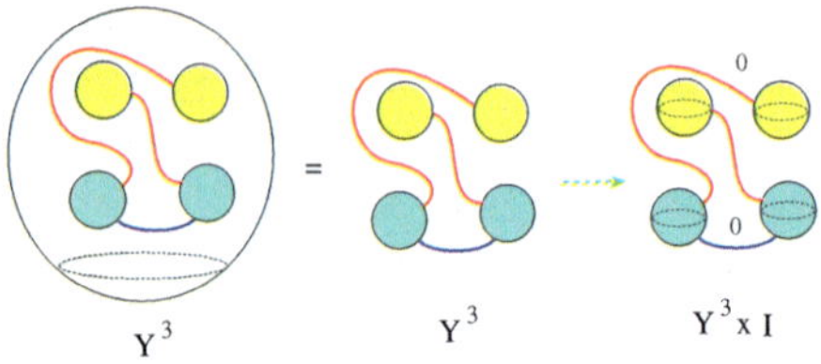

Figure 2.2: $Y^3 = S^3$

After this, by changing the 1-handle notations to the circle with dot notation, in Figure 2.3 we get a handlebody picture of $T^2 \times B^2$, and proceeding in a similarly way we get a handlebody of $F_g \times B^2$, where F_g denotes the surface of genus g.

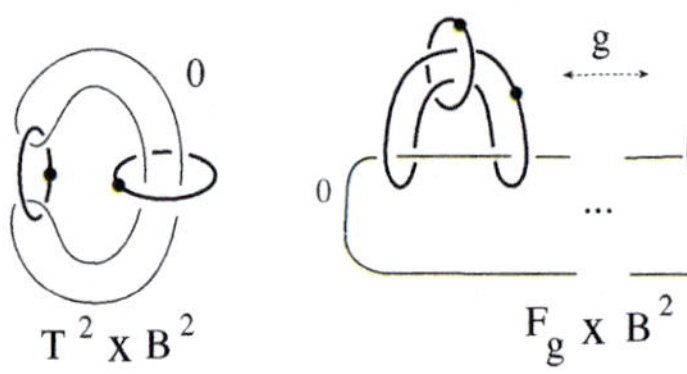

Figure 2.3

Exercise 2.1. *By using this method, verify Exercise 1.11, and also show that the handlebody pictures in Figure 2.4 are twisted D^2 bundles over S^2, $\mathbb{RP}^2$, and F_g, respectively (more specifically they are all Euler class k D^2-bundles).*

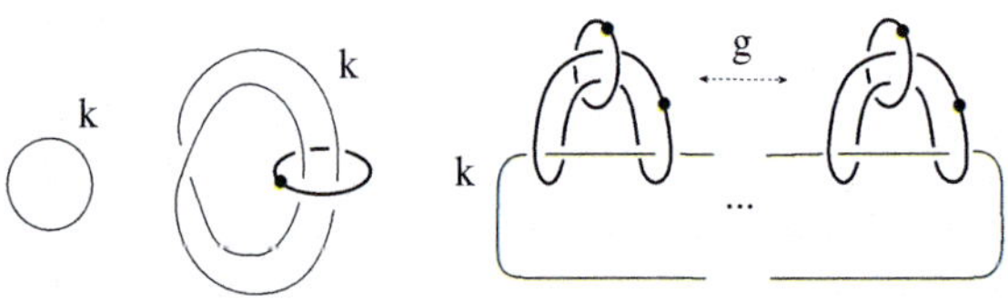

Figure 2.4

2.1 Plumbing

For $i = 1, 2$ let $D_i \hookrightarrow E_i \to S_i$ be two Euler class k_i disk bundles over the 2-spheres S_i, and let $B_i \subset S_i$ be the disks giving the trivializations $E_i|_{B_i} \overset{\approx}{\to} B_i \times D_i$. Plumbing E_1 and E_2 is the process of gluing them together by identifying $D_1 \times B_1$ with $B_2 \times D_2$ (by switching base and fiber directions). Hence, in this manifold if we locate ourselves on the boundary of $D_1 \times B_1 = B^4$, we will see a two 2-handles attached to B^4 along the link of Hopf circles $\{\partial D_1, \partial B_1\}$ with framings k_1, k_2. This plumbed manifold is symbolically denoted by a graph with one edge and two vertices weighted by integers k_1 and k_2 (Figure 2.5).

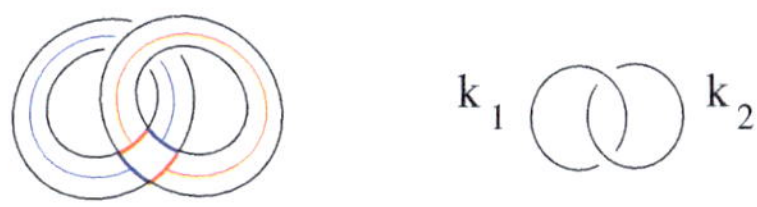

Figure 2.5

By iterating this process we can similarly construct 4-manifolds corresponding to any graph whose verticies are weighted with integers. Also in this construction S_i' s don't have to be spheres, they can be any surface (orientable or not), but in this case when we abbreviate these plumbings using graphs, to each vertex we need to specify a specific surface alongside an integer weight. The reader can check that the manifolds corresponding to the graphs in Figure 2.6 are given by the handlebodies on the right.

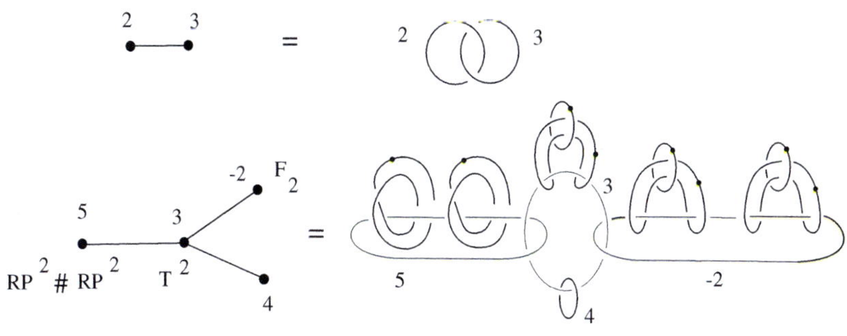

Figure 2.6

The following plumbed manifolds (Figure 2.7) have special names: E_8 and E_{10} .

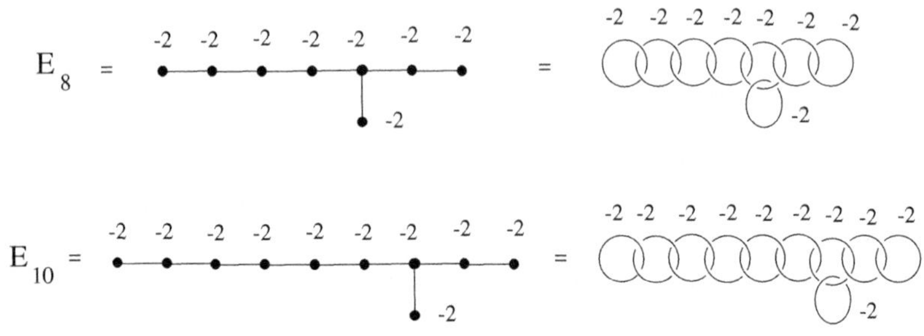

Figure 2.7

An alternative way to denote a plumbing diagram of 2-spheres is to draw the dual of the plumbing graph, where the weighted vertices are replaced by weighted intersecting arc segments, intersecting according to the graph (e.g. Figure 2.8).

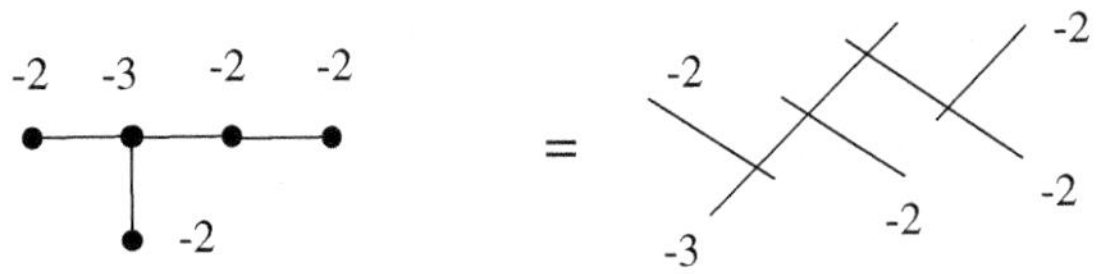

Figure 2.8

2.2 Self plumbing

Using a method similar to that one above, we can plumb a disk bundle $D \hookrightarrow E \to S$ to itself. So the only difference is, when we locate ourselves on the boundary of $D_1 \times B_1 = B^4$, we will see a link of thickened Hopf circles, identified with each other by a cylinder $(S^1 \times [0,1]) \times D^2$. This is equivalent to first attaching a 1-handle to B^4 (thickened point $\times [0,1]$), followed by attaching a 2-handle to the connected sum of the two circles over the 1-handle (a tunnel). Figure 2.9 gives self plumbings of Euler class k bundles over S^2 and over T^2 (see the "round handle" discussions in Definition 7.8 and Figure 3.10).

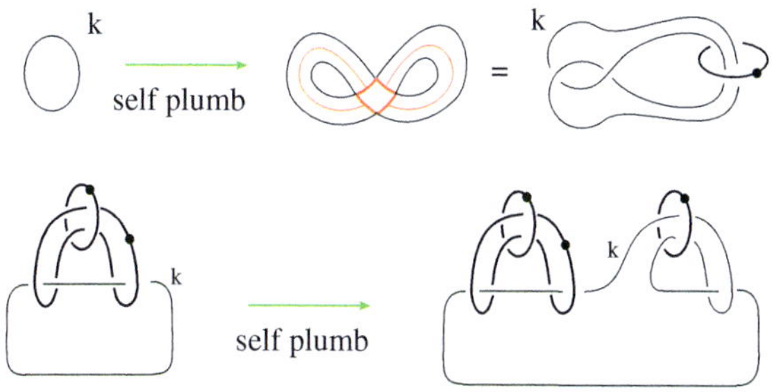

Figure 2.9

2.3 Some useful diffeomorphisms

As shown in Figure 2.10, by introducing a canceling pair of 1- and 2-handles, and sliding the new 1-handle over the existing 1-handle, and then canceling the resulting 1-and 2-handle pair, we get an interesting diffeomorphism between the first and last handlebodies of this figure. These diffeomorphisms are used in [AK2], identifying various 3-manifolds.

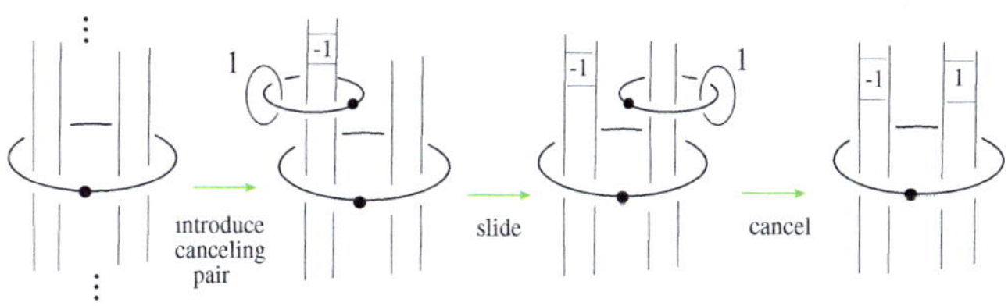

Figure 2.10

Exercise 2.2. *Let $W^{\pm}(l,k)$ be the contractible manifolds given in Figure 2.11 (where the integer l denotes l-full twist between the stands) and K is the knot in Figure 2.12. By using the above diffeomorphisms construct the following diffeomorphisms:*

- $W^{\pm}(l,k) \approx W^{\pm}(l+1,k-1)$
- $W^{-}(l,k) \approx -W^{+}(-l,-k+3)$
- $W^{\pm}(l,k) \# (\mp\mathbb{CP}^2) \approx K^{\mp 1}$ *(Hint: Figure 2.13)*

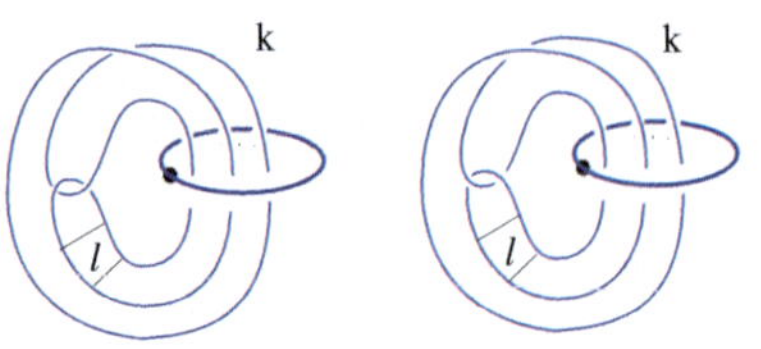

Figure 2.11: $W^-(l,k)$ and $W^+(l,k)$

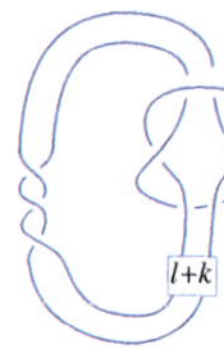

Figure 2.12: K

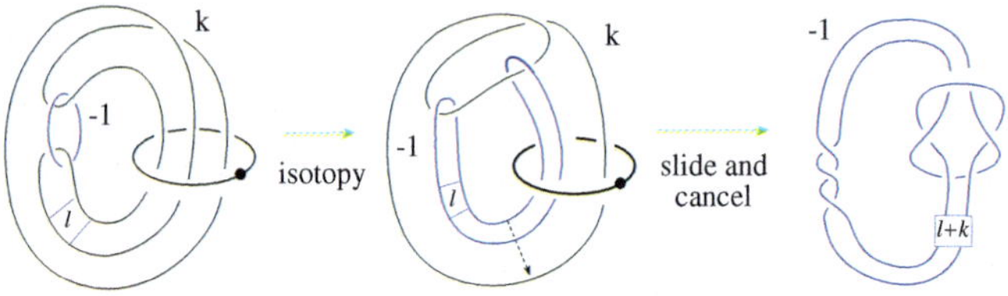

Figure 2.13: $W^+(l,k)\#\bar{\mathbb{CP}}^2 \approx K^{-1}$

2.4 Examples

Links of hypersurface singularities provide a rich class of 3-manifolds (Section 12.4); they are obtained by intersecting hypersurfaces in $\mathbb{C}^3$ with isolated singularities at the origin with a sphere S^5_ϵ of small radius ϵ, centered at the origin. The *Brieskorn* 3-manifolds

$$\Sigma(a,b,c) := \{(x,y,z) \in \mathbb{C}^3 \mid x^a + y^b + z^c = 0\} \cap S^5_\epsilon$$

defined this way, are boundaries of some interesting 4-manifolds of Figure 2.14 (Exercise 12.4). $\Sigma(2,3,5)$ is called the *Poincare homology sphere.* There is a very useful boundary identification $f : \Sigma(2,3,5) \overset{\approx}{\to} \partial E_8$ (Fig 2.7) given by the steps in Figure 2.15.

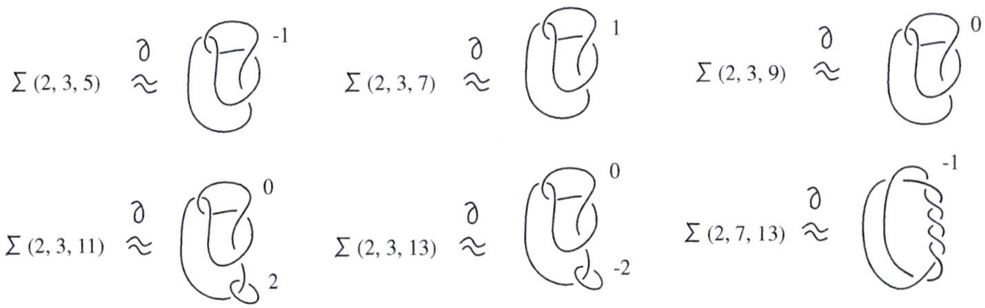

Figure 2.14: Various Brieskorn homology spheres

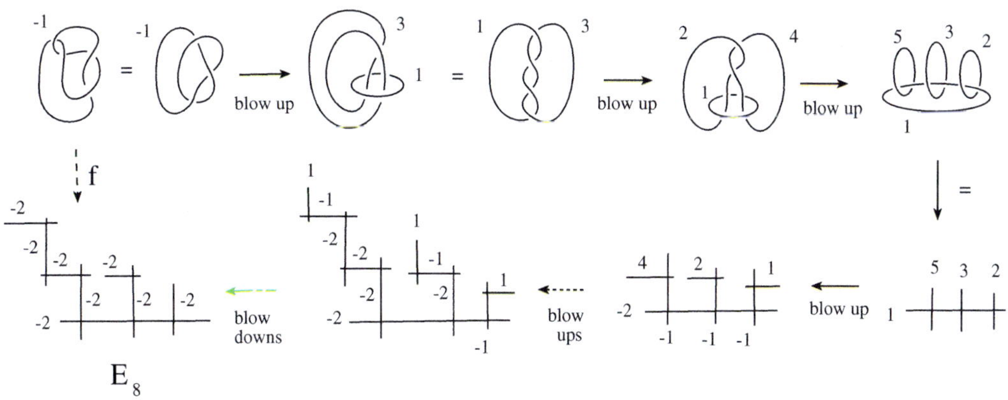

Figure 2.15: The diffeomorphism $f : \Sigma(2, 3, 5) \to \partial E_8$

Exercise 2.3. *By imitating the steps in Figure 2.15, show that* $\Sigma(2,3,7) \approx \partial E_{10}$*. Also by justifying the diffeomorphisms in Figure 2.16, show that* $\partial E_{10} \approx M^4$*, where* M^4 *is a manifold obtained from* E_8 *by attaching a pair of 2-handles, and* M *has the intersection form of* $E_8 \# (S^2 \times S^2)$*.*

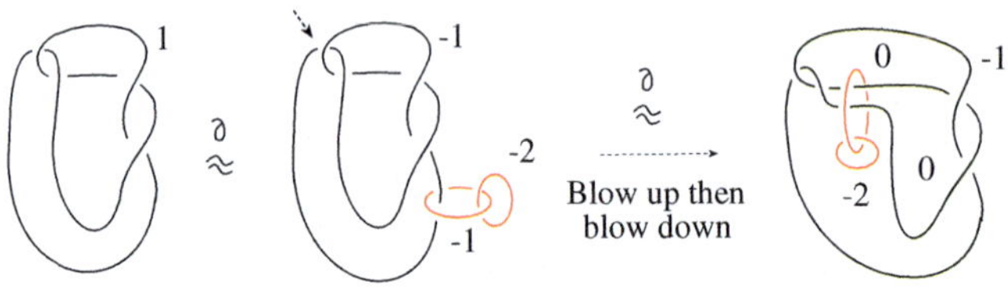

Figure 2.16

Exercise 2.4. *By justifying the steps in Figure 2.17, construct a closed, simply connected, smooth manifold with signature* -16 *and the second Betti number* 22*. (In the figure* $M = A + B$ *means* M *contains each of the handlebodies* A *and* B*, and the handles of* A *and* B *have zero algebraic linking number, in which case we can write* $M = A + handles = B + handles$*, and* M *is homology equivalent to* $A \# B$*.)*

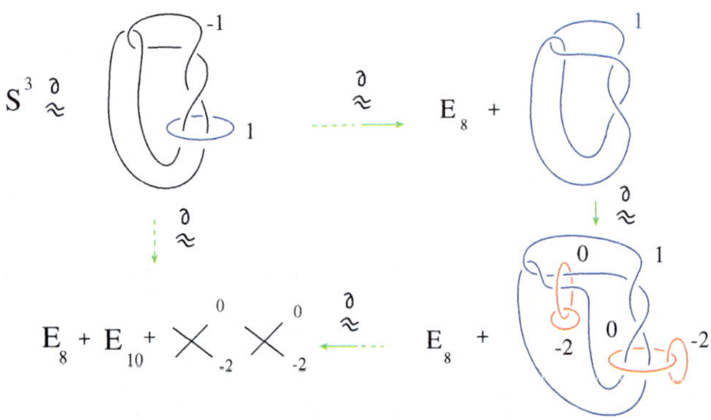

Figure 2.17

$\Sigma(2,3,13)$ bounds a contractible manifold, which can be seen by the identification $\partial\Sigma(2,3,13) \approx \partial W^{+}(1,0)$, as indicated in Figure 2.18.

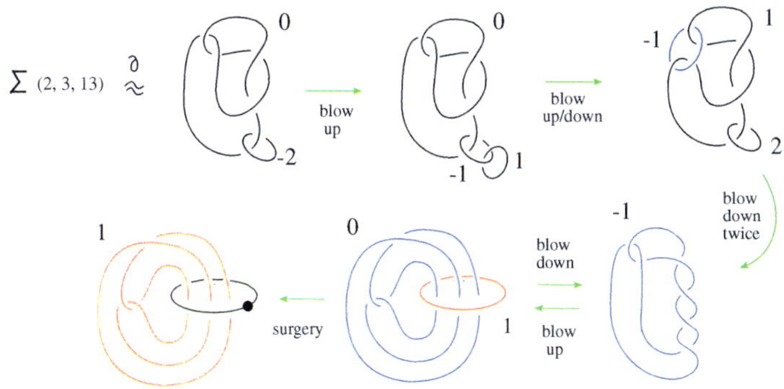

Figure 2.18

Exercise 2.5. *By blowing up and blowing down operations, verify the diffeomorphism of Figure 2.19 (Hint: imitate the initial steps of Figure 2.18).*

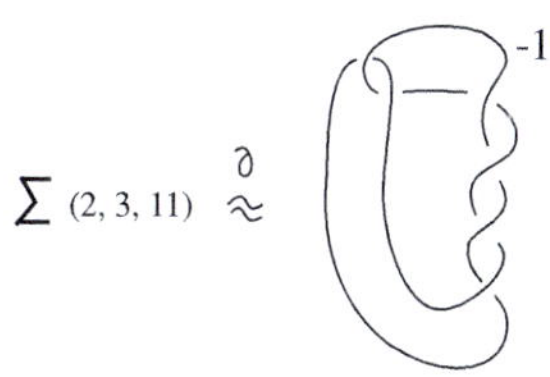

Figure 2.19

Exercise 2.6. *By verifying the identifications of Figure 2.20, show that* $\Sigma(2,3,11)$ *bounds a definite manifold with intersection form* $E_8 \oplus (-1)$*, and also bounds a smooth simply connected manifold with signature* -16 *and the second Betti number* 20.

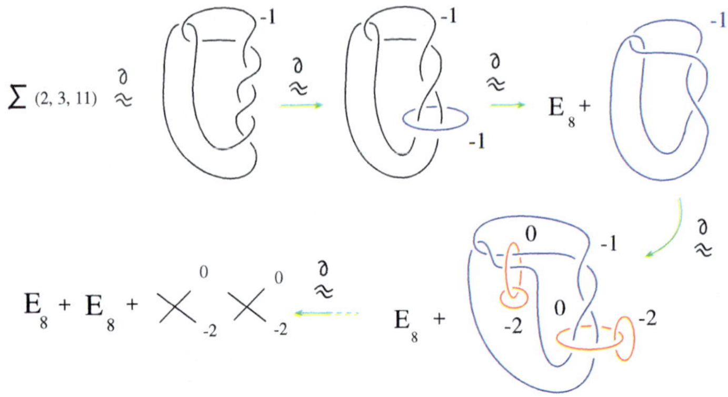

Figure 2.20

Exercise 2.7. *By thickening the handlebody of $\mathbb{RP}^2$ in two different ways, obtain two different D^2 bundles over $\mathbb{RP}^2$: one is the trivial bundle with non-orientable total space, the other is the twisted bundle with orientable total space (Figure 2.21). The second one is the normal bundle of an imbedding $\mathbb{RP}^2 \subset \mathbb{R}^4$.*

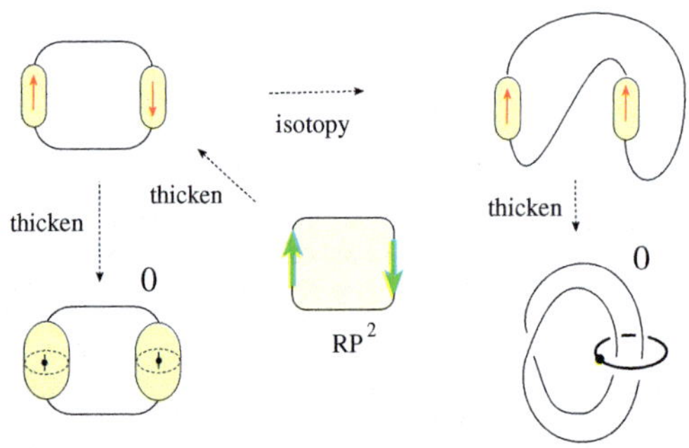

Figure 2.21

2.5 Constructing diffeomorphisms by carving

Given a diffeomorphism $f : \partial M \xrightarrow{\approx} \partial N$, when does f extend to a diffeomorphism inside $F : M \to N$? Some instances of carving can provide a solution. One necessary condition is that f must extend to a homotopy equivalence inside, so let us assume this as a hypothesis. Now let us start with $M = M_\Lambda$, where $\Lambda = \{K_1^{r_n}, \dots, K_n^{r_n}, C_1, \dots, C_s\}$, and let $\{\gamma_1, \dots, \gamma_k\}$ be the dual circles of the 2-handles, i.e. $\gamma_j = \partial B_j$, where B_j is the co-core of the dual 2-handle of $K_j^{r_j}$. Then if $\{f(\gamma_1), \dots, f(\gamma_k)\}$ is a slice link in N, that is, if each $f(\gamma_j) = \partial D_j$ where $D_j \subset N$ are properly imbedded disjoint disks. We can then extend f to a diffeomorphism:

$$f' : M' := \partial M_\epsilon \smile \cup_j \nu(B_j) \to \partial N_\epsilon \smile \cup_j \nu(D_j) := N'$$

where ∂M_ϵ and ∂N_ϵ are the collar neighborhoods of the boundaries, and $\nu(B_j)$, $\nu(D_j)$ are the tubular neighborhoods of the disks B_j, D_j (Figure 2.22).

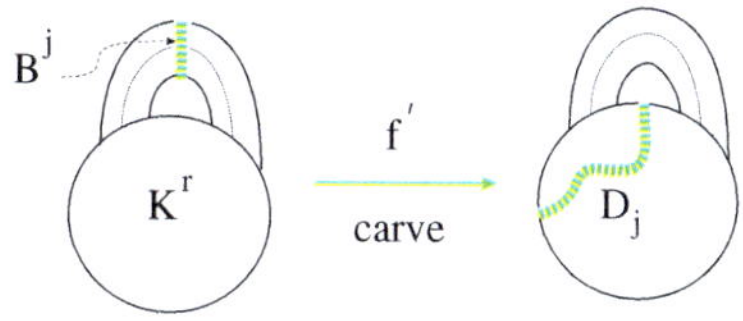

Figure 2.22

This reduces the extension problem to the problem of extending f' to complements $M - M' \to N - N'$. Notice that $M - M' = \#_s(S^1 \times B^3)$ and $N - N'$ is a homotopy equivalent to $\#_s(S^1 \times B^3)$. Since every self diffeomorphism $\#_s(S^1 \times S^2)$ extends to a unique self diffeomorphism of $\#_s(S^1 \times B^3)$, the only way f' doesn't extend to a diffeomorphism $M - M' \to N - N'$ is when $N - N'$ is an exotic copy of $\#_s(S^1 \times B^3)$. The case of $s = 0$ is particularly interesting, since the only way the last extension problem $M - M' \to N - N'$ fails is when the 4-dimensional smooth Poincare conjecture fails (in examples this step usually goes through). Even in the cases of single 2-handle $M = K^r$ and $N = L^r$, extending a diffeomorphism $f : \partial(K^r) \to \partial(L^r)$ can be a difficult task without carving. Because there are no other handles to slide over to construct diffeomorphisms in a conventional way, Cerf theory says that if they are diffeomorphic there must be canceling handle pairs; but we don't know where they are. Carving at least gives us a way to start. For example, Figure 2.23 describes a diffeomorphism between the boundaries of two 4-manifolds ([A1]).

$$f : \partial(K^1) \xrightarrow{\approx} \partial(L^1)$$

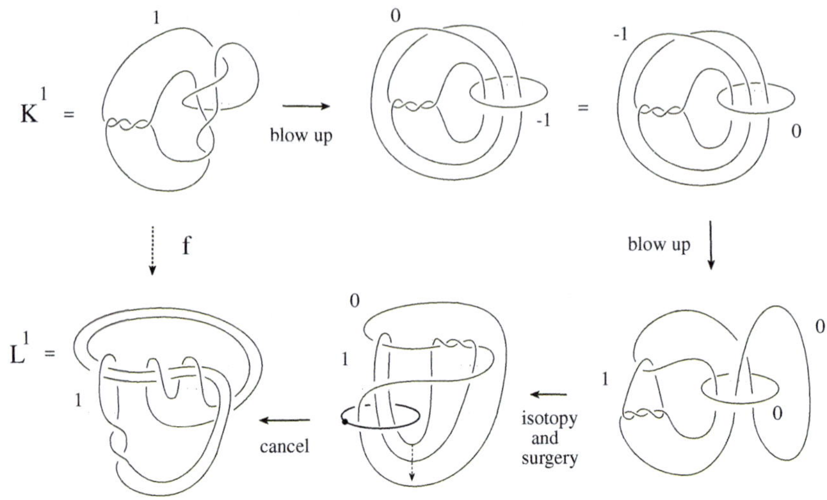

Figure 2.23

Figure 2.24 shows the dual circle γ and its image $f(\gamma)$ under this diffeomorphism. By carving along γ and $f(\gamma)$ we can extend f to a diffeomorphism. To do this we put dots on these circles (making 1-handles). Then we only have to check that the picture on the right is B^4. For this we slide two stands of the 2-handle over the 1-handle, as indicated in the figure, and then check that we get B^4.

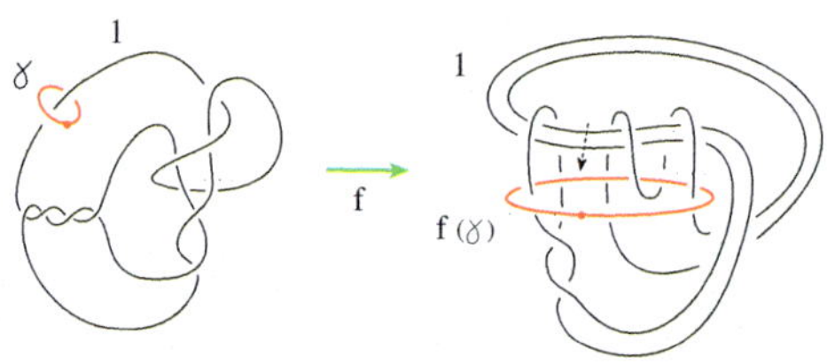

Figure 2.24

The following is a generalization of the previous example of [A1].

Theorem 2.1. *([A4]) For each $r \in \mathbb{Z} - 0$, there are of distinct knot pairs, K and L_r (one is slice the others are non-slice knots) such that for all $r \neq 0$*

$$K^r \approx L_r^r \tag{2.1}$$

Proof. The previous example gives a hint of how to see this by Cerf theory: for this we introduce and then cancel 1- and 2-handle pairs, as shown in Figure 2.25.

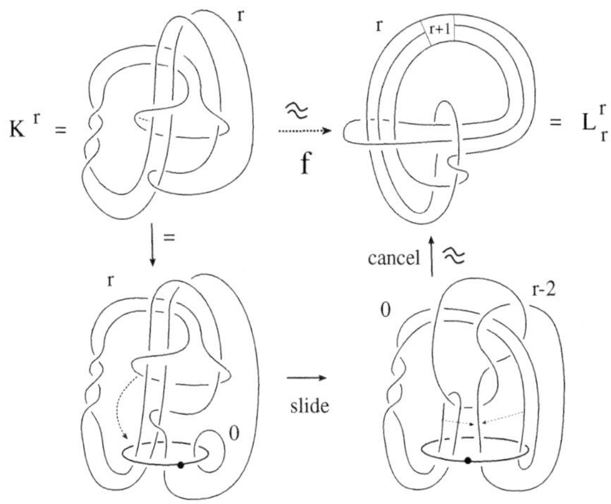

Figure 2.25

Exercise 2.8. *Show K is slice (Hint: perform the slice move along the dotted line indicated in the figure). Show L_r are non-slice for $r > 1$(compute the p-signatures, see Definition 2.3 and Exercise 2.11 (c)).*

Exercise 2.9. *(Case of 0-surgeries, [O], [T]) Show that in the boundaries ∂M and ∂N of the manifolds of Figure 2.26 the loops a and b are isotopic to each other (Hint: slide them over the 2-handle). From this, produce distinct knots K_r and L_r with $K_r^0 \approx K_s^0$ and $L_r^4 \approx L_s^4$ for all $r \neq s$ (Hint: consider repeated ± 1 surgeries to a and b).*

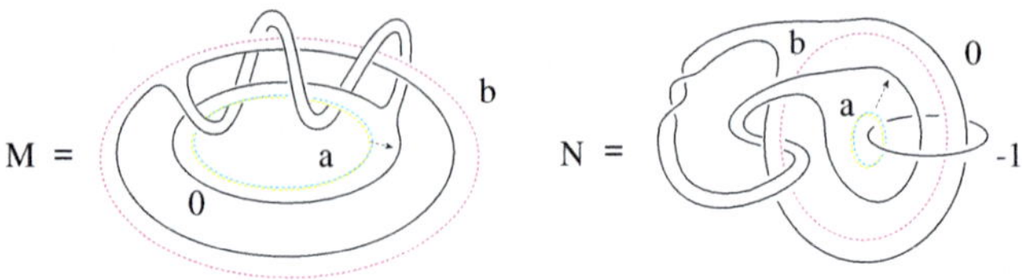

Figure 2.26

2.6 Shake slice knots

Definition 2.2. *We call a link $\mathcal{L} = \{K, K_+, \dots, K_+, K_-, \dots, K_-\}$, consisting of K and an even number of oppositely oriented parallel copies of K (pushed off by the framing r), an r-shaking of K. We call a knot $K \subset S^3$ an r-shake slice if an r-shaking of K bounds a disk with holes in B^4. The r-shake genus of a knot K is the smallest integer g where some r-shaking of K occurs as a boundary of a genus g surface with holes in B^4.*

As in the (2.1) examples, if K is a knot and L is a slice knot, such that the 4-manifolds L^r and K^r are diffeomorphic, then K is an r-shake slice. This is because with L being slice there is a smooth 2-sphere $S \subset L^r$ representing the generator of the homology group $H_2(L^r) \cong \mathbb{Z}$ (namely the slice disk which L bounds in B^4 together with the core of the handle), then if $F : L^r \to K^r$ is a diffeomorphism, by making F transverse to the co-core of the 2-handle of K^r and by an isotopy, we can make the link $F(S) \cap S^3$ an r-shaking of the knot K; clearly this link bounds a disk with holes (Figure 2.27). Theorem 9.6 gives an example of pair of knots K_1 and K_2, such that one is slice the other is not, and K_1^{-1} is homeomorphic to K_2^{-1}, even though they are not diffeomorphic to each other. So not all knots are shake slice.

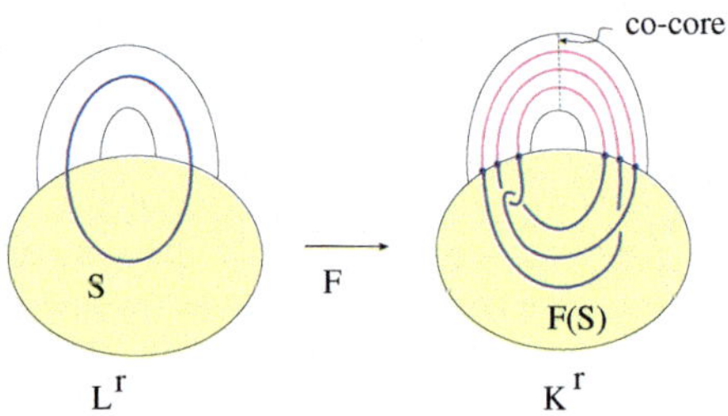

Figure 2.27

Exercise 2.10. *Tristram inequality ([Tr]) states that if an oriented link $\mathcal{L}$ bounds a disk with holes in B^4, and $\sigma_p(\mathcal{L})$, $n_p(\mathcal{L})$ and $\mu(\mathcal{L})$ denote p-signature, p-nullity, and the number of components of the link $\mathcal{L}$ (Section 2.7), then for any prime p:*

$$|\sigma_p(\mathcal{L})| + n_p(\mathcal{L}) \le \mu(\mathcal{L})$$

Prove that if $|\sigma_p(K)| \ge 2$ and p divides r, then K is not r-shake slice (Hint: show that r-shaking doesn't change the p-signature, whereas it increases the nullity by 2m where m is the number of pairs of oppositely oriented copies to K ([A1]).

2.7 Some classical invariants

Let $K \subset S^3$ be a link. Any oriented surface $F \subset S^3$ bounding K is called a Seifert surface of K. An *Alexander matrix* of F is a matrix for the linking form $L(a,b) = Link(a, v^*b)$, where v^*b is a push-off of b along a normal vector field v to F (e.g. [Ro], [AMc], [Liv]),

$$L : H_1(F;\mathbb{Z}) \times H_1(F;\mathbb{Z}) \to \mathbb{Z}$$

Then define an *Alexander polynomial* by $\mathbb{A}_F(t) = det(V - tV^T)$. For example, when K is the trefoil knot and F is as in Figure 2.28, we get $V = \begin{pmatrix} -1 & 1 \\ 0 & -1 \end{pmatrix}$ and $A_F(t) = t^2 - t + 1$.

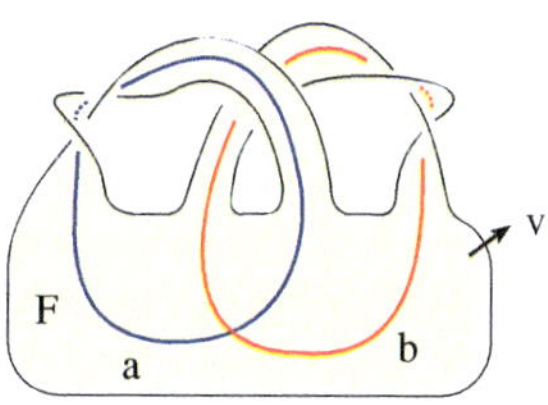

Figure 2.28

Definition 2.3. *Let $\omega_p = exp(2\pi i/p)$. Then the p-signature $\sigma_p(K)$, and the p-nullity $n_p(K)$ of K are defined to be the signature and nullity of the skew Hermitian matrix:*

$$(1 - \bar{\omega}_p)A_F(\omega_p) = (1 - \bar{\omega}_p)(V - \omega_p V^T)$$

Definitions of $\sigma_p(K)$, $n_p(K)$ are independent of the choice of the Seifert surface F. The following exercise gives a 4-dimensional proof of this, at least for the case $p = 2$.

Exercise 2.11. *(Well definedness of signatures) By pushing the interior of F into B^4, we get a proper imbedding $F \subset B^4$; let B^4_F denote the 2-fold branched covering space of B^4 branched along F. (a) By the handlebody description of B_F (Section 11.2, Figure 11.11), show that $\sigma_2(K) = \sigma(B_F)$, where σ denotes the signature. (b) By using Remark 1.3, show that $\sigma(B_F)$ does not depend on F. (How can you generalize this process to other $\sigma_p(K)$?). (c) Show that if K is slice then $\sigma_2(K) = 0$ (Hint: Any closed oriented surface $\Sigma^2 \subset S^4$ bounds a 3-manifold in $Y^3 \subset B^5$, then the corresponding 2-fold branched covering space B^5_Y provides a 5-manifold which S^4_Σ bounds; then use the fact that if a 4-manifold bounds it must have zero signature). Prove similar statements for p-signatures.*

It is known that $A_F(t)$ is defined only up to a multiplication by a unit, that is if F' is another Seifert surface then $A_F(t) = t^n A_{F'}(t)$, for some $n \in \mathbb{Z}$.

Definition 2.4. *The symmetrized Alexander polynomial is defined to be*

$$\Delta_K(t) = t^{-r/2} A_F(t)$$

where $r = b_1(F)$ is the first betti number of F. It satisfies the relation $\Delta_K(t^{-1}) = \Delta_K(t)$.

The symmetrized Alexander polynomial satisfies a nice skein relation: Suppose K_-, K_+, and K_0 be the links in S^3, with projections differing from each other by a single crossing, as shown in Figure 2.29 (part of their Seifert surfaces are shaded). Then

$$\Delta_{K_+}(t) - \Delta_{K_-}(t) = (t^{1/2} - t^{-1/2})\Delta_{K_0}(t) \qquad (2.2)$$

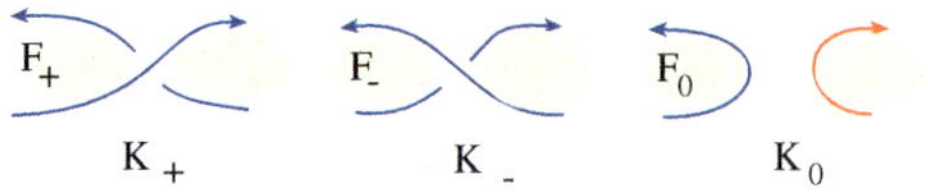

Figure 2.29

This is because the corresponding Alexander matrices V_+, V_-, and V_0 of the Seifert surfaces F_+, F_- and F_0 (shaded regions of Figure 2.29) can be computed from:

$$V_\pm = \left(\begin{array}{c|c} a_\pm & * \\ \hline * & V_0 \end{array} \right)$$

where $a_- = a_+ + 1$ (one circle generator is going across the shaded region of $K_\pm$). from which we get $A_{F_+} - A_{F_-}(t) = (t-1)A_{F_0}$, and from this the result follows.

The skein relation (2.2) with the additional axiom $\Delta_{\mathbf{O}}(t) = 1$, where $\mathbf{O}$ is the unknot, completely determines $\Delta_K(t)$, so it can be used as a definition.

Remark 2.5. *The Alexander polynomial of a slice knot K factors as a product $\Delta_K(t) = f(t)f(t^{-1})$, where $f(t)$ is some integral Laurent polynomial (e.g.[Ro]), this gives an obstruction to sliceness. By Exercise 2.11, knot signatures are also obstruction to sliceness, More subtle obstruction sliceness comes from the adjunction formula, Theorem 9.1.*

Next, we discuss torsion invariants (e.g. [Ni2]): Let V be a finite-dimensional vector space over a field F. A *volume* element of V is a choice of a nonzero element v of the 1-dimensional vector space $\Lambda^{top}(V)$ (i.e. top exterior power). For example, any choice of an orientable basis of V gives a volume element. Given two volume elements w and w' let $[w\ ;\ w']$ denote the scalar c so that $w = cw'$. Let $\mathbf{C}_* = \{C_i, \partial\}$ be an acyclic complex

$$C_m \xrightarrow{\partial} C_{m-1} \xrightarrow{\partial} \dots \xrightarrow{\partial} C_0$$

of finite-dimensional vector spaces (acyclic means $H_*(\mathbf{C}_*) = 0$). Now choose a volume element v_i for each C_i, and choose elements $\omega_i \in \Lambda^* C_i$ so that, $0 \neq \partial\omega_{i+1} \wedge \omega_i \in \Lambda^{top} C_i$, then *Reidemeister torsion* $\tau(\mathbf{C}_*) \in F$ is defined as in (2.3). Here, $\partial^2 = 0$ implies that this definition of $\tau(\mathbf{C}_*)$ is independent of the choice of $\omega_i's$.

$$\tau(\mathbf{C}_*) = \prod_{i=0}^{m} [\partial(\omega_{i+1}) \wedge \omega_i\ ;\ v_i]^{(-1)^{i+1}} \tag{2.3}$$

We extend this definition to non-acyclic chain complexes $\mathbf{C}_*$ simply by setting $\tau(\mathbf{C}_*) = 0$. Let Y be a finite CW complex and $\mathbf{C}_*(Y) = \{C_i(Y), \partial\}$ be its chain complex, clearly i-cells of Y define volume on each $C_i(Y)$. Usually $\mathbf{C}_*(Y)$ is not acyclic (e.g. closed manifolds are never acyclic). To define torsion nontrivially we twist the coefficients of $\mathbf{C}(Y)$ so that it will have more chance to be acyclic: For example, we take the universal cover $\tilde{Y} \to Y$, which gives $\mathbf{C}_*(\tilde{Y})$ a $\mathbb{Z}[\pi_1(Y)]$-module structure, then we pick a homomorphism to a field $\lambda : \mathbb{Z}[\pi_1(Y)] \to F$ and form a chain complex $\mathbf{C}_*(Y; F) := \mathbf{C}_*(\tilde{Y}) \otimes_{\mathbb{Z}[\pi_1(Y)]} F$ over the field F; we then define $\tau_\lambda(Y) = \tau(\mathbf{C}_*(Y, F))$.

Example 2.1. *Let Y be the lens space $L(p,q)$ (Section 5.3) obtained by taking the quotient of $S^3 = \{(z_1, z_2)|\ |z_1|^2 + |z_2|^2 = 1\}$ by the $\mathbb{Z}_p$ action $(z_1, z_2) \mapsto (tz_1, t^q z_2)$, where $t = e^{2\pi i/p}$ and where p, q are relatively prime integers. Then $R := \mathbb{Z}[\mathbb{Z}_p] = \mathbb{Z}[t, t^{-1}]/(1 - t^p)$. Write a cell decomposition $Y = \sum_{j=0}^{3} e^j$ coming from its handlebody structure induced from the union of two solid tori, and lift these cells to get an equivariant cell decomposition of $\tilde{Y} = S^3 = \sum_{j=0}^{3} \sum_{k \in \mathbb{Z}_p} e_k^j$. By squeezing the first solid torus (where the action is $z_1 \mapsto tz_1$) to the unit circle (1-skeleton) we get $te_k^j = e_{k+1}^j$ for $j = 0, 1$. The second solid torus (where the action is $z_2 \mapsto t^q z_2$) contains the 2 and 3 cells with $te_k^j = e_{k+q}^j$ for $j = 2, 3$ (Figure 2.30).*

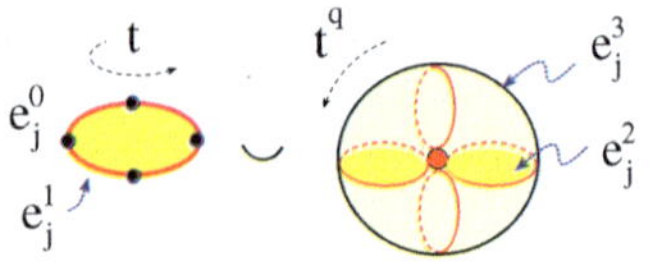

Figure 2.30

Here $\lambda : R \to \mathbb{C}$ *identifies* $t \mapsto e^{2\pi i/p}$, *so we have* $\partial e_j^1 = (t-1)e_j^0$, $\partial e_j^2 = (\sum_{j=0}^{p-1} t^j)e_0^1 = 0$, *and* $\partial e_j^3 = e_{j+1}^2 - e_j^2 = (t^a - 1)e_j^2$ *where* $aq = 1 \; mod(p)$. *Then the chain complex* $\mathbf{C}_*(Y;\mathbb{C})$ *is given by (2.4); hence* $\tau_\lambda(Y) = (t^a - 1)(t-1) = (s-1)(s^q - 1)$, *where* $t = s^q$.

$$0 \to \mathbb{C} \xrightarrow{t^a - 1} \mathbb{C} \xrightarrow{0} \mathbb{C} \xrightarrow{t-1} \mathbb{C} \tag{2.4}$$

Remark 2.6. *Torsion is a combinatorial invariant, that is, finite cell complexes with isomorphic subdivisions have the same torsion. Torsion is not a homotopy invariant, it is only a simple-homotopy invariant (e.g.[Co]). For example* $L(7,1)$ *and* $L(7,2)$ *are homotopy equivalent manifolds with different torsions. Compact smooth manifolds* Y *have unique PL-structure, so torsion is a diffeomorphism invariant.*

There is an other notion of torsion, defined by Milnor, for manifolds with $b_1(Y) > 0$, as follows. Let $\hat{Y} \to Y$ be the maximal abelian covering of Y, corresponding to the kernel of the natural homomorphism $\pi_1(X) \to H$, where H is the free abelian group which is the quotient of $H_1(Y)$ by its torsion subgroup. H acts on $\tilde{Y}$ by deck transformations making $\mathbf{C}_*(\tilde{Y})$ a $\mathbb{Z}[H]$-module. $\mathbb{Z}[H]$ is an integral domain (because it is UFD ([Tan]) so we can take its field of fractions $\mathcal{F} = \mathbb{Q}[H]$, which we can consider a $\mathbb{Z}[H]$-module. Hence we can form a chain complex $\mathbf{C}(Y,\mathcal{F}) := \mathbf{C}_*(\tilde{Y}) \otimes_{\mathbb{Z}[\pi_1(Y)]} \mathcal{F}$ over the field $\mathcal{F}$. *Milnor torsion* is defined to be as the torsion of this chain complex $\nu(Y) \in \mathcal{F}/H$.

Remark 2.7. *If* $K \subset S^3$ *is a knot, and* $Y = S_K^3$ *is the 3-manifold obtained by surgering* S^3 *along* K *with framing* 0, *then* $H = \mathbb{Z}$ *and* $\mathbb{Z}[H] = \mathbb{Z}[t, t^{-1}]$. *By [M5] and [Tu] there is the following identification:*

$$\nu(S_K^3) = \frac{\Delta_K(t)}{(t^{1/2} - t^{-1/2})^2} \tag{2.5}$$

Chapter 3

Gluing 4-manifolds along their boundaries

Given two connected smooth 4-manifolds with boundary M, N and an orientation preserving diffeomorphism $f : \partial M \to \partial N$, we can ask how to draw a handlebody of the oriented manifold obtained by gluing with this map:

$$-M \smile_f N$$

Also we can ask when given two disjoint copies of codimension zero submanifolds $L \sqcup -L \subset \partial M$, how can we draw the manifold $M(f)$ obtained from M by identifying these two copies by a diffemorphism $f : L \to L$?

$$M(f) = M \sqcup (L \times [0,1]) \Big/ \begin{matrix} (x,0) & \sim x \in L \\ (f(x),1) & \sim x \in -L \end{matrix}$$

There are two ways of doing this: the upside down method, and the cylinder method.

3.1 Constructing $-M \smile_f N$ by the upside down method

Let $M = M_\Lambda$ with $\Lambda = \{K_1^{r_1}, \ldots, K_n^{r_n}, C_1, \ldots, C_s\}$ and $N = N_{\Lambda'}$. Let $\{\gamma_1, \ldots, \gamma_n\}$ be the dual zero-framed circles of the 2-handles $K_1, .., K_n$. Then $-M \smile_f N$ is represented by $Q_{\Lambda''}$ where $\Lambda'' = \Lambda' \smile \{f(\gamma_1), \ldots, f(\gamma_n)\}$. In particular, if we apply this process to $M^0 := M - \{$3- and 4-handles$\}$ and any diffeomorphism $f : \partial(M^0) \to \partial \#_p(S^1 \times B^3)$ (p is the number of 3-handles), we get the *upside down handlebody* of M (Figure 3.1). In the special case of when M has no 3-handles, then clearly the framed link $\{f(\gamma_1), \ldots, f(\gamma_n)\}$ in ∂B^4 gives its upside down handlebody of M.

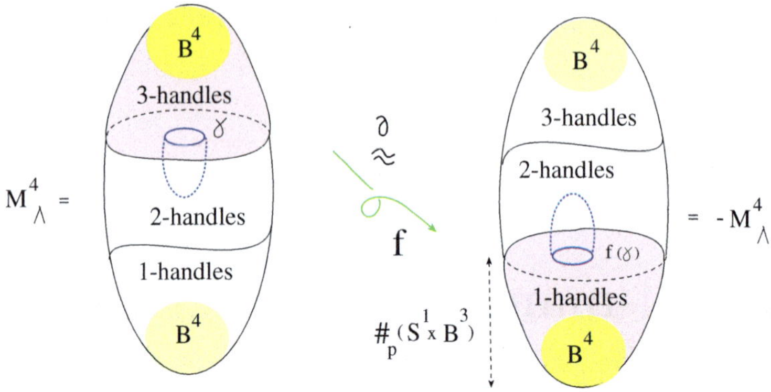

Figure 3.1

The manifold $-M \smile_{id} M$ is called *the double* of M and denoted by $D(M)$. So by the above, $D(M)$ is a handlebody obtained from M by attaching 2-handles along the zero-framed dual circles of the 2-handles of M. This gives $D(T^2 \times B^2) = T^2 \times S^2$. The *cusp* is defined to be the 4-manifold K^0, where K is the right handed trefoil knot. By using this method, we see that the double $D(C)$ of the cusp C is diffeomorphic to $S^2 \times S^2$ (to see this slide the 2-handle in Figure 3.2 over its dual to simplify).

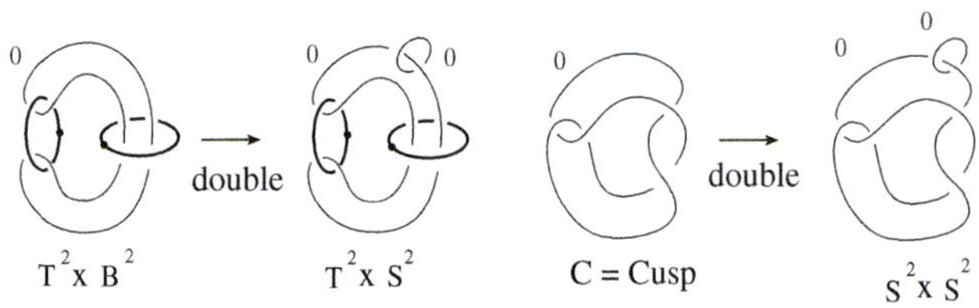

Figure 3.2

Exercise 3.1. *(a) Show that $D(W) = S^4 = W \cup_f -W$, where W is the Mazur manifold and f is the involution, as shown in Figure 1.8. (b) The fishtail is defined to be the manifold of the first picture of Figure 3.3. Let $f : \partial F \to \partial F$ be the diffeomorphism on the boundary of the fishtail described in Figure 3.3 (obtained by the dot and* 0 *exchange). Show that $-F \smile_f F = S^4$ (given by a canceling pair of* 2- *and* 3-*handles).*

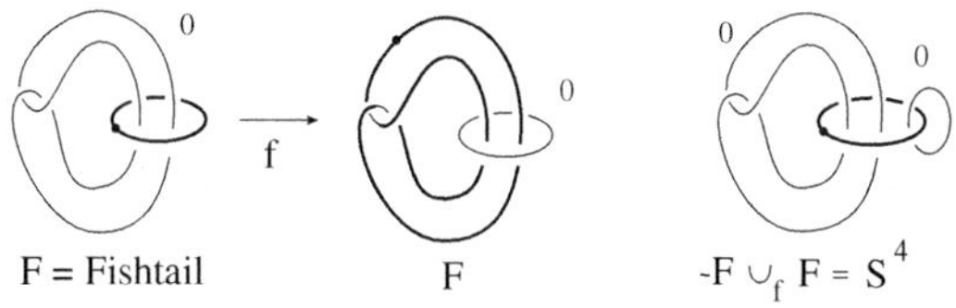

Figure 3.3

Exercise 3.2. *Let M be B^4 with a 2-handle attached to the left handed trefoil knot with -1 framing, and $f : \partial M \to \partial E_8$ be the diffeomorphism described in Figure 2.15, show that $-M \smile_f E_8 = \mathbb{CP}^2 \# 8\bar{\mathbb{CP}}^2$ (Hint Figure 3.4).*

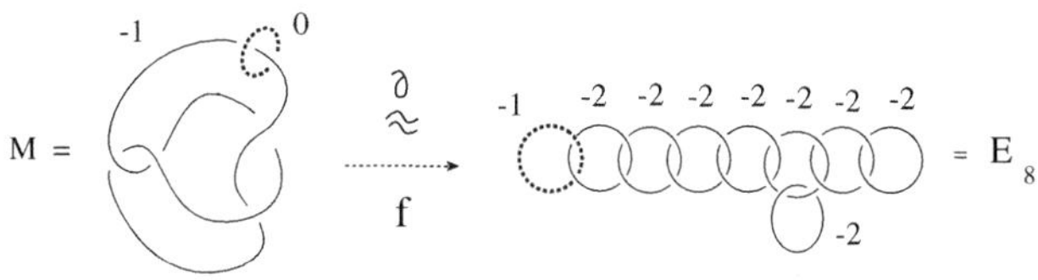

Figure 3.4

3.2 Constructing $-M \smile_f N$ and $M(f)$ by the cylinder method (roping)

For this we attach a cylinder $\partial M \times [0,1]$ to disjoint union $-M \sqcup N$, where one end of the cylinder is attached by the identity, the other end is attached by the diffeomorphism $f : \partial M \to \partial N$.

$$-M \smile_f N = -M \smile_{id} (\partial M \times [0,1]) \smile_f N$$

Similarly, for $M(f)$ we glue a cylinder to M running from $-L$ to L

$$M(f) = M \smile_{id \sqcup f} (\partial L \times [0,1])$$

Think of f as a force field hovering over $-M \sqcup N$ carrying points of ∂M to ∂N. To observe the effects of f, we lower ropes (from a central point) with hooks tied at their ends, and the hooks go through the cores γ_j of the 1-handles of the 3-manifold ∂M; then we watch where f takes them, as indicated in Figure 3.5.

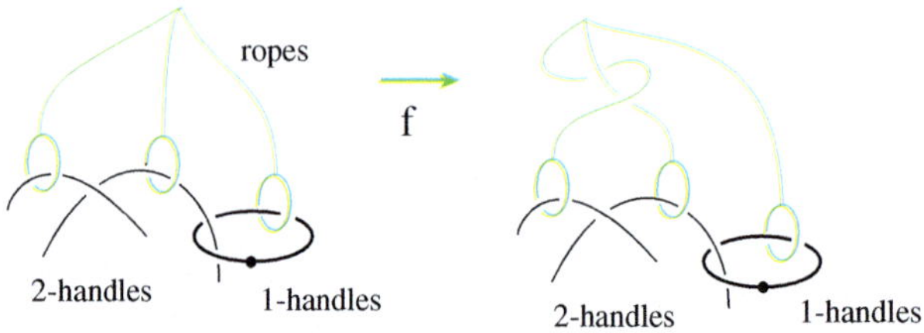

Figure 3.5: The effect of f

To describes the process of gluing the two boundary components by f, we attach 2-handles to $\gamma_j \# f(\gamma_j)$ by using the ropes as a guide, as shown in Figure 3.6, which is $-M \smile_f N$. Put another way, f connects the 0-handle of M with the 0-handle of N by a 1-handle (so this 1-handle cancels one of the two 0-handles), then by going over this 1-handle it identifies the neighborhoods of γ_j with $f(\gamma_j)$, which amounts to attaching 2-handles to $\gamma_j \# f(\gamma_j)$ (i.e. creating tunnels). We don't need to describe how f identifies the 2- and 3-handles of M with that of N, since by the similar description, they amount to adding 3- and 4- handles to $-M \sqcup N$. Recall that describing 4-manifold handlebodies we don't need to specify 3- and 4-handles, because they are always attached canonically.

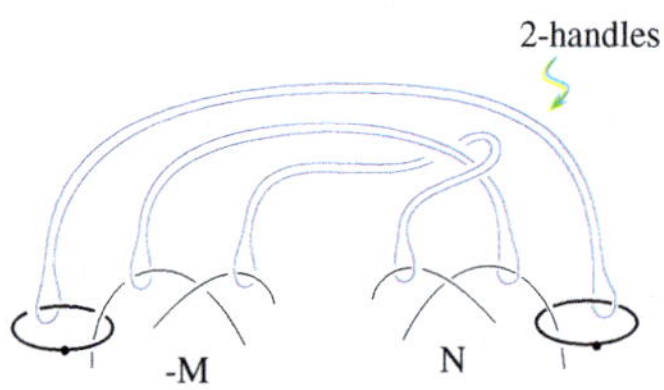

Figure 3.6: $-M \smile_f N$

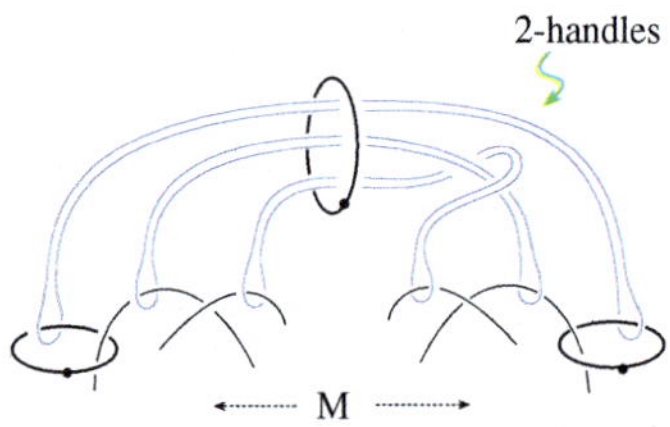

Figure 3.7: $M(f)$

The handlebody of $M(f)$ can be constructed almost the same way, except that we must have an extra 1-handle, as shown in Figure 3.7, because in this case the identifying cylinder $L \times [0,1]$ is attached to a single connected manifold M, as opposed to between two disjoint manifolds $-M \sqcup N$. In the former case, this 1-handle was canceled by one of the 0-handles of $-M \sqcup N$. Figure 3.8 and Figure 3.9 are depicting $-M \smile_f N$ and $M(f)$.

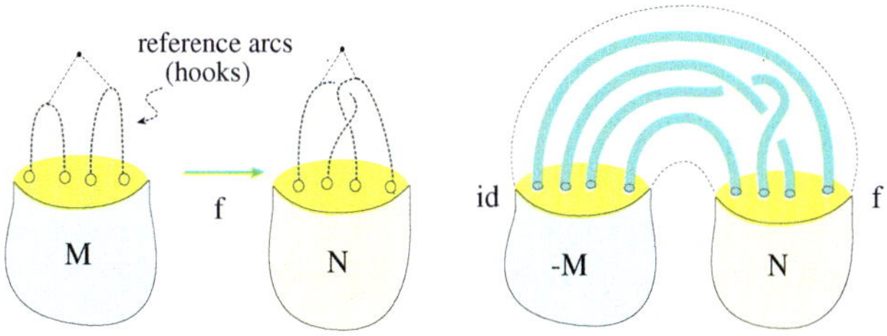

Figure 3.8: $-M \smile_f N$

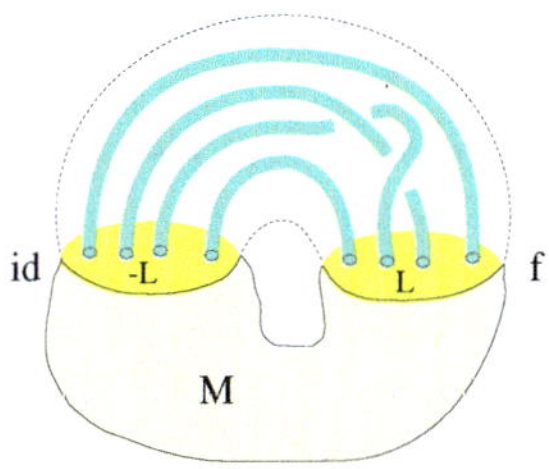

Figure 3.9: $M(f)$

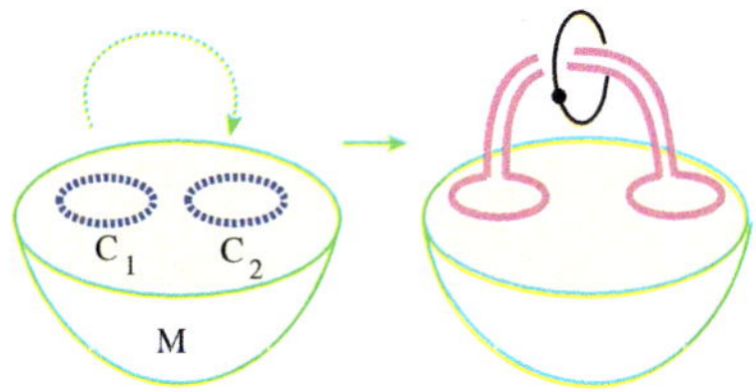

Figure 3.10: Round 1-handle

Remark 3.1. *A simple but useful example of this process is so called round 1-handle attachment (Definition 7.8), which is obtained by identifying the tubular neighborhoods of two disjoint framed circles $\{C_1, C_2\}$ lying in ∂M^4, hence it can be described by attaching a 1- and 2- handle pair, where the 2-handle is attached to the curve obtained by connected summing C_1 and C_2 along the 1-handle, as shown in Figure 3.10 by the framing induced from the framings of C_1 and C_2. On the boundary 3-manifold ∂M^4 this process corresponds to doing 0-surgery to the connected summed knot $C_1 \# C_2$, then doing 0-surgery to the linking circle of the connected summing band (the meridian of the band). Here 0-surgery means surgery by framing induced from ∂M. We will call this operation a round surgery operation to $Y = \partial M^4$ along $\{C_1, C_2\}$ and denote it by $Y(C_1, C_2)$.*

3.3 Codimension zero surgery $M \mapsto M'$

Let $N \subset M$ be a codimension zero submanifold giving the decomposition $M = N \smile_\partial C$, where N is the complement of C, and let $f : \partial N \to \partial N'$ be a diffeomorphism. We call the process of cutting N out and gluing N' by f a *codimension zero surgery.*

$$M \mapsto M' := N' \smile_f C$$

To visualize this process think of all the handles of C as placed at the top of N. Notice that only the 2-handles of C interact with the handles of N. So think of N as hanging from the 2-handles of C (Figure 3.11) like hangers in a dress closet, where f shuffles the hangers below. So while the handles of N are hanging from the top, they look different according to how f has rearranged them below.

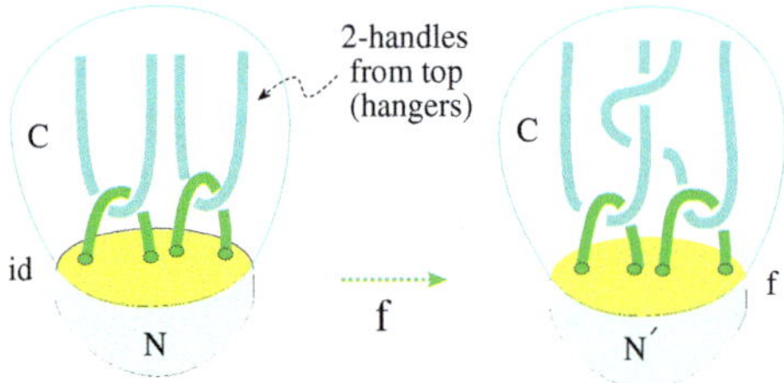

Figure 3.11: Codimension zero surgery $M \mapsto M'$

More precisely, we apply diffeomorphism $f : \partial N \to \partial N'$ keeping track of where f throws the 2-handles of C in N'. That is, if $N = N_\Lambda$ and $N' = N'_{\Lambda'}$ such that $M = M_{\mathbf{\Lambda}}$ with $\mathbf{\Lambda} = \Lambda \smile \{K_1^{r_1}, \dots, K_p^{r_p}\}$, then $M' = M'_{\mathbf{\Lambda}'}$, where $\mathbf{\Lambda}' = \Lambda' \smile \{f(K_1^{r_1}), \dots, f(K_p^{r_p})\}$.

In this case we say that M *is obtained from* M' *by cutting out* N'*, and gluing* N *via* $f : \partial N \to \partial N'$. So if $M' = M'_{\mathbf{\Lambda}'}$ with $\mathbf{\Lambda}' = \Lambda' \smile \{L_1^{r_1}, \dots, L_p^{r_p}\}$, then $M = M_{\mathbf{\Lambda}}$ where $\mathbf{\Lambda} = \Lambda \smile \{f^{-1}(L_1^{r_1}), \dots, f^{-1}(L_p^{r_p})\}$.

Chapter 4

Bundles

Here we describe handlebodies of 4-manifolds which are bundles. First a simple example:

4.1 $\mathbf{T}^4 = T^2 \times T^2$

We start with T^2 (Figure 2.1) then thicken it to $T^2 \times [0,1]$, then by identifying the front and the back faces of $T^2 \times [0,1]$ we form T^3 (Figure 4.1 -recall the recipe of Section 3.2), we then construct T^4 by identifying the front and back faces of $T^3 \times [0,1]$ (Figure 4.2).

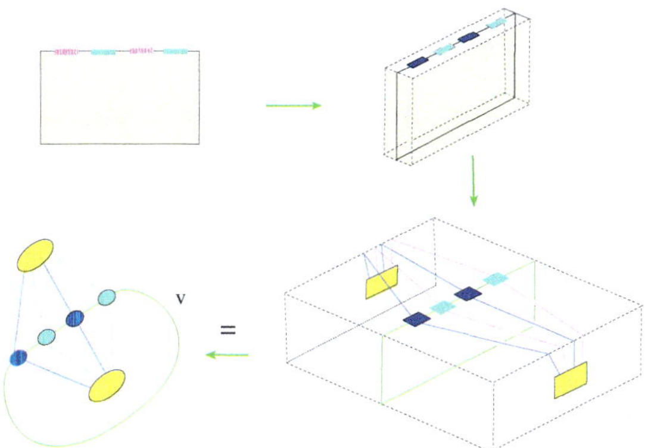

Figure 4.1: T^3

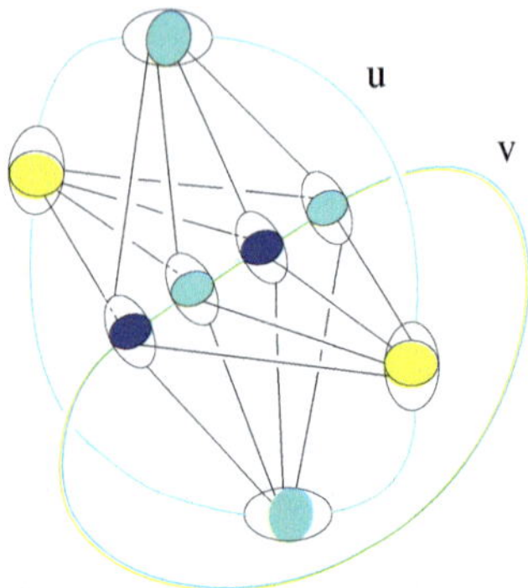

Figure 4.2: T^4

By converting the 1-handle notation to the circle-with-dot notation of Section 1.1, in Figure 4.5 we get another handlebody picture of T^4. For the benefit of the reader we do this conversion gradually. For example, we first flatten the middle 1-handle balls as in Figure 4.3, obtaining Figure 4.4, then convert them to the circle-with-dot notation with ease, obtaining the final Figure 4.5.

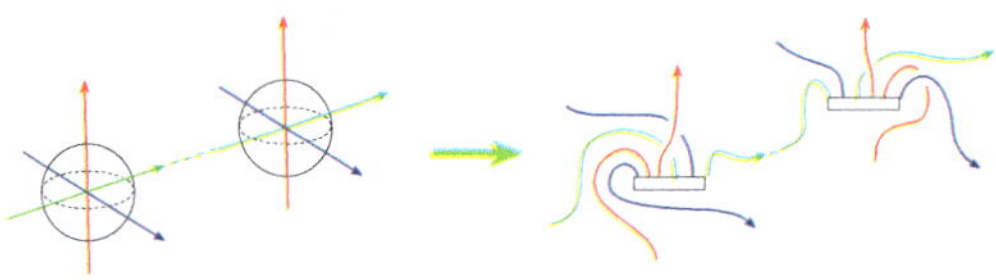

Figure 4.3

Exercise 4.1. *Show that without the 2-handle v, Figure 4.2 describes $T_0^2 \times T^2$, and deleting two 3-handles from $T_0^2 \times T^2$ gives just $T_0^2 \times T_0^2$, where $T_0^2 = T^2 - D^2$ is the punctured T^2.*

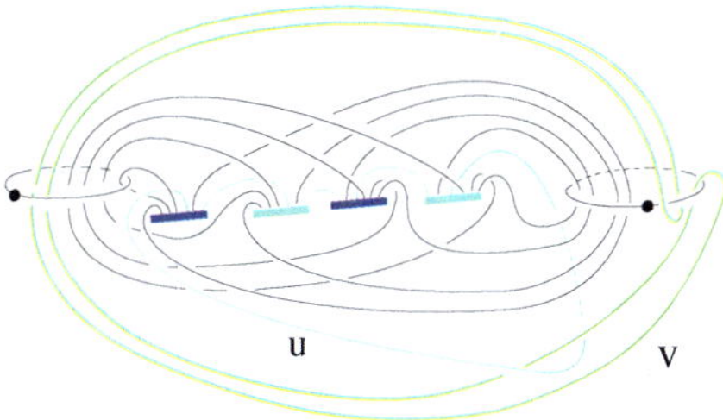

Figure 4.4: T^4

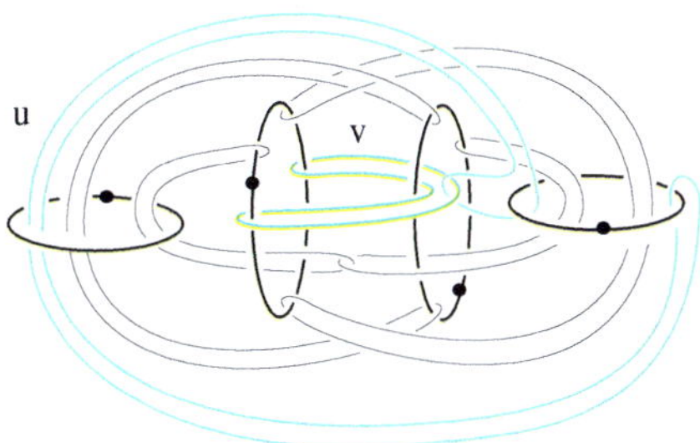

Figure 4.5: T^4

4.2 Cacime surface

Cacime is a particular surface bundle over a surface, which appears naturally in complex surface theory [CCM]. An understanding this manifold is instructive, because it is a good test case for understanding many of the difficulties one encounters in constructing handlebodies of surface bundles over surfaces. We will first draw a handlebody of Cacime; from this derive a recipe for drawing surface bundles over surfaces in general.

Let F_g denote the surface of genus g. Let $\tau_2 : F_2 \to F_2$ be the elliptic involution with two fixed points, and $\tau_3 : F_3 \to F_3$ be the free involution induced by 180^0 rotation (Figure 4.6). The Cacime surface M is the complex surface obtained by taking the quotient of $F_2 \times F_3$ by the product involution:

$$M = (F_2 \times F_3)/\tau_2 \times \tau_3$$

By projecting to the second factor we can describe M as an F_2-bundle over $F_2 = F_3/\tau_3$. Let A denote the twice punctured 2-torus $A = T^2 - D^2_- \sqcup D^2_+$. Clearly, M is obtained

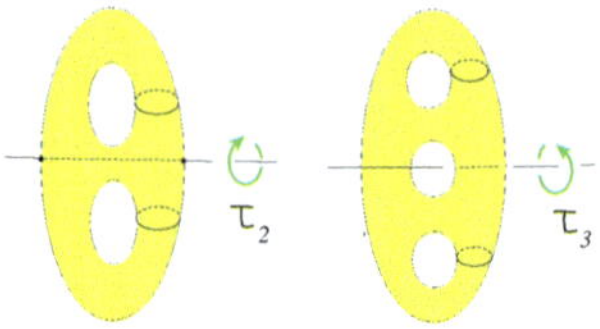

Figure 4.6: The involutions τ_2 and τ_3

by identifying the two boundary components of $F_2 \times A$ by the involution induced by τ_2 (notice that A is a fundamental domain of the action τ_3).

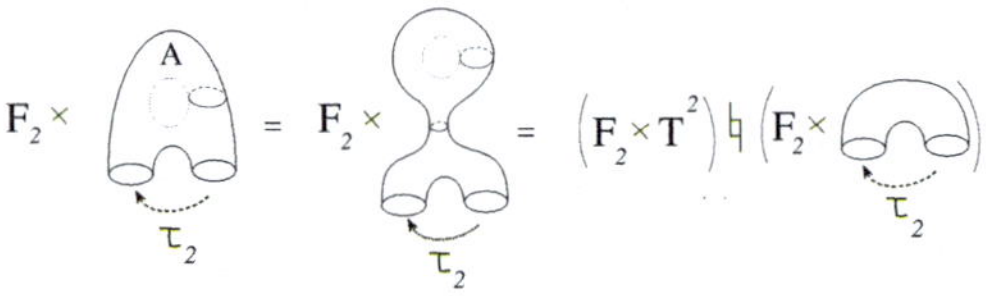

Figure 4.7: M

By deforming A, as in Figure 4.7, we see that $M = E \natural E'$ is the fiber sum of two F_2 bundles over T^2, where $E = F_2 \times T^2 \to T^2$ is the trivial bundle and E' is the bundle:

$$E' = F_2 \times S^1 \times [0,1]/(x,y,0) \sim (\tau_2(x), y, 1) \to T^2$$

Now by using the techniques developed in Chapter 3, step by step we will construct the following manifolds and diffeomorphisms:

(a) $E_0 := E - F_2 \times D^2 = F_2 \times T_0^2$

(b) $f_1 : \partial E_0 \xrightarrow{\approx} F_2 \times S^1$

(c) $E_0' = E' - F_2 \times D^2$

(d) $f_2 : \partial E_0' \xrightarrow{\approx} F_2 \times S^1$

(e) $M = -E_0 \smile_{f_2^{-1} \circ f_1} E_0'$

(a) By imitating the construction of T^4 in Figures 4.1 through 4.5, in Figures 4.8 through 4.9 we build $F \times T^2$.

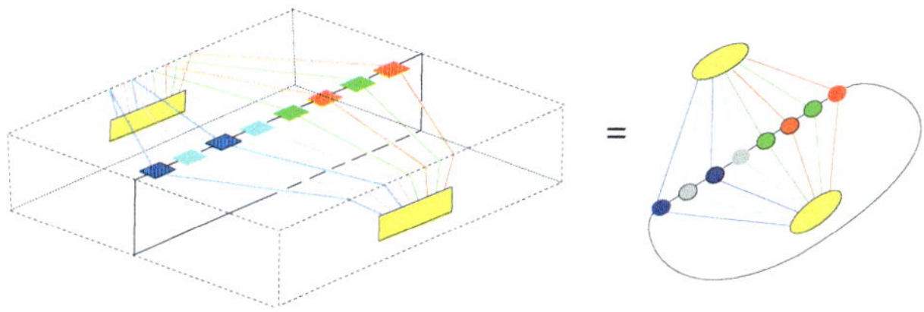

Figure 4.8: $F_2 \times S^1$

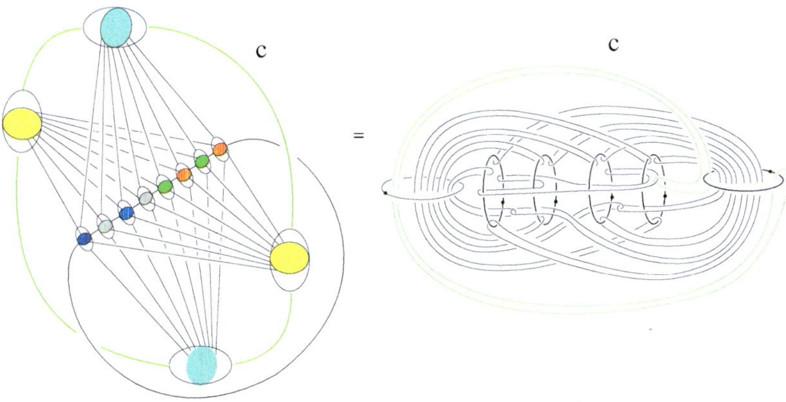

Figure 4.9: $F_2 \times T^2$

Exercise 4.2. *Show that if we remove the 2-handle c from the Figure 4.9, we get a handlebody picture for* $E_0 = F_2 \times T_0^2$.

(b) We claim that there is a diffeomorphism, $f_1 : \partial E_0 \xrightarrow{\approx} F_2 \times S^1$, which takes the ropes with hooks of Figure 4.9 to the corresponding ropes, as indicated in Figure 4.10 (see Section 3.2 for discussion of ropes).

Exercise 4.3. *Show that this diffeomorphism* f_1 *can be obtained by the handle slides on the boundary, as indicated in Figure 4.11. (Hint: just perform the indicated handle slides while keeping track of the ropes.)*

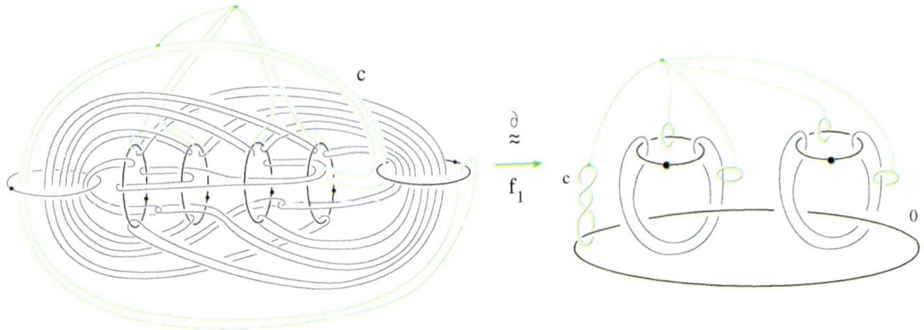

Figure 4.10: Diffeomorphism $\partial E_0 \approx F_2 \times S^1$ made concrete

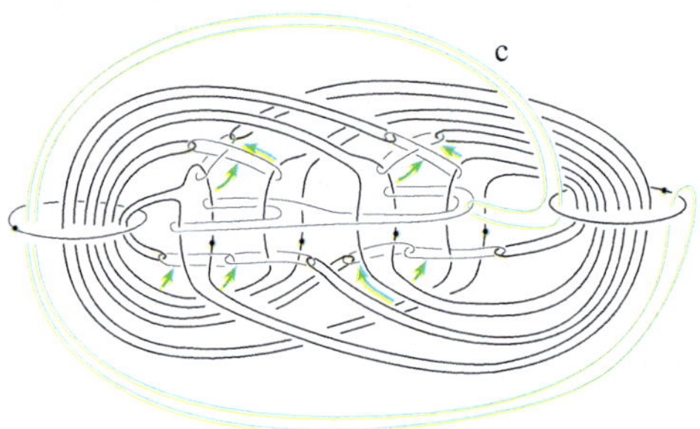

Figure 4.11: Describing f_1

(c) $E_0' = E' - F_2 \times D^2$ is a twisted version of E_0; we proceed as in Figure 4.8 except that we identify the front and back faces of $F_2 \times [0,1]$ by $\tau : F_2 \to F_2$ (Figure 4.12 depicts the induced map on F_2 by τ) and get Figure 4.13, which is a twisted version of Figure 4.8. To construct E' we simply cross this twisted F_2 bundle over the circle $F_2 \times_\tau S^1$ with S^1. This gives Figure 4.14 (drawn in two different 1-handle notations), which is the analogous version of Figure 4.9.

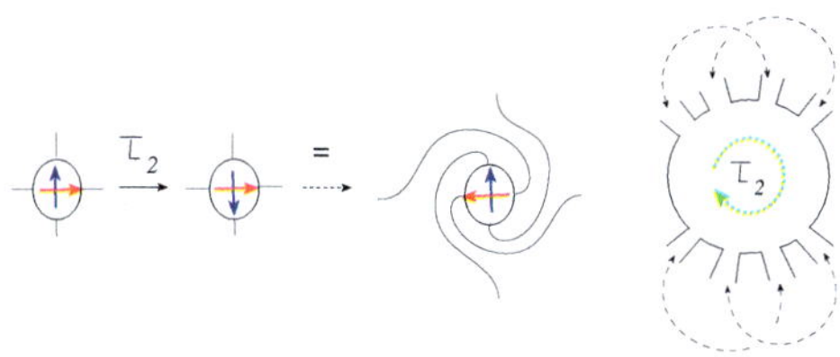

Figure 4.12: The action of $\tau : F_2 \to F_2$

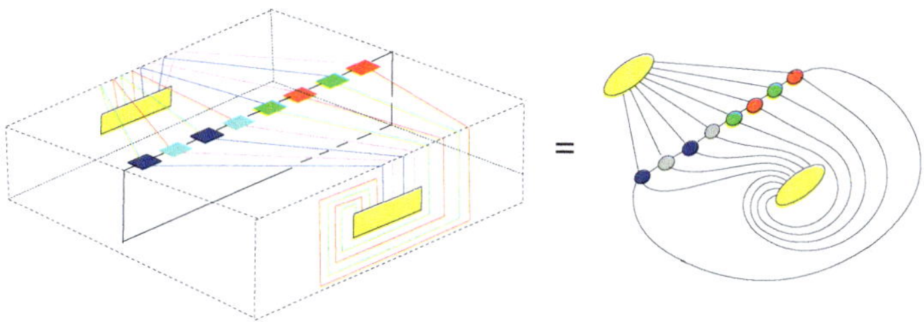

Figure 4.13: $F_2 \times_\tau S^1$

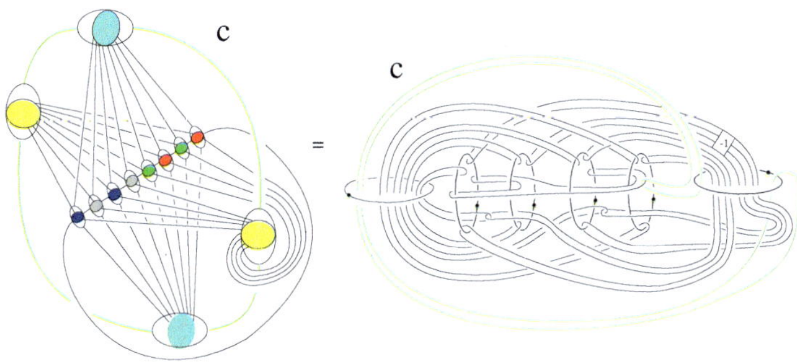

Figure 4.14

Exercise 4.4. *Show that without the 2-handle c, Figure 4.14 describes* $E_0' = E' - F_2 \times D^2$.

(d) Similar to (b) there is a diffeomorphism, $f_2 : \partial E_0' \xrightarrow{\approx} F_2 \times S^1$, obtained by the handle slides described in Figure 4.15. As before, to describe this diffeomorphism geometrically in pictures, we lower ropes (with hooks) and trace out what f does to these ropes during the diffeomorphism of Figure 4.15. This gives Figure 4.16.

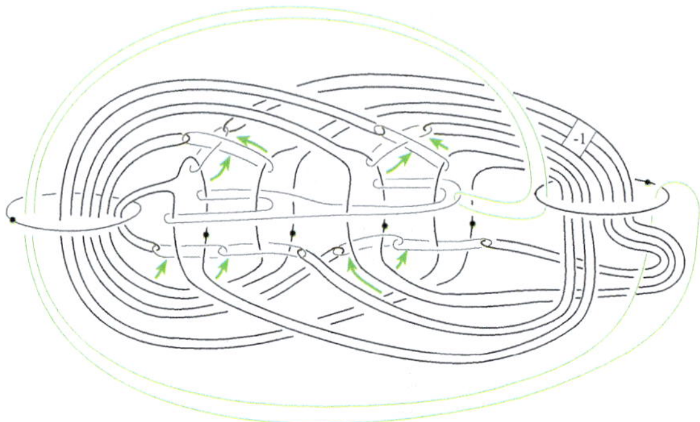

Figure 4.15: Describing f_2

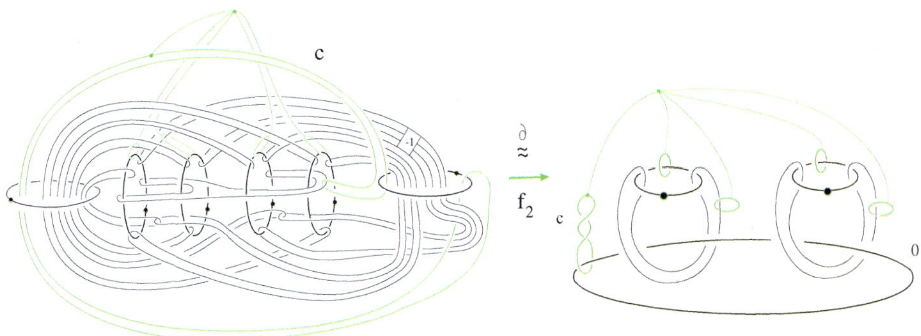

Figure 4.16: Diffeomorphism $\partial E_0' \approx F_2 \times S^1$ made concrete

(e): Finally, by applying the recipe of Section 3.2 we construct the manifold $M = -E_0 \smile_{f_2^{-1}\circ f_1} E_0'$, which is Figure 4.17. Notice that the ropes in ∂E_0 in the left picture of Figure 4.10 are mapped to the ropes in the left hand picture of Figure 4.16 by $f_2^{-1} \circ f_1$.

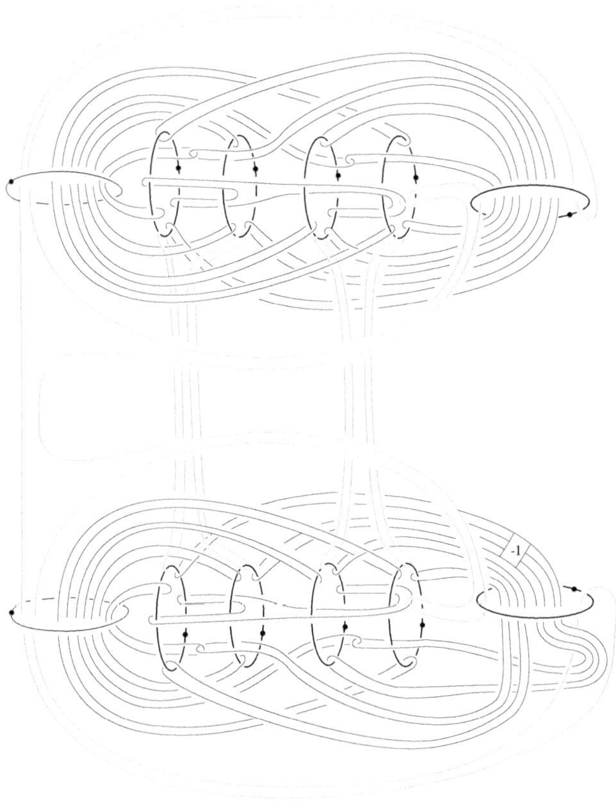

Figure 4.17: Cacime surface

4.3 General surface bundles over surfaces

Now it is clear how to proceed in drawing a handlebody picture of a general F_g bundle over F_p, denoted by $M = F_g \tilde{\times} F_p$. We first decompose M as a fiber sum of F_g bundles over T^2, with monodromies $\tau_j : F_g \to F_g$ $(j = 1, .., p)$, as shown in Figure 4.18. By removing $F_g \times D^2$ from each, we write $M = \smile_\partial E_j$, where each E_j is a F_g bundle over T_0^2, then we perform gluing operations along the boundaries, as described in Chapter 3.

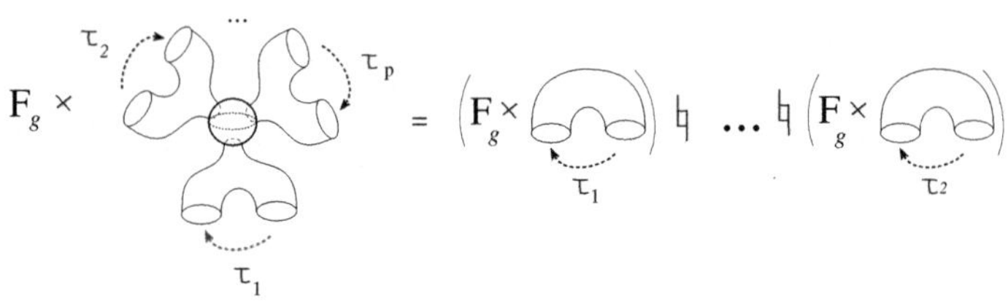

Figure 4.18

4.4 Circle bundles over 3-manifolds

Let $Y^3 = H_g \smile_f H_g$ be a 3-manifold given by its Heegaard decomposition, i.e. H_g is the solid handlebody bounding F_g, and $f : F_g \to F_g$ is a diffeomorphism (see Chapter 5).

Exercise 4.5. *Show that any oriented circle bundle M^4 over Y^3 is a union of two copies of $S^1 \times H_g$ glued along their boundaries by a diffeomorphism. A handlebody for M can be constructed by applying the gluing techniques of Chapter 3 to two copies of the handlebody of $S^1 \times H_g$ (Figure 4.19).*

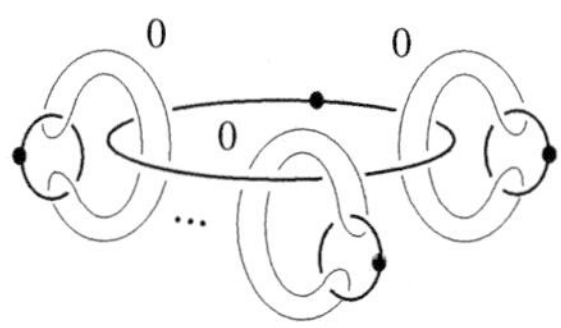

Figure 4.19: $S^1 \times H_g$

4.5 3-manifold bundles over the circle

Section 3.2 essentially gave a recipe for drawing the handlebody picture of a 3-manifold bundle over the circle with a monodromy $f : Y^3 \to Y^3$:

$$M^4 = M(f) = Y \times [0,1]/(x,0) \sim (f(x),1)$$

To put this technique into practice, we will apply it to a very interesting example Q^4 of [CS], which is an exotic copy of $\mathbb{RP}^4$. Recall that $\mathbb{RP}^4 = N \smile_\partial C$ is a union of codimension zero submanifolds N and C, glued along their common boundaries; where N is a twisted B^2-bundle over $\mathbb{RP}^2$, and C is the non-orientable B^3- bundle over $\mathbb{RP}^1$ (see Section 1.5). Similarly $Q^4 = N \smile_\partial C_A$ is the union of codimension zero submanifolds N and C_A, glued along their common boundaries, where C_A is the mapping torus of the diffeomorphism, $f_A : T^3_0 \to T^3_0$, induced by the following integral matrix $A : \mathbb{R}^3/\mathbb{Z}^3 \to \mathbb{R}^3/\mathbb{Z}^3$ (4.1). It is easy to see that C_A is homology equivalent to C, and $\pi_1(Q) = \mathbb{Z}_2$.

$$A = \begin{pmatrix} 0 & 1 & 0 \\ 0 & 0 & 1 \\ -1 & 1 & 0 \end{pmatrix} \tag{4.1}$$

The exotic manifold of Figure 1.29 [A3] was derived from Q^4. Also, the double covering of Q is the homotopy sphere $\Sigma = D^2 \times S^2 \smile C_B$, where C_B is the mapping torus of the diffeomorphism $f_B : T^3_0 \to T^3_0$, induced from $B = A^2$. Σ is the first member Σ_0 of a similarly defined infinite family of homotopy spheres Σ_m, which turned out to be diffeomorphic to the standard 4-sphere ([AK1], [G3], [A5]).

Let us draw a handlebody picture of Σ (following [AK1]). Notice that $C_B = M(f)$, where $M = T^3 \times [0,1]$ and f is the diffeomorphism f_B. By applying the technique of Section 3.2 we draw $M(f)$. For this we start with a Heegaard picture of T^3 (Figure 4.20), as drawn with 1- and 2-handles on S^2 (Figure 4.1).

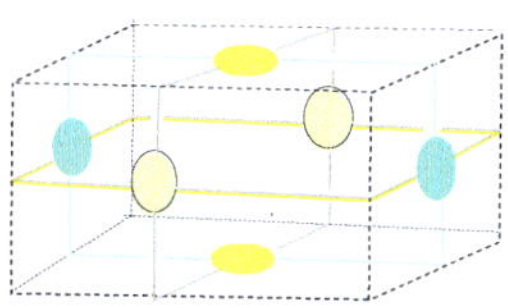

Figure 4.20: T^3

We then study $f_B : T^3 \to T^3$ by first checking what it does on the coordinate axis, i.e. the cores of the 1-handles corresponding to the hooks of the "ropes with hooks" technique of Section 3.2 (Figure 4.21). Then Figure 4.22 gives $M(f)$. For simplicity, we put one of the attaching balls of the 1-handle (the companion of the center ball) at ∞.

$$B = \begin{pmatrix} 0 & 0 & 1 \\ -1 & 1 & 0 \\ 0 & -1 & 1 \end{pmatrix} \tag{4.2}$$

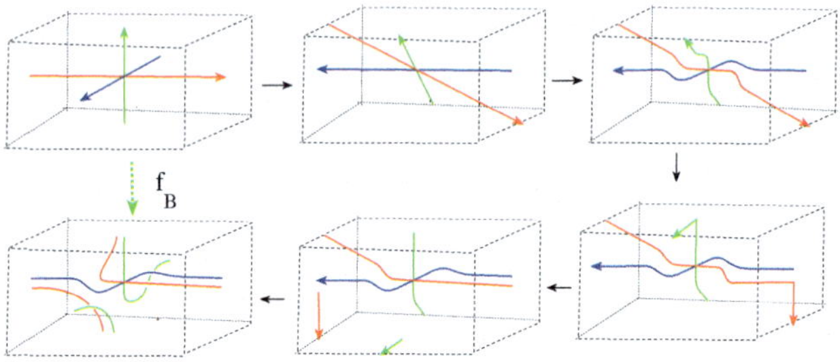

Figure 4.21: Isotoping f_B so that it takes the coordinate axis to the coordinate axis

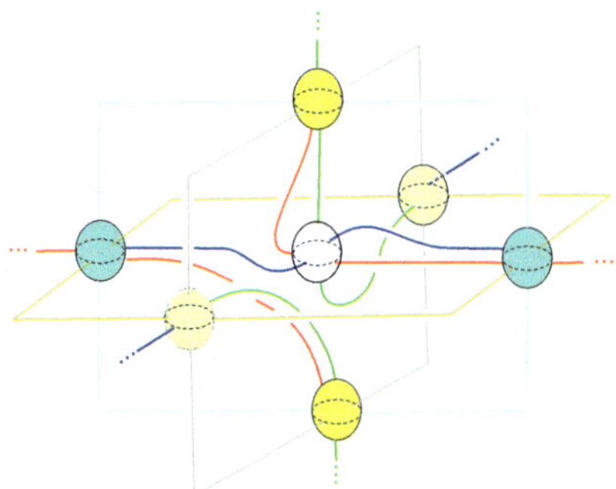

Figure 4.22

Exercise 4.6. *Convert the* 1*-handles of Figure 4.22 to the circle-with-dot notation (this is done in Figure 14.1), and locate the framed circle of the* 2*-handle corresponding to* $D^2 \times S^2$ *of* Σ*, and decide the parity of its framing (in [AK1] this was mistakenly assumed to be even, but in [ARu] it was shown to be odd).*

Chapter 5

3-manifolds

Every closed orientable 3-manifold Y^3 bounds an closed orientable 4-manifold (e.g.[Li], [W]), in fact it bounds a 4-manifold M_Λ consisting of 0- and 2-handles. One can see this from the *Heegaard decomposition* of $Y^3 = H_g \cup_\tau -H_g$, which is basically a union of two copies of a solid handlebody (i.e. the handles of index ≤ 1, and the handles of index ≥ 2 respectively) glued along their boundary $\partial H_g := F_g$ by an orientation preserving diffeomorphism $\tau : F_g \to F_g$. It is known that any surface diffeomorphism can be written as a composition of Dehn twists, $\tau = \tau_{C_1} ... \tau_{C_n}$, along imbedded curves, $C_j \subset F_g$.

Definition 5.1. *Dehn twist $\tau_C : F \to F$ along C is a diffeomorphism, which is identity outside of a cylinder $C \times [-1,1] \subset F$ centered at C, and a (right or left) full rotation inside the cylinder, as in Figure 5.1, so that $\tau_C| : C \to C$ is the antipodal map (c.f. [Iv]).*

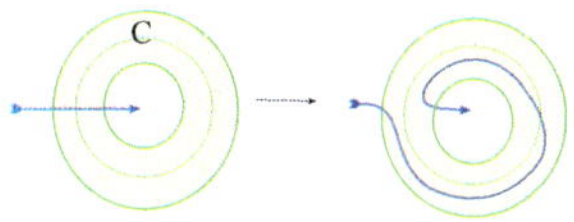

Figure 5.1: τ_C

Repeated application of the following implies that any closed oriented Y^3 is obtained by an integral surgery to a framed link, i.e. $Y^3 = \partial M^4_\Lambda$ for some Λ.

Exercise 5.1. *For $0 \leq k \leq n$, let $Y_k = H_g \cup_{\tau_k} H_g$, where $\tau_k = \tau_{C_1} ... \tau_{C_k}$ with $\tau_0 = id$. Show that $Y_0 = \#_g S^1 \times S^2$, and Y_k is obtained from Y_{k-1} by an integral surgery to a knot in Y_{k-1} istopic to C_k (Hint: Push the last curve C_k to the interior of the solid handlebody H_g, then compare the manifolds Y_{k-1} and Y_k, cf. [Sa]).*

5.1 Dehn surgery

Dehn surgery of a 3-manifold Y^3 is an operation taking out an imbedded copy of $S^1 \times B^2$ from Y^3 and regluing it by some diffeomorphism of its boundary. Let $K \subset Y^3$ be a null homologous knot, and $N(K)$ be its tubular neighborhood. Let μ and λ be the meridian and longitude of $\partial N(K)$. This means λ is the parallel copy of K which is null homologous in $Y - K$. Let p, q be coprime integers. Fix a trivialization:

$$\phi : B^2 \times S^1 \xrightarrow{\approx} N(K)$$

such that $\phi(1,0) = \mu$ and $\phi(0,1) = \lambda$, where $(1,0)$ and $(0,1)$ denotes the two generators of $S^1 \times S^1$. For simplicity we will view ϕ as an identification, and parametrize the curves on $\partial N(K)$ by $(p,q) \leftrightarrow p\mu + q\lambda$. Then $r = p/q$ surgery to Y^3 is the manifold,

$$Y(K,r) = [\, Y - N(K)\,] \smile_{\phi_r} (B^2 \times S^1)$$

where $\phi_r : \partial B^2 \times S^1 \to \partial N(K)$ is the unique diffeomorphism with $\phi_r(1,0) = (p,q)$. With the above identifications we can express the gluing map by the matrix (cf. [Ro])

$$\phi_r = \begin{pmatrix} p & r \\ q & s \end{pmatrix}$$

where $ps - qr = 1$. In particular $Y(K,\infty) = Y$. For brevity, when there is no danger of confusion we will abbreviate $K^r = Y(K,r)$, even though previously when r is an integer, we used this notation for the 4-manifold obtained B^4 by attaching a 2-handle to K.

Exercise 5.2. *Let $K \subset S^3$ be a knot. By using the identity (5.1), justify the 3-manifold diffeomorphism in Figure 5.2 (in [GS] this identification is called "slam-dunk" and attributed to T. Cochran). Note that by the diffeomorphism of Figure 5.3 we can reduce this to the case $\{K,\mu\}$, which is the Hopf link, and also note that in $S^3(K,n)$, the circle μ isotopes to the center circle in the surgery solid torus $B^2 \times S^1$,*

$$\begin{pmatrix} n & -1 \\ 1 & 0 \end{pmatrix}\begin{pmatrix} p & r \\ q & s \end{pmatrix} = \begin{pmatrix} np-q & nr-s \\ p & r \end{pmatrix} \tag{5.1}$$

Exercise 5.3. *By iterating the diffeomorphism of Exercise 5.2, justify the 3-manifold diffeomorphisms of Figure 5.4, where $p/q = [a_1, a_2, .., a_k]$ is the continued fraction expansion (when $a_i \geq 2$ the expression is unique) (see [HNK], [Sa] for more in-depth discussion),*

$$p/q = a_1 - \cfrac{1}{a_2 - \cfrac{1}{\ddots - \cfrac{1}{a_k}}}$$

$$\begin{pmatrix} p & r \\ q & s \end{pmatrix} = \begin{pmatrix} a_1 & -1 \\ 1 & 0 \end{pmatrix} \begin{pmatrix} a_2 & -1 \\ 1 & 0 \end{pmatrix} .. \begin{pmatrix} a_k & -1 \\ 1 & 0 \end{pmatrix}$$

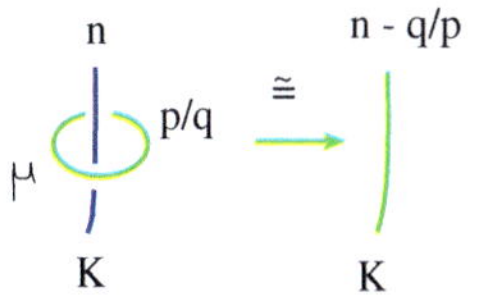

Figure 5.2

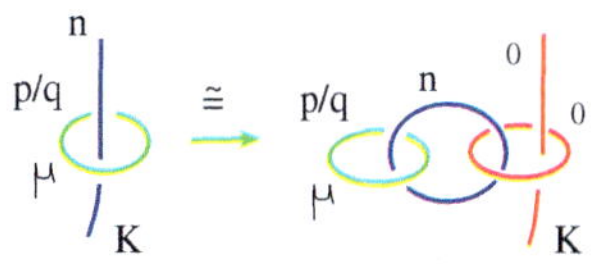

Figure 5.3

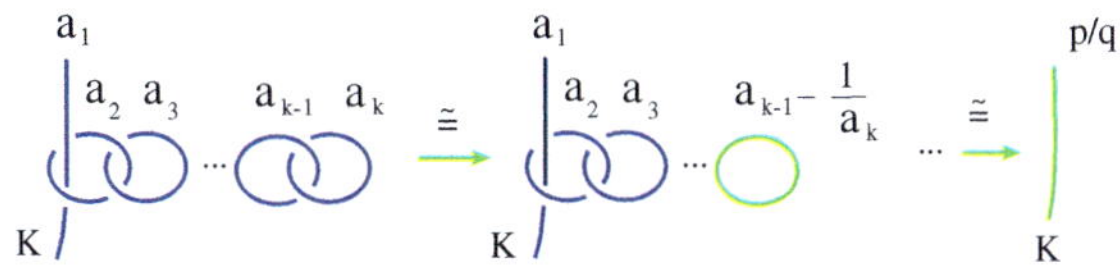

Figure 5.4

Exercise 5.4. *Let $K \subset S^3$ be a knot, and $K_n \subset S^3$ be a framed link consisting of n parallel copies of K (pushed off by the zero framing), each component assigned -1 framing. Show that as 3-manifolds we have the identification $K^{1/n} \approx K_n$ (Hint: use $1/n = [1, 2, 2, .., 2]$).*

5.2 From framed links to Heegaard diagrams

Given a 4-manifold $M^4 = M_\Lambda$, with $\Lambda = \{K_1^{r_1}, \dots, K_n^{r_n}\}$, how can we construct the Heegaard diagram of its boundary 3-manifold $Y_\Lambda^3 := \partial M_\Lambda$? First, note that if the complement $C(\Lambda) := S^3 - \cup_j N(K_j)$ of the tubular neighborhood of the framed link Λ is a 1-handlebody (i.e. B^3 with 1-handles), then $(C(\Lambda), K_1, \dots, K_n)$ gives a Heegaard decomposition of Y_Λ. To use this fact, we first remove solid pipes (i.e. properly imbedded thickened arcs) from $C(\Lambda)$ to turn it into a 1-handlebody, $C'(\Lambda) = C(\Lambda) - \cup N(I_j)$. To undo the damage we cause by removing pipes I_j from the complement, we attach 2-handles γ_j to the meridians of I_j's (Figure 5.5). Then the end result is a Heegaard decomposition, given by the 3-dimensional 1-handlebody $C'(\Lambda)$, along with simple disjoint simple closed curves $\{K_1, \dots, K_n, \gamma_1, \dots, \gamma_s\}$ lying on its boundary, signifying the 2-handles. Any Heegaard diagram can be described by a genus g surface F with two sets of disjoint simple closed curves α and β (compressing curves), each describing a solid handlebody. Here in our case $F = \partial C'(\Lambda)$ and $\beta = \{K_1, \dots, K_n, \gamma_1, \dots, \gamma_s\}$, and $\alpha = \{\alpha_1, \dots, \alpha_r\}$, $(r = n + s)$ are the obvious collection of curves on F compressing in $C'(\Lambda)$. For example, by applying this process to $\Lambda = \{K^2\}$, where K is the trefoil knot, we get Figure 5.6.

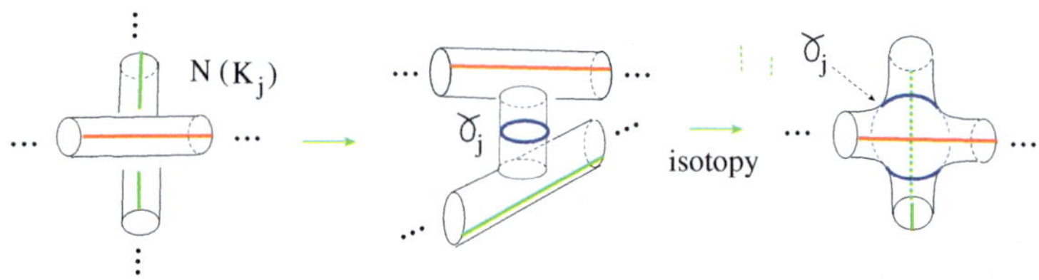

Figure 5.5: Adding solid pipes to the complement

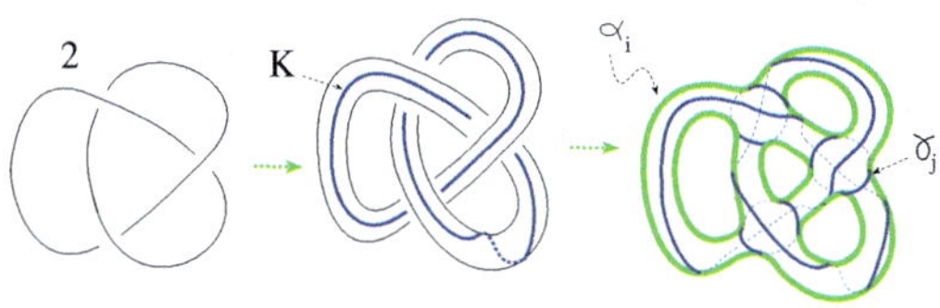

Figure 5.6: Constructing a Heegaard picture of $\partial(K^2)$

Exercise 5.5. *By considering moving pictures (and the above process), identify the complement of a properly imbedded 2-disk in B^4, with a single transverse self intersection, with the fishtail (Hint: Consider the isotopies of Figure 5.7.) Compare this with the self plumbing operation of Figure 2.9.*

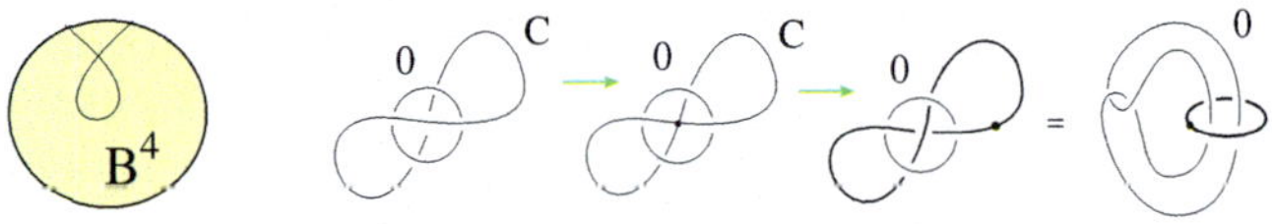

Figure 5.7

5.3 Gluing knot complements

Let $K, L \subset S^3$ be knots, and let $\phi : \partial N(K) \xrightarrow{\approx} \partial N(L)$ be a diffeomorphism, between the boundaries of the tubular neighborhoods of these knots. We want to find a 4-manifold $M_\varphi(K,L)$ whose boundary is the 3-manifold formed by gluing the knot complements:

$$\partial M_\varphi(K,L) \approx (S^3 - N(K)) \smile_\phi (S^3 - N(L))$$

We do this by the method of Section 3.2. First by using $\{\mu, \lambda\}$ coordinates, express

$$\phi = \begin{pmatrix} p & r \\ q & s \end{pmatrix}$$

When K and L are the unknots U, we get $\partial M_\varphi(U,U) = L(p,q)$ (Example 2.1). Therefore we can take the plumbing $M_\varphi(U,U) = C_{p,q}$ described in Figure 5.8, where $p/q = [a_1, a_2, \ldots, a_k]$ (e.g. [Ro]). Write $C = S^3 - N(U) \approx S^1 \times B^2$ so that $L(p,q) = C \smile_\phi C$,

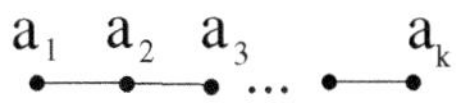

Figure 5.8: $C_{p,q}$

Next we will extend this process to any knots. Let γ and γ' be the dual circles of K and L in S^3, respectively. Let $N(\gamma)$ denote the tubular neighborhood of γ in $\partial(K^0)$. Notice that we have $\partial(K^0) - N(\gamma) \approx S^3 - N(K)$. We need to identify boundaries of two such knot complements with a cylinder $(S^1 \times S^1) \times [0,1]$ via ϕ, which is the boundary of the manifold obtained from the disjoint union $K^0 \sqcup C_{p,q} \sqcup L^0$, by identifying $N(\gamma)$ with one copy of C in $L(p,q)$, and identifying the other copy of C with $N(\gamma')$ (Figure 5.9):

$$M_\varphi(K,L) = K^0 \smile U \smile C_{p,q} \smile U' \smile L^0$$

.

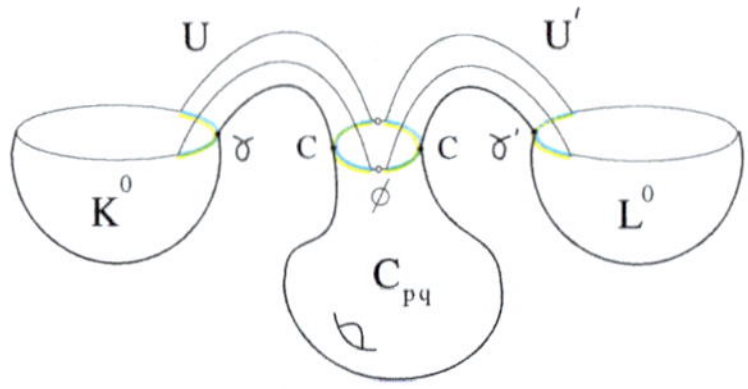

Figure 5.9: $M_\varphi(K, L)$

Exercise 5.6. *Show that Figure 5.10 gives a handlebody for $M_\phi(K, L)$.*

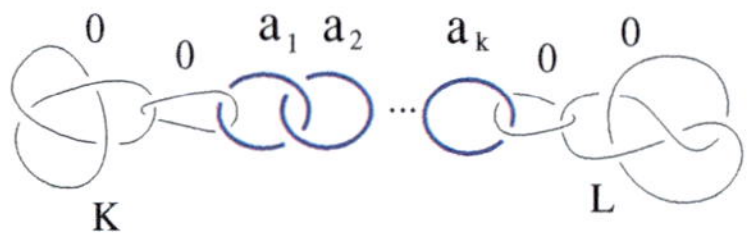

Figure 5.10

Exercise 5.7. *Show that Figure 5.11 gives $\partial M_\phi(K, L)$ when* $\phi = \begin{pmatrix} p & -1 \\ 1 & 0 \end{pmatrix}$.

Exercise 5.8. *By modifying the above construction, verify that Figure 5.12 also describes $\partial M_\phi(K, L)$ for some ϕ, where K is the trefoil knot, and L is the figure-eight knot.*

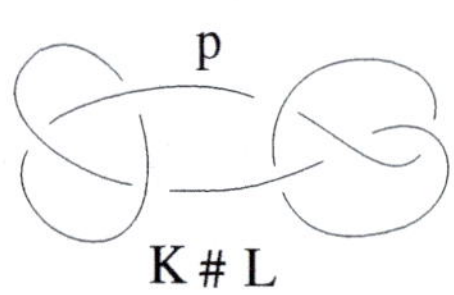

Figure 5.11

Figure 5.12

5.4 Carving 3-manifolds

By applying the carving principle of Section 1.1 to 3-manifolds, we can view any 1-handle of a 3-manifold as a carved out (complement of) properly imbedded unknotted thickened arc in B^3; for this we can introduce a 3-manifold analogue of "circle with dot" notation. We denote this by a pair of thick dots and a dotted line connecting them, signifying the 1-handle membrane (any arc intersecting this is going over the 1-handle) (Figure 5.13).

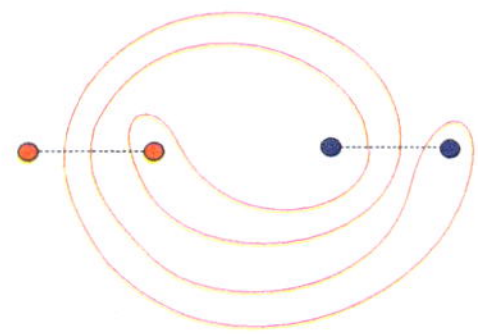

Figure 5.13

By this we can give a 3-dimensional analogue of the "carving ribbons from B^4" operation discussed in Section 1.4. Here we are carving knotted arcs in B^3. The analogy is completely similar. For example, in Figure 5.14 we start with S^0 at top, then by moving pictures introduce a 1- and 2- canceling handle pair, and end up with a pair of 1-handles (represented by $S^0 \sqcup S^0$) and a 2- handle (the view from the bottom). In this case we get the manifold of Figure 5.13.

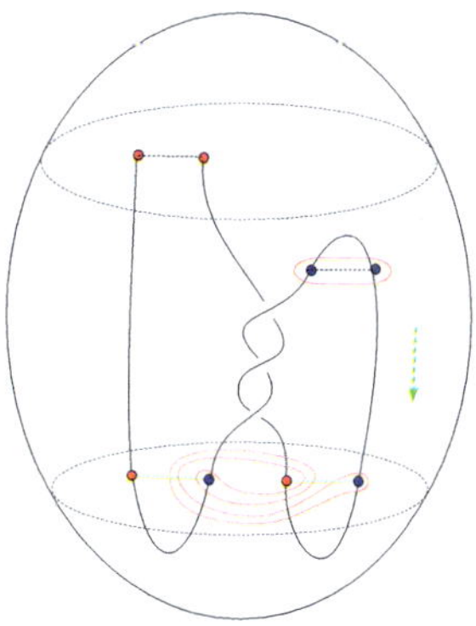

Figure 5.14

5.5 Rohlin invariant

Definition 5.2. *Let X^4 be a closed smooth manifold. A homology class $\alpha \in H_2(X;\mathbb{Z})$ is called characteristic if for all $x \in H_2(X;\mathbb{Z})$ the following congruence holds:*

$$\alpha.x = x.x \ (mod\, 2) \tag{5.2}$$

(5.2) is called the "Wu relation", it says that α intersects odd classes odd times, and even classes even times. The mod 2 reduction of any characteristic classes corresponds to the second Steifel–Whitney class $w_2(X) \in H^2(X;\mathbb{Z}_2)$ under the Poincare duality (e.g.[MS]). Also any characteristic class α satisfies the following congruence ([Bl], [MH]):

$$\sigma(M) = \alpha.\alpha \ \ (mod\, 8) \tag{5.3}$$

Let Y^3 be an orientable closed 3-manifod with $H^1(Y;\mathbb{Z}_2) = 0$. We know that every 3-manifold bounds an oriented 4-manifold, in fact $Y^3 = \partial M_\Lambda$ for some framed link $\Lambda = \{K_1^{r_1}, \dots, K_n^{r_n}\}$ (Chapter 5). In fact we can assume that Λ is an even framed link, i.e. all $r_1, \dots, r_n$ are even integers. This can be seen by first choosing a framed knot

$$K = \lambda_1 K_1^{r_1} + \ldots + \lambda_n K_n^{r_n} \tag{5.4}$$

representing a homology class $\alpha \in H_2(M_\Lambda)$ (as discussed in Section 1.6) satisfying:

$$\lambda_1 a_{1j} + \ldots + \lambda_n a_{nj} = r_j \ \ (\text{mod } 2) \quad , \text{ for all } j = 1, \dots, n \tag{5.5}$$

where $a_{ij} = L(K_i, K_j)$ (so $a_{jj} = r_j$). These equations can easily be solved over $\mathbb{Z}_2$ (why), but solutions are not unique. For example, when Λ is an even link $K = 0$ is a solution.

Exercise 5.9. *Show that $\alpha = [K]$, above, satisfies the Wu relation, hence is characteristic.*

Exercise 5.10. *Show that, given Y, we can arrange Λ with $\partial M_\Lambda = Y$, so that $\alpha = [K]$, where K is the connected sums of trefoil knots. Hint: Write a Seifert surface F of K as a disk with handles, then by blowing up M respecting the Wu property (± 1 blowup circles should always link K odd times) make the handles unknotted and unlinked from each other, and then by blowings up across three stands, as in the next to last step of Figure 2.18, change the twisting on each handle by 2, so as to end up with 1 twisting. After this, show that by blowups we can transform K to a 1-framed unknot, and by blowing it down we transform Λ to an even framed link.*

So, by Exercise 5.10 we can make K the empty knot, i.e. we can choose the bounding 4-manifold M_Λ to be even, this means that M_Λ has a spin structure (i.e. $w_2(M_\Lambda) = 0$) inducing a spin structure on the boundary (Y, s), which in turn means a trivialization of the tangent bundle TY. A theorem of Rohlin states that the signature of a closed spin 4-manifold satisfies the following congruence ([Roh], [K3]).

Theorem 5.3. *([Roh]) If X^4 is a closed smooth spin 4-manifold , then $\sigma(X) = 0$ (mod 16).*

Therefore the following is a well defined definition:

Definition 5.4. *(Rohlin invariant) Let (Y^3, s) be a spin 3-manifold, let (X^4, s) be a spin 4-manifold bounding (Y, s), then define the Rohlin invariant of (Y, s) to be:*

$$\mu(Y, s) = \sigma(X) \ (mod\ 16)$$

Note that when Y is a homology sphere, it has a unique spin structure, so from the definition we can drop the dependency to spin structure s and write $\mu(Y, s) = \mu(Y)$. For example, by using Section 2.4 we compute $\mu(\Sigma(2,3,5)) = \mu(\Sigma(2,3,7)) = -8$.

Exercise 5.11. *Prove a stronger version of (5.3) with mod (16) congruence, by using Theorem 5.3. (Hint: represent α by an imbedded surface $\Sigma \subset X^4$, then after blowing up X make the normal bundle $N(\Sigma)$ of Σ trivial $\Sigma \times D^2$, then replace $N(\Sigma)$ by $Y^3 \times S^1$, where Y is a solid handlebody which Σ bounds. Then record the change of the signatures in the blowing up and regluing processes.)*

Exercise 5.12. *Prove that every closed smooth orientable 3-manifold Y^3 is parallelizable, and imbeds into $\mathbb{R}^5$. (Hint: First pick an even X^4 as above, consisting of only one 0-handle and 2-handles, which Y bounds. Then the double of X is $\#_k S^2 \times S^2$ which imbeds into $\mathbb{R}^5$ as the boundary of $\#_k S^2 \times B^3$.)*

Chapter 6

Operations

6.1 Gluck twisting

It is known that any diffeomorphism $S^2 \times S^1 \xrightarrow{\approx} S^2 \times S^1$ is either isotopic to the identity, or isotopic to the map $\varphi : S^2 \times S^1 \to S^2 \times S^1$ defined by (e.g. [Wa2])

$$\varphi(x,t) = (\varphi_t(x), t)$$

where $S^1 \ni t \mapsto \varphi_t \in SO(3)$ is the nontrivial element of $\pi_1 SO(3) = \mathbb{Z}_2$. Let X be a smooth 4-manifold, and $S \subset X$ be a copy of S^2 imbedded with trivial normal bundle $S^2 \times B^2 \subset X$ (i.e. homological self intersection is zero: $S.S = 0$). We call the operation (Figure 6.1)

$$X \mapsto X_S := (X - S^2 \times B^2) \smile_{\varphi} (S^2 \times B^2)$$

Gluck twisting. The best way to understand how a Gluck twisting operation alters the handles of X is by drawing a handlebody of X as handles attached to the top of this particular $S^2 \times B^2$ (an unknot with 0-framing) (Figure 6.1). Then the Gluck twisting corresponds to the operation of putting one twist to all the strands going through the 0-framed circle:

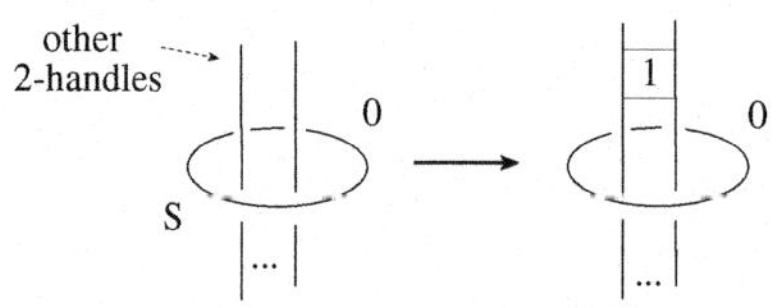

Figure 6.1: Gluck twisting

Exercise 6.1. *Show that the Gluck twisting operation is equivalent to one of the following operations:*

(i) The operation described in Figure 6.2 (where n, m are arbitrary).

(ii) The zero and dot exchange operation in Figure 6.3.

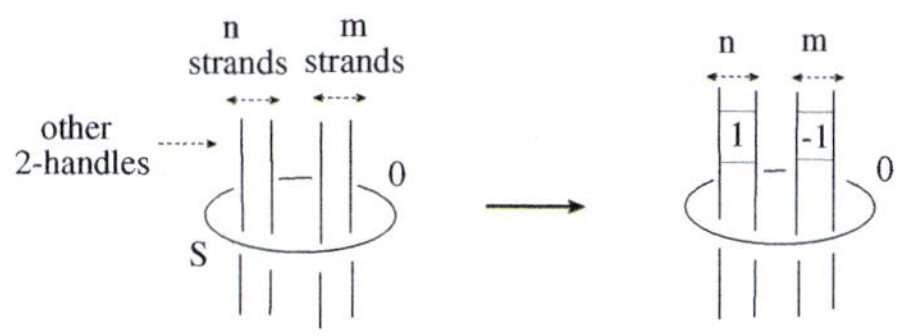

Figure 6.2

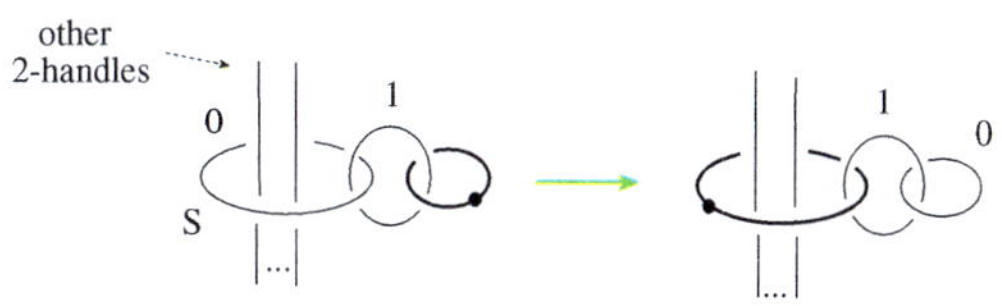

Figure 6.3

If X is simply connected and $S \subset X$ is null homologous, then it is clear that the operation $X \mapsto X_S$ does not change the homology groups and the intersection form of X, and also keeps X simply connected, hence X_S is h-cobordant to X and hence it is homeomorphic to X by [F]. Also, when the intersection form of X is odd, the Gluck twisting does not change the smooth structure of X under some conditions (e.g. [AY1]).

Theorem 6.1. *Let X be a simply connected smooth 4-manifold, and $S \subset X$ be a 2-sphere with trivial normal bundle, then $X_S \approx X$, provided that in $X - S$ there is a 2-dimensional spherical homology class with odd self intersection.*

Proof. Let $X^{(i)}$ denote the union of handles index $\le i$. Let $\alpha \in H_2(X, \mathbb{Z})$ be a spherical homology class represented by $f : S^2 \to X - S$ with $\alpha.\alpha$ odd. First we claim α can be represented by a 2-handle in $X - S$ whose attaching circle is null homotopic in $X^{(1)}$. This is because after a homotopy we may assume that f is an immersion, and its image is contained in $X^{(2)}$ (since the cocores of a 3-handles are codimension 3). Let $p_1, p_2, \dots, p_n$ be the transverse intersection points of $f(S^2)$ and the cocores of 2-handles of X. Let D_i $(1 \le i \le n)$ be a small 2-disk neighborhoods of $f^{-1}(p_i)$ in S^2. We may assume that each $f(D_i)$ is a core of a 2-handle of X and that $f(S^2 - \coprod_{i=1}^n D_i)$ is contained in $X^{(1)}$. For each i, take an arc $\gamma_i \subset S^2$ connecting D_i and D_{i+1}, with $\nu(\gamma_i)$ its tubular neighborhood, so that $\Sigma := (\cup_{i=1}^n D_i) \cup (\cup_{i=1}^{n-1} \nu(\gamma_i))$ is a 2-disk in S^2. We can assume $f|_{\nu(\gamma_i)}$ are embeddings into $\partial X^{(1)}$. This gives a 2-handle h^2 of X representing α, whose attaching circle $f(\partial\Sigma)$ is null homotopic in $X^{(1)}$ (introduce a canceling 2/3-handle pair and slide its 2-handle over the 2-handles of X until it becomes $f(\partial\Sigma)$).

The Gluck twisting operation corresponds to zero and dot exchanges between the zero framed 2-handle S and the dotted circle T, as in Figure 6.4. Let K be the attaching circle of h. By first sliding the middle 1-framed handle over K we can make its framing even, and then by sliding over T make it zero, and finally by using the "null-homotopy" assumption we can slide it over T, as shown in Figure 6.5, to make it unknotted and then get the third picture of Figure 6.4. Now it is clear that the zero and dot exchange between S and T does not change the diffeomorphism type, i.e. $X_S \approx X$. □

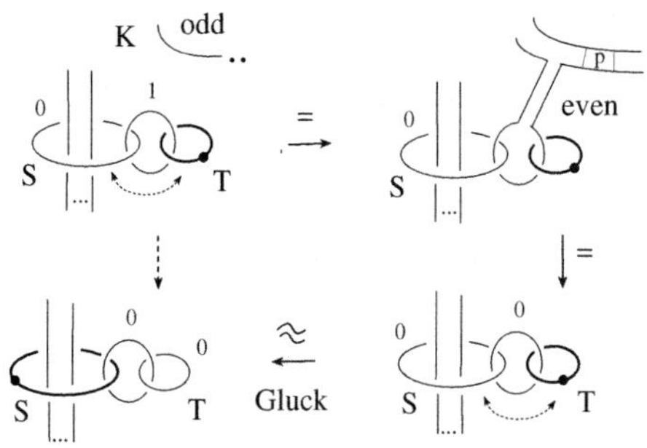

Figure 6.4

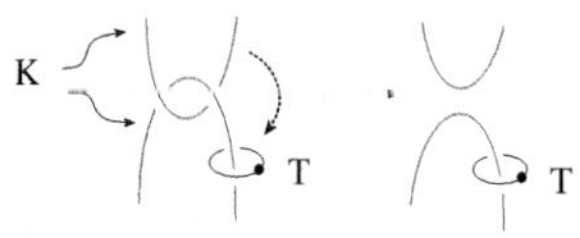

Figure 6.5

6.2 Blowing down ribbons

Let us decompose the 2-sphere as a union of upper and lower hemispheres $S^2 = D^2_- \cup_\partial D^2_+$, then since the diffeomorphism $t \mapsto \varphi_t(x)$ is an isotopy of (doner kebab) rotation of S^2 by the north-south-pole axis, along the circle; we can restrict it to pieces $S^1 \times D_\pm$

$$\text{Diff}(S^1 \times S^2) \to \text{Diff}(S^1 \times D^\pm)$$

Also we can arrange any imbedding $S^2 \subset X^4$ so that D_+ lies in the top 4-handle B^4 as the trivial disk, and D_- as a slice disk in $X_0 := X - B^4$ bounding the unknot ∂D_+. Then we can perform the Gluck twisting operation on two parts separately and take their union: (1) Twist X_0 along D_-, and (2) Twist B^4 along D_+ which gives back B^4 (Figure 6.6).

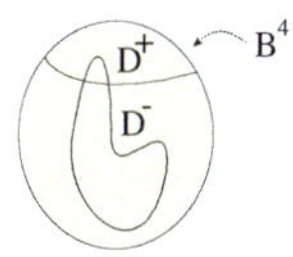

Figure 6.6

It is particularly interesting when D_- is a ribbon in X_0, since we know how to draw pictures of ribbons (Section 1.4). For example, a typical ribbon in X_0 will look like Figure 6.7 (plus the handles of X_0, not drawn in the figure), then the Gluck twisting operation to X_0 along D_- corresponds to attaching a ± 1 framed 2-handle to the trivially linking circle C of K (why?). When $X_0 = B^4$ this gives a contractible manifold.

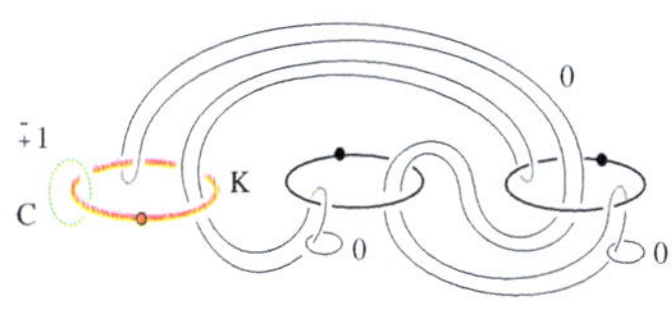

Figure 6.7: Ribbon complement $B^4 - D_-$

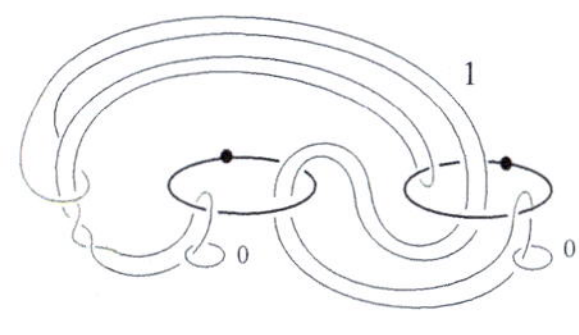

Figure 6.8: B^4 twisted along D_-

Exercise 6.2. *Show that Figure 6.8 is the picture of S^4 Gluck twisted by the 2-sphere, which is the double of the ribbon surface described in Figure 6.7. Conclude that Gluck twisting S^4 along a 2-knot $S^2 \subset S^4$ which is the double of a ribbon, gives S^4 (Hint: By sliding 2-handles over the small 0-framed linking circles move them to trivial position).*

6.3 Logarithmic transform

Given a smooth 4-manifold X^4 and an imbedding $T^2 \times B^2 \subset X^4$, the operation of removing this $T^2 \times B^2$ and then gluing it back by a nontrivial diffeomorphism of its boundary $\varphi : T^3 \to T^3$ is called T^2-surgery operation. It is known that any closed smooth 4-manifold can be obtained from a connected sum of number of copies of $S^1 \times S^3$ and $\pm\mathbb{CP}^2$ by a sequence of these operations, [I]. A special case of this operation is called *p-log transform operation*, when the map $\phi_p : T^2 \times S^1 \to T^2 \times S^1$ $(p \in \mathbb{Z})$ corresponds to:

$$\phi_p = \begin{pmatrix} 1 & 0 & 0 \\ 0 & 0 & 1 \\ 0 & -1 & p \end{pmatrix}$$

This operation will be denoted by $X \mapsto X_p$, and it is called a *logarithmic transform of order* p, because the degree of the composition of the maps below has order p.

$$S^1 \xrightarrow{\text{inc}} T^2 \times S^1 \xrightarrow{\phi_p} T^2 \times S^1 \xrightarrow{\text{proj}} S^1 \tag{6.1}$$

In the literature any ϕ with this property is called a *log*-transformation; for simplicity, here we have adapt this more restricted definition. By the techniques of Section 3.3 we can draw a handlebody picture of this operation (e.g. [A23], [AY2], [GS]). The recipe from Section 3.3 says that for a given imbedding $T^2 \times D^2 \subset X$, write $X = T^2 \times D^2 \smile$ (other handles), then carry the other handles by the inverse diffeomorphism ϕ_p^{-1} to the top of $T^2 \times D^2$,

$$\phi_p^{-1} = \begin{pmatrix} 1 & 0 & 0 \\ 0 & p & -1 \\ 0 & 1 & 0 \end{pmatrix}$$

Figure 6.9 describes this handlebody operation with pictures (check that the three circles of T^3 are mapped by ϕ_p^{-1}). This figure gives a picture recipe of how to modify a 4-manifold handlebody, containing a framed torus $T^2 \times D^2$, in order to get the handlebody of the p-log transformed 4-manifold along this torus. For example, Figure 6.9 describes how the linking loop B is changed by this operation. To emphasize the core torus, this operation is referred as the p-log transformation operation done along $(S^1 \times S^1, S^1)$. In short when (a, b, c) are the homology generators of $T^2 \times S^1$, we call it a $(a \times b, c)$ operation. The formula (13.45) computes the change of Seiberg–Witten invariant under a $p - log$ transformation $X \mapsto X_p$, under some conditions (Chapter 13).

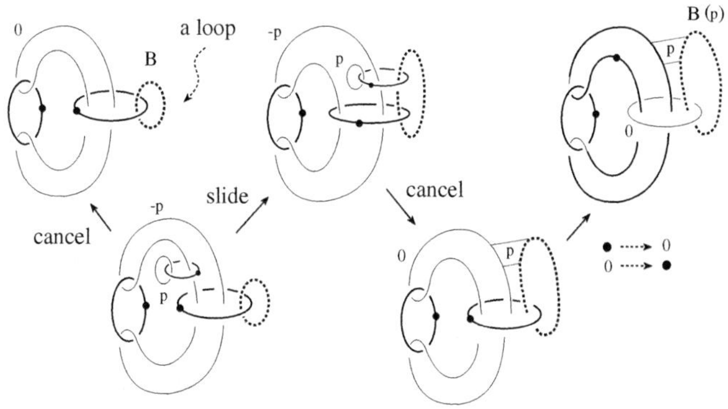

Figure 6.9: p-log transform operation $\phi_p^{-1} : X \mapsto X_p$

6.4 Luttinger surgery

In case (X^4, ω) is a symplectic manifold (Chapter 8) and $T^2 \subset X$ is a Lagrangian torus (i.e. $\omega|_{T^2} = 0$), there is a special T^2-surgery operation called a *Luttinger surgery* ([Lu],[EP],[ADK1]), which preserves the symplectic structure of X. The Luttinger surgery $X \mapsto X_{m,n}$ corresponds to the diffeomorphism $\phi(x,y,\theta) = (x + m\theta, y + n\theta, \theta)$,

$$\varphi_{m,n} = \begin{pmatrix} 1 & 0 & m \\ 0 & 1 & n \\ 0 & 0 & 1 \end{pmatrix}$$

Since T^2 is Lagrangian, in an open neighborhood T^2 the symplectic structure of (X, ω) can be identified by the symplectic structure $T^*(T^2) \cong T^2 \times \mathbb{R}^2$ (e.g. [MSa]) and

$$\omega = d\left(r[(cos2\pi\theta)dx + (sin2\pi\theta)dy]\right)$$

where $[(x,y),(r,\theta)]$ are the coordinates of $T^2 \times \mathbb{R}^2$, and the boundary of a closed tubular neighborhood $T^2 \subset N$ is represented by the hypersurface $r = 1$. Clearly on ∂N we have $\phi_{m,n}^* \omega = \omega$, so the symplectic structure of N extends to a symplectic structure on $X_{m,n}$. As before, to perform this operation we carry other handles to the top of $T^2 \times D^2$ by the diffeomorphism $\varphi_{m,n}^{-1} = \varphi_{-m,-n}$, as in the example of Figure 6.10. In the literature, sometimes Luttinger surgery just refers to the operation $X \mapsto X_{0,\pm 1}$ and it is related to the (± 1)-log transformation by the following exercise.

Exercise 6.3. *For simplicity, abbreviate* $\varphi_\pm := \varphi_{0,\pm 1}$ *and* $\phi_\pm := \phi_{\pm 1}$. *Then show that:* *(a)* $\varphi_\pm = \phi_\pm \circ \gamma_\pm$, *where* $\gamma_\pm$ *is the restriction of the self diffeomorphism of* $T^2 \times D^2$, *e.g.*

$$\phi_+ = \begin{pmatrix} 1 & 0 & 0 \\ 0 & 0 & 1 \\ 0 & -1 & 1 \end{pmatrix} = \begin{pmatrix} 1 & 0 & 0 \\ 0 & 1 & 1 \\ 0 & 0 & 1 \end{pmatrix} \begin{pmatrix} 1 & 0 & 0 \\ 0 & 1 & 0 \\ 0 & -1 & 1 \end{pmatrix} = \varphi_+ \circ \gamma_-$$

More specifically, $\gamma_\pm$ *is the map* $(x, y, \theta) \mapsto (x, y, \pm y + \theta)$; *(b)* $\varphi_+^{-1} = \varphi_-$ *and* $\gamma_+^{-1} = \gamma_-$, *(c) There is a diffeomorphism* $X_{\pm 1} \approx X_{0,\pm 1}$, *in particular if* X *is symplectic and* $T^2 \subset X$ *Lagrangian, then log transformation* $X \mapsto X_{\pm 1}$ *produces a symplectic manifold.*

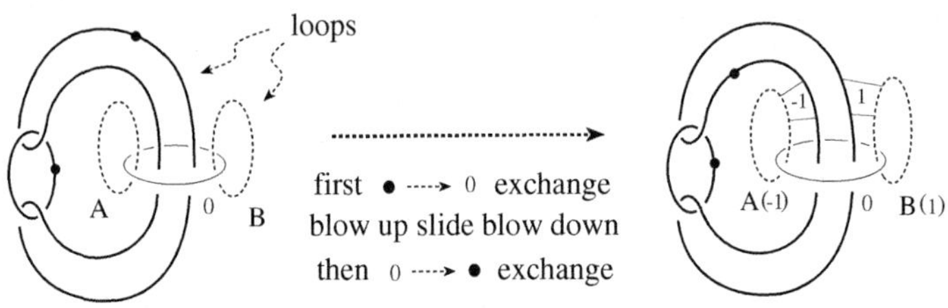

Figure 6.10: Luttinger operation $\varphi_+^{-1} : X \mapsto X_{0,1}$

Remark 6.2. *By comparing Figures 6.10 and 6.2, we see that the Luttinger surgery operation is a certain generalization of the Gluck twisting operation (where the* 0*-framed* 2*-handle need not represent an imbedded* 2*-sphere).*

Exercise 6.4. *Verify that Figure 6.11 describes* 1*-log transformation operation* ϕ_+.

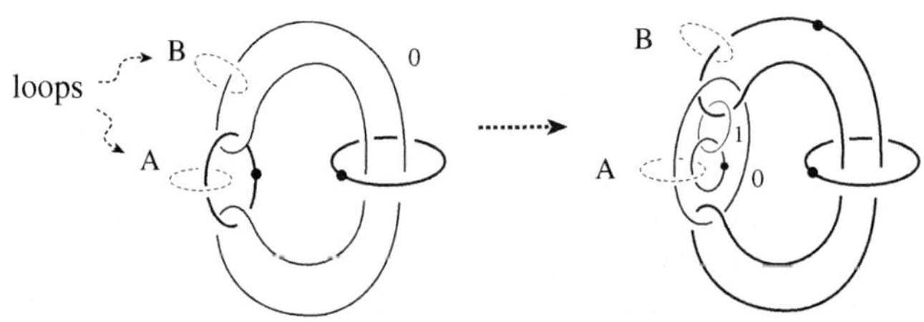

Figure 6.11: 1-log transformation operation $\phi_+^{-1} : X \to X_1$

6.5 Knot surgery

Let X be a smooth 4-manifold, $T^2 \times B^2 \subset X$ an imbedded torus with trivial normal bundle, $K \subset S^3$ a knot and $N(K)$ its tubular neighborhood. The operation of replacing $T^2 \times B^2$ with $(S^3 - N(K)) \times S^1$, so that the meridian $p \times \partial B^2$ of the torus coincides with the longitude of K, is called a Fintushel–Stern *knot surgery operation* ([FS1]),

$$X \rightsquigarrow X_K = (X - T^2 \times B^2) \cup [S^3 - N(K)] \times S^1$$

A handlebody of this operation was constructed in [A6]. Notice $(S^3 - N(K)) \times S^1$ is obtained by identifying two ends of $(S^3 - N(K)) \times [0,1]$. Since 4-manifolds are determined by their 1- and 2- handles, it suffices to draw $(B^3 - N(K_0)) \times [0,1]$ with its ends identified, where $K_0 \subset B^3$ is a properly imbedded arc with the knot K tied onto it (the rest is a 3-handle). The second picture of Figure 6.12 describes $(B^3 - N(K_0)) \times [0,1]$; identifying its two ends (up to 3-handles) corresponds to attaching a new 1-handle, and attaching 2-handles to the knots obtained by connected summing (over the new 1-handle) of the core circles of the 1-handles of two boundary components of $(B^3 - N(K_0)) \times [0,1]$, as shown in the third picture of Figure 6.12 (this is the $M(f)$ technique of Section 3.2, here we can use Section 5.2 for locating the 1-handle core circles of 3-manifolds).

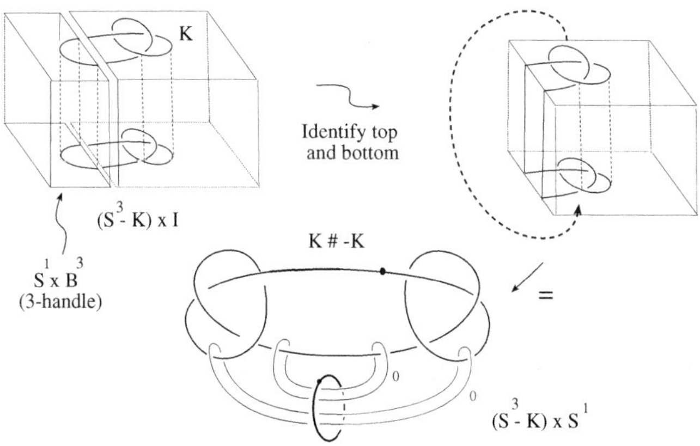

Figure 6.12: $(S^3 - K) \times S^1$, where K is the trefoil knot

Summarizing: Figure 6.13 gives a recipe of how to modify a 4-manifold handlebody, containing a framed torus $T^2 \times B^2$ to get the handlebody of the knot surgered 4-manifold along this torus, by using a knot K (in the figure K is taken to be a trefoil knot). Figure 6.13 describes how the linking loops a, b, c change by this operation. For example in this figure, if we attach a -1 framed 2-handle to either (both) of the loops a, c we get a fishtail (cusp) on the left, and the knot surgered fishtail (cusp) on the right (see Section 3.1).

Figure 6.14 describes the slow evolution of the two pictures in Figure 6.13, from left to right (i.e. description of the boundary diffeomorphism). In Figure 6.14, first we introduce a canceling 2/3-handle pair, then replace the dot with zero in the middle 1-handle turning it into a 2-handle; we then do the indicated handle slides, and put back the dot on the resulting ribbon 2-handle. To go in reverse, from right to left, we perform the same operation to $(T^2 \times B^2)_K$, this time by using the dotted arrows in Figure 6.13.

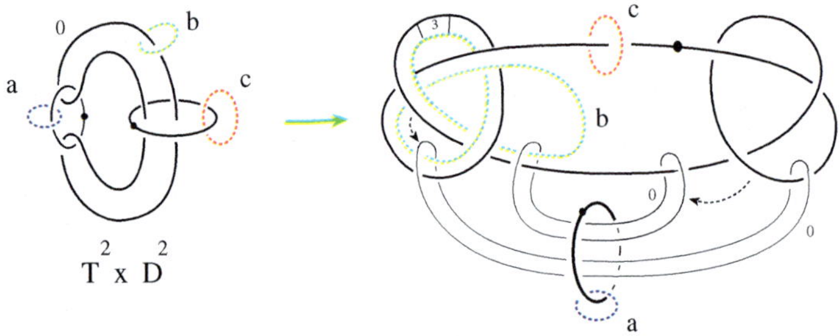

Figure 6.13: The operation $T^2 \times B^2 \mapsto (T^2 \times B^2)_K$

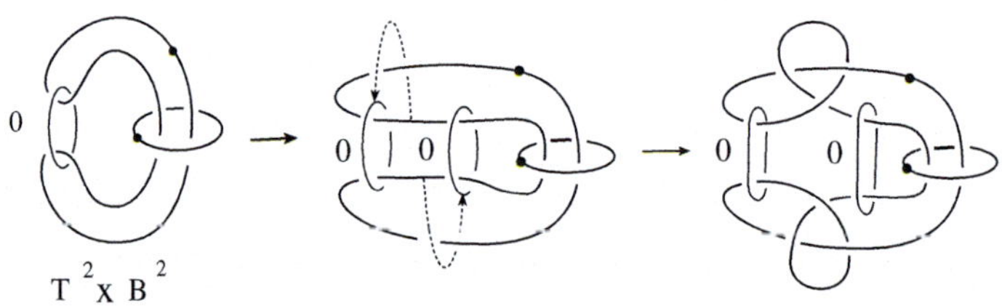

Figure 6.14

As an example, let us take the elliptic surface $E(1)$ (see Chapter 7) and apply a knot surgery operation to the cusp inside. The first picture of Figure 6.15 is a handlebody of $E(1)$ (from [A7], which reader can identify with Figure 7.5); the second picture of this figure is $E(1)_K$, which is obtained by applying the the previous algorithm.

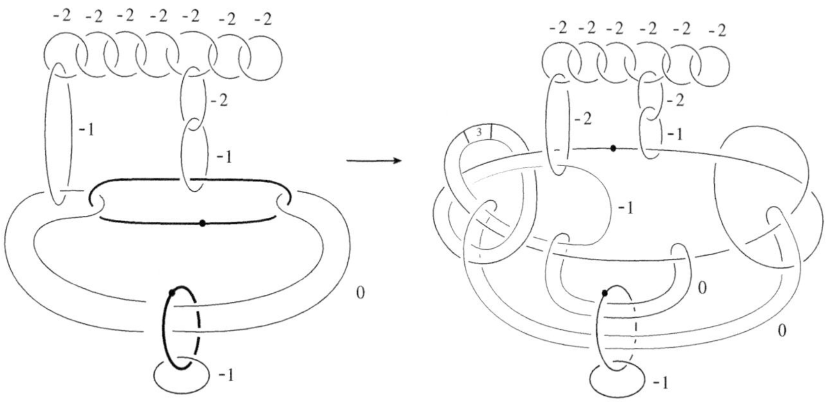

Figure 6.15

Theorem 6.3. *([FS1]) Let X be a closed smooth 4-manifold with $b_2^+(X) > 1$, and let $T \subset X$ be a smoothly imbedded torus, which is contained in a node neighborhood, and $0 \neq [T] \in H_2(X;\mathbb{Z})$, and X and $X - T$ are simply connected. Then X_K is homeomorphic to X and the Seiberg–Witten invariant (Chapter 13) of X_K is given by*

$$SW_{X_K} = SW_X \,.\, \Delta_K(t^2)$$

where $\Delta_K(t)$ is the (symmetrized) Alexander polynomial of the knot $K \subset S^3$ (see Section 2.7), and $t = t^T$ (13.33).

Proofs of this theorem will be discussed in Sections 13.11 and 13.12.

Remark 6.4. *The handlebody construction technique of Figure 6.12 is an essential tool in the solution of the celebrated Scharlemann problem [A19].*

6.6 Rational blowdowns

The *rational blowing down* operation was introduced by Fintushel and Stern as a useful tool in gauge theory to construct exotic manifolds ([FS2], [P]). More recently J. Park has used this technique to find new small closed exotic 4-manifolds (e.g. [P2]). Let $C \subset X^4$ be a negative definite plumbing in a smooth 4-manifold, such that $\partial C \approx \partial B$ for some rational ball B. Call $X' = X - int(C)$, then the operation of replacing C by B

$$X \mapsto X_{(p)} := X' \smile B$$

is called a *rational blowing down* operation. An important special case is when $C_{p,q}$ is the 4-manifold given by the plumbing Figure 6.16 ([P]), where each $b_i \geq 2$, and

$$p^2/pq - 1 = [b_k, b_{k-1}, .., b_1]$$

Figure 6.16: $C_{p,q}$

For $p > q \geq 1$, relatively prime, the boundary is the lens space: $\partial C_{p,q} = L(p^2, pq - 1)$, which by [CH] bounds a rational ball. In [LM] it was shown that in fact $L(p^2, pq - 1)$ bounds the very simple rational ball $B_{p,q}$ shown in Figure 6.17.

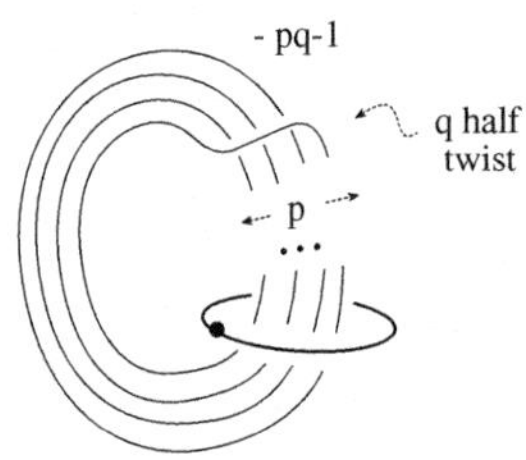

Figure 6.17: $B_{p,q}$

Exercise 6.5. *Prove that $C_{p,q}$ is a Stein manifold (Theorem 8.11), and there is a diffeomorphism $\partial C_{p,q} \approx \partial B_{p,q}$; describe this diffeomorphism (see [Wil], [Go]).*

In the special case of $q = 1$ the manifold $C_p := C_{p,1}$ is the plumbing of a chain of 2-spheres $\{u_0, u_1, \dots, u_{p-1}\}$ with Euler numbers $-p-2, -2, -2, \dots, -2$. Figure 6.18 demonstrates a concrete diffeomorphism $\partial B_{p,1} \approx \partial C_p$, which allows us to perform the blowing down operation concretely on handlebodies ([GS]).

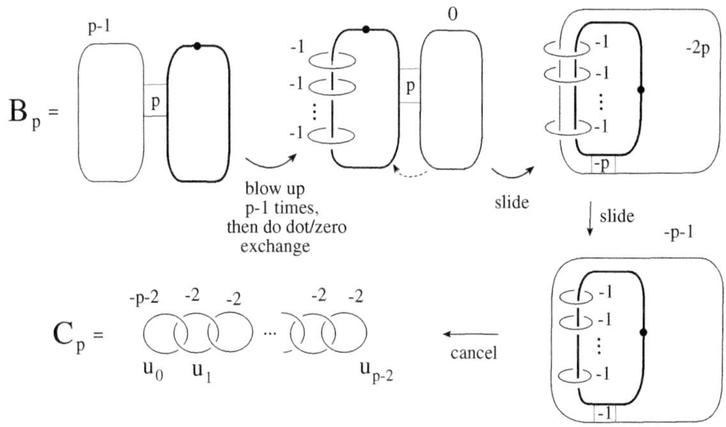

Figure 6.18: Describing a diffeomorphism $\partial B_p \approx \partial C_p$

Exercise 6.6. *Call $L := L(p^2, p-1)$ and $B_p = B_{p,1}$. Let δ to the dual circle of the 2-handle of B_p, and let γ_j denote the dual circles of the 2-spheres u_j of C_p, $0 \le j \le p-2$. We can consider these dual circles as elements of the 2-dimensional cohomology or (by Poincare duality) 2-dimensional relative homology classes of B_p and C_p. Show that:*

(a) In $\pi_1(L) = \mathbb{Z}_{p^2}$ the loop γ_j represents $[1 + j(p+1)]\gamma_0$, where γ_0 is a generator.

(b) The map induced by restriction $H^2(B_p) \to H^2(L)$ is the "multiplication by p" map $\mathbb{Z}_p \to \mathbb{Z}_{p^2}$, sending $\delta \longmapsto p\gamma_0$ (Hint: trace the dual circles in Figure 6.18).

Theorem 6.5. *([FS2]) Let $k \in H^2(X_{(p)}; \mathbb{Z})$ be a characteristic class (Section 5.5) then the restriction $k|_{X'}$ lifts to a characteristic class $\bar{k} \in H^2(X; \mathbb{Z})$. Furthermore, for any lifting $\bar{k}$ of k with $d(\bar{k}) \ge 0$ (see 13.24) we have:*

$$SW_{X_{(p)}}(k) = SW_X(\bar{k}) \tag{6.2}$$

Proof. : Let $i^*(k|_{X'}) = k_L \in H^2(L)$ be the restriction induced by inclusion. Since the map $j^* : H^2(C_p) \to H^2(L)$, induced by inclusion, is onto, we can always choose a class $k_{C_p} \in H^2(C_p)$ with $j^*(k_{C_p}) = k_L$. Then $\bar{k} = k|_{X'} \oplus k_{C_p}$ defines a class in $H^2(X)$ by the exactness of the Mayer–Vietoris sequence

$$0 \to H^2(X) \to H^2(X') \oplus H^2(C_p) \xrightarrow{i^*-j^*} H^2(L)$$

We need to choose k_{C_p} carefully so that $\bar{k}$ is characteristic in X, i.e. it intersects even classes even times and odd classes odd times. Next we show how to do this. By Exercise 6.6 we have $k_L = m(p\gamma_0)$ for some m. Since B_p is an even manifold when p is odd, and since k is characteristic, m and p must have different parity. It is easy to see that

$$(p-1)\bar{\mathbb{CP}}^2 = \bar{B}_p \smile_\partial C_p$$

Let $E_1, .., E_{p-1}$ be the 2-sphere generators of $(p-1)\bar{\mathbb{CP}}^2 = \bar{\mathbb{CP}}^2\#..\#\bar{\mathbb{CP}}^2$, then we see that

$$u_j = \begin{cases} -2E_1 - E_2 - .. - E_{p-1} & \text{if } j = 0 \\ E_j - E_{j+1} & \text{if } j \geq 1 \end{cases}$$

Now choose $k_{C_p} = -E'_1 - .. - E'_{(m+p-1)/2} + E'_{(m+p+1)/2} + .. + E'_{p-1} \in H^2(C_p)$

where $E'_j = E_j|_{C_p} = \sum(E_j.u_j)u_j$, i.e $E'_1 = 2u_0 - u_1$, $E'_2 = u_0 + u_1 - u_2$ .. etc. Furthermore, by Exercise 6.6 *(b)* each E'_j restricts to $-p\gamma_0 \in H^2(L)$ hence k_{C_p} restricts to $m(p\gamma_0)$.

Since $\bar{k}^2 = k^2 + k_{C_p}^2 = k^2 - (p-1)$, the dimensions of Seiberg–Witten moduli spaces coincide $d_X(\bar{k}) = d_{X_{(p)}}(k)$ (13.24). For the part $SW_{X_{(p)}}(k) = SW_X(\tilde{k})$, see [FS3] □

The rational blowing down process $X \rightsquigarrow X_{(p)}$ reduces $b_2(X)$, so it is used to obtain smaller exotic manifolds from larger ones. For this we need to first locate $C_p \subset X$, and find a characteristic class $\bar{k}$ of X, which descends to characteristic classes k of X.

Remark 6.6. *$C_p \subset X$ is called a taut imbedding if each basic class $\bar{k}$ of X satisfies $|\bar{k}.u_0| \leq p$, and $\bar{k}.u_i = 0$ for all other $i \geq 1$ (recall the adjunction inequality Section 13.17). If $X \mapsto X_{(p)}$ is a rational blowdown of a simply connected manifold along a taut imbedding, then the lifting $\bar{k}$ of any basic class k of X_p satisfies $|\bar{k}.u_0| = p$ ([FS2]).*

Remark 6.7. *(Relating p-log transform to a rational blowdown [FS2]) Figure 6.18 shows that the rational blowdown operation is just the operation of Figure 6.19. As a corollary, we can check the claim of [FS2] that, in presence of a "vanishing cycle" (the -1 curve in the first picture of Figure 6.20) the p-log transformation (Section 6.3) of X^4 is equivalent to a rational blowing down operation performed to $X\#(p-1)\mathbb{CP}^2$. This is explained in Figure 6.20.*

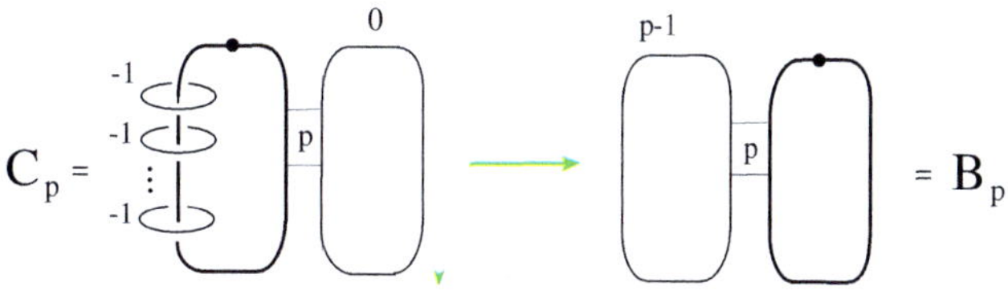

Figure 6.19: Rational blowdown

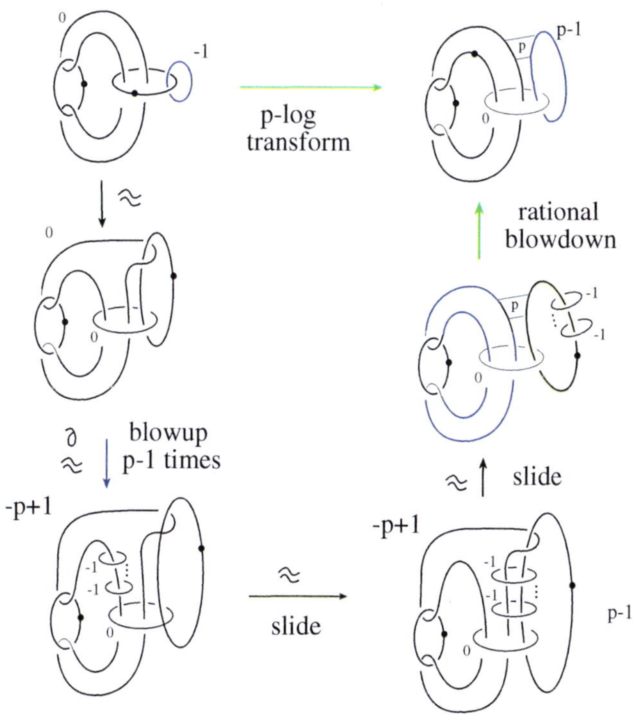

Figure 6.20

Chapter 7

Lefschetz fibrations

Let $\pi : X \to \Sigma$ be a map from an oriented 4-manifold to a surface, we say $p \in X$ is a *Lefschetz singularity* of π, if we can find charts $(\mathbb{C}^2, 0) \hookrightarrow (X, p)$ and $\mathbb{C} \hookrightarrow \Sigma$, on which π is given by $(z, w) \mapsto zw$. A *Lefschetz fibration* (LF in short) is a map $\pi : X \to \Sigma$ which is a submersion in the complement of a finitely many Lefschetz singularities $\mathcal{P}$, and an injection on $\mathcal{P}$. Here we will only consider the case $\Sigma = S^2$ (and later $\Sigma = D^2$). The set of finite points $\pi(\mathcal{P}) = \mathcal{C}$ is called the *singular values* of π. Let $\mathcal{C} = \{p_1, .., p_k\}$. Clearly, over $S^2 - \mathcal{C}$ the map π is a fiber bundle map, with fiber a smooth surface F (regular fiber). Pick $p_0 \in D \subset X - \mathcal{C}$ where D is a disk, then we have can identify $\pi^{-1}(D) \approx D \times F$. Now connect p_0 to p_j with a path γ_j in the complement of $\mathcal{C}$. Let $D_j = D \smile N(\gamma_j)$, where $N(\gamma_j)$ is the tubular neighborhood of γ_j (Figure 7.1).

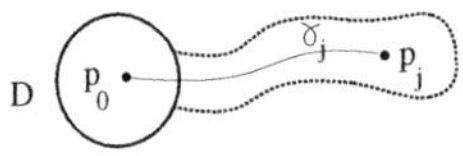

Figure 7.1: D_j

In [Ka] Kas sowed that $\pi^{-1}(D_j)$ is obtained by attaching a 2-handle h_j^2 to $\pi^{-1}(D)$, along a circle $C_j \subset p \times F \subset S^1 \times F$ (for some $p \in S^1$) with framing one less than the framing induced from the surface F,

$$\pi^{-1}(D_j) = \pi^{-1}(D) \smile h_j^2 \tag{7.1}$$

h_j^2 is called a *Lefschetz handle*. After attaching h_j^2, the trivial fibration $\pi| : S^1 \times F \to \partial D$ becomes a nontrivial fibration over ∂D_j, with monodromy given by the right handed Dehn twist $\tau_j : F \to F$ along C_j. By continuing with this process we obtain Lefschetz handles $h_1^2, \dots, h_k^2$, which extends the Lefschetz fibration over $D_I = D \smile \cup_{j=1}^k N(\gamma_j)$ with

$$\pi^{-1}(D_I) = \pi^{-1}(D) \smile \cup h_j^2 \tag{7.2}$$

so that the compositions of the mondromies is identity $\tau_k \tau_{k-1} \dots \tau_1 = id$. Since $S^2 - D_I = D_\infty$ is a disk, we can extend the fibration over D_∞ trivially $D_\infty \times F \to D_\infty$, since the monodromy on the boundary is identity.

Conversely, any collection of imbedded curves $\{C_j\}_{j=1}^k$ in F describes a Lefschetz fibration over D^2 with vanishing cyles $\{C_j\}$; furthermore if we have $\tau_{C_k}\tau_{C_{k-1}} \dots \tau_{C_1} = id$, then they describe a Lefschetz fibration $\pi : X \to S^2$.

Exercise 7.1. *Let a, b be the two circle generators of $T^2 = S^1 \times S^1$, then show that the vanishing cycles $\{a\}$ and $\{a, b\}$ on T^2 give Lefschetz fibration structures (over D^2) on the fishtail and the cusp, respectively (Figure 7.2). (Hint: Change the 1-handles in the middle of Figure 7.2 to the circle with dot notation.)*

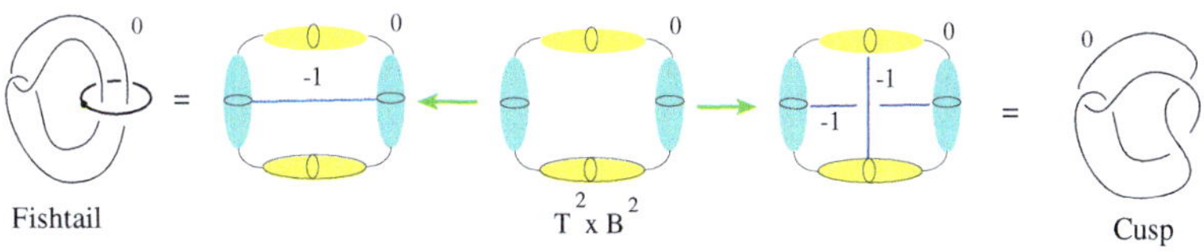

Figure 7.2: Making fishtail and cusp Lefschetz fibrations over D^2

Example 7.1. *Let $P_0(z)$ and $P_1(z)$ be a generic pair of homogenious cubic polynomials in $\mathbb{C}^3$. For each $t = [t_0, t_1] \in \mathbb{CP}^1$ the sets $Z_t = \{z \in \mathbb{CP}^2 \mid t_0 P_0(z) + t_1 P_1(z) = 0\}$ fill $\mathbb{CP}^2$ (Z_t is a torus for generic t). Blowing up $\mathbb{CP}^2$ along nine common zeros gives a Lefschetz fibration $\pi : \mathbb{CP}^2 \# 9\bar{\mathbb{CP}}^2 \to \mathbb{CP}^1$ with regular fiber T^2, which is called $E(1)$ (Figure 7.3).*

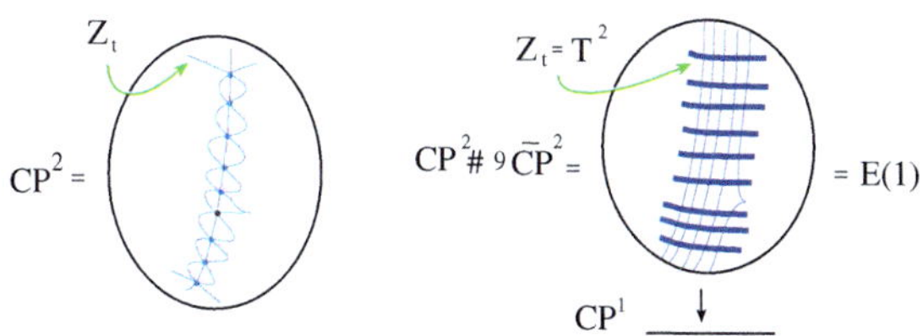

Figure 7.3: $E(1)$

7.1 Elliptic surface E(n)

Let a, b be the two circle generators of $T^2 = S^1 \times S^1$. It is known that $(\tau_a \tau_b)^6 = id$. Let $E(1)$ be the corresponding Lefschetz fibration. Figure 7.4 describes how to construct a handlebody of $E(1)$ starting with $D^2 \times T^2$ (framings of all the middle 2-handles are -1). In [A7] this handlebody was identified with the handlebody of Figure 7.5, from which it follows that $E(1) \approx \mathbb{CP}^2 \# 9\bar{\mathbb{CP}}^2$. More generally, the Lefschetz fibration corresponding to the relation $(\tau_a \tau_b)^{6n} = id$ is called $E(n)$, which has 12n handles (vanishing cycles), each one after the other attached to $D^2 \times T^2$ with the same weaving pattern of Figure 7.4. $E(2)$ is known to be diffeomorphic to the hypersurface $V_4 \subset \mathbb{CP}^3$ (Section 12.2).

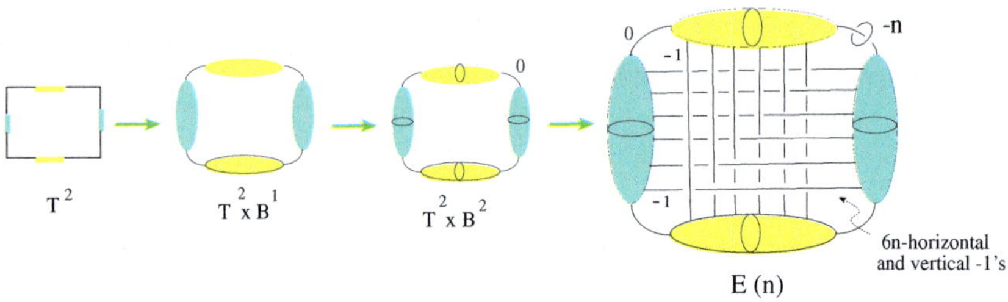

Figure 7.4: $E(n)$, drawn when $n = 1$

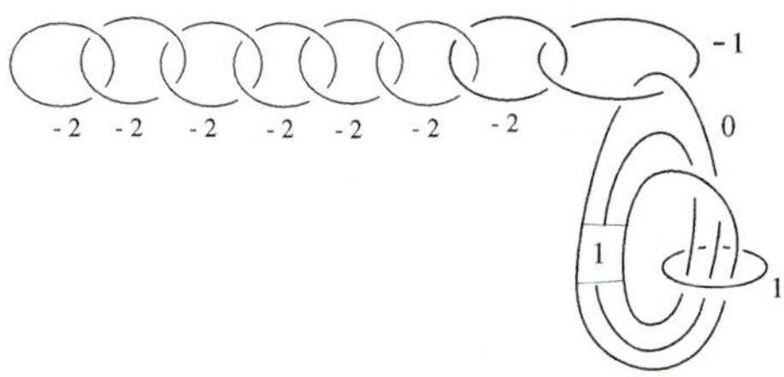

Figure 7.5: E(1)

Exercise 7.2. *(a) Show that in Figure 7.4 the small circle with $-n$ framing corresponds to the 2-handle extending the fibration over $D_\infty \times T^2 \to D_\infty$. (b) Show that $E(n)$ is the fiber sum $E(n-1) \natural_F E(1)$; more specifically, it is obtained by removing a tubular neighborhood $D^2 \times F$ of the regular fiber $F = T^2$ from each $E(n-1)$ and $E(1)$, and then by identifying them along the common boundary T^3, respecting the projection.*

7.2 Dolgachev surfaces

Let $E(1)_{p,q}$ denote the manifold obtained from the Lefschetz fibration, $\pi : E(1) \to S^2$, by performing log transformations of order p and q to two parallel T^2-fibers. These manifolds are known as *Dolgachev surfaces.* In [D1] Donaldson proved that, when p, q are relatively prime integers all $E(1)_{p,q}$ are distinct exotic copies of $E(1)$. Following [A13], here we draw a handlebody for $E(1)_{p,q}$. In Figure 7.6 we start with the handlebody of $T^2 \times [0,1]$ (cf. Figure 4.1), and then by gluing two copies, $T^2 \times [0,1]$, along one of their faces (by Section 3.2) we get a Heegaard picture of $T^2 \times [0,1]$ so that two parallel copies of T^2 inside are visible.

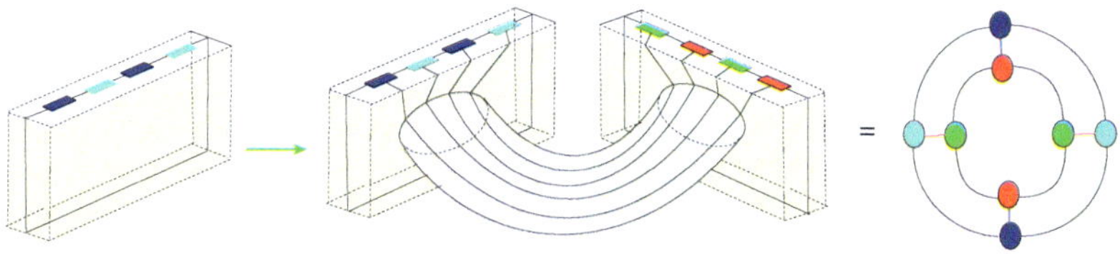

Figure 7.6

By thickening this Heegaard picture, in Figure 7.7, we obtain a handle picture for $T^2 \times D^2$ where two parallel $T^2 \times D^2$'s inside are visible. By applying the same process to the cusp C (which is $T^2 \times D^2$ with two vanishing cycles) we get Figure 7.8. Now by Section 6.3, we apply log transforms of order p and q to these parallel tori and get $(T^2 \times D^2)_{p,q}$, and also $C_{p,q}$ (Figure 7.9). To apply this to the cusp inside $E(1)$, we first push two parallel tori from the cusp inside $E(1)$ in Figure 7.5 and get Figure 7.10 (where, for empasis the 1-handles are drawn in pair-of-balls notation), and we then apply the log transforms as in Figure 7.9 to obtain Figure 7.11, which is $E(1)_{p,q}$.

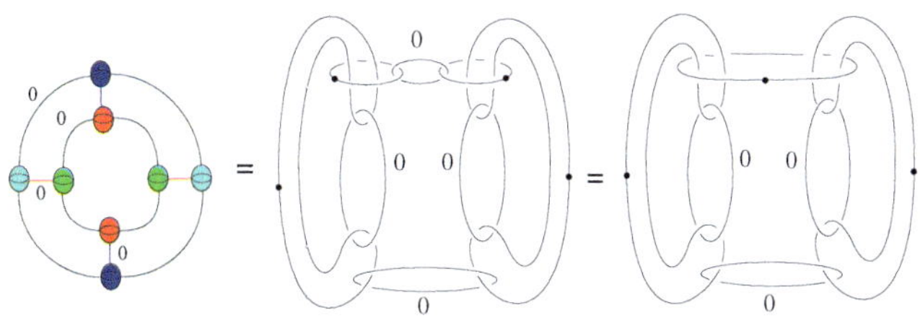

Figure 7.7: $T^2 \times D^2$ with two disjoint tori visible inside

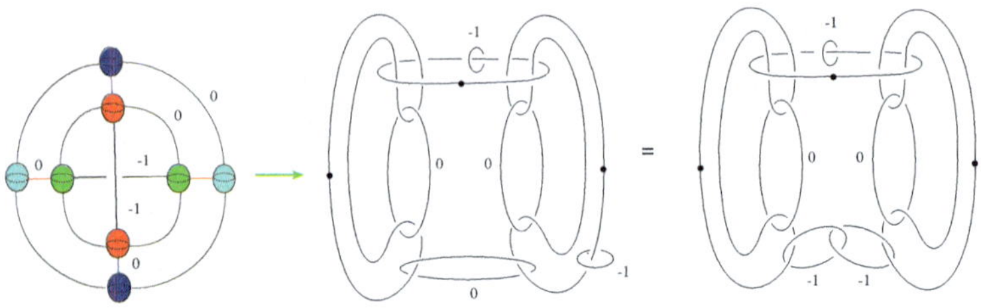

Figure 7.8: The cusp C with two disjoint tori visible inside

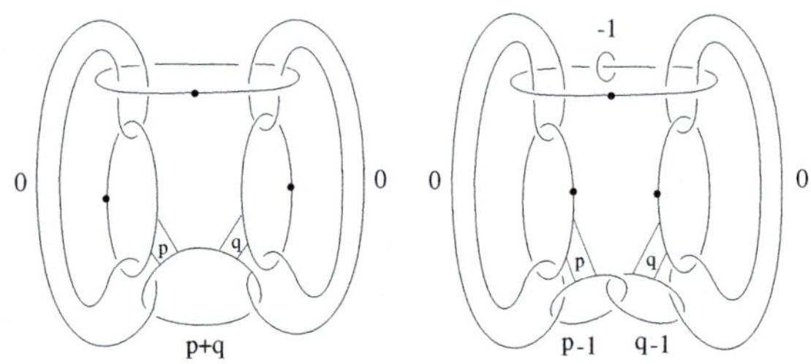

Figure 7.9: $(T^2 \times D^2)_{p,q}$ and $C_{p,q}$

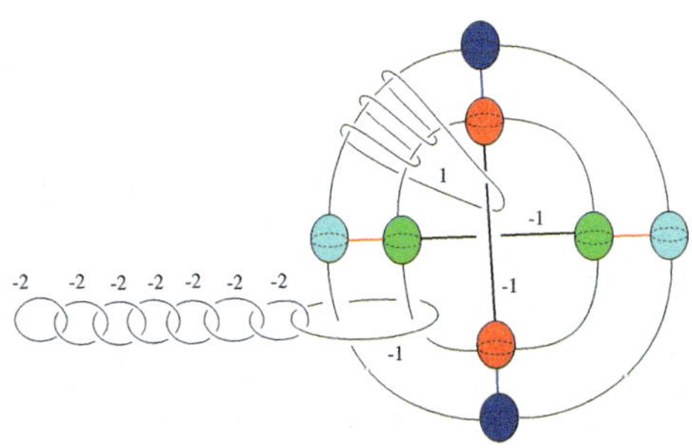

Figure 7.10: $E(1)$ with two copies of T^2 inside its cusp

Exercise 7.3. *Verify that converting* 1*-handles of Figure 7.10 to circle-with-dot notation, and applying the log transforms as described in Figure 7.9 gives Figure 7.11.*

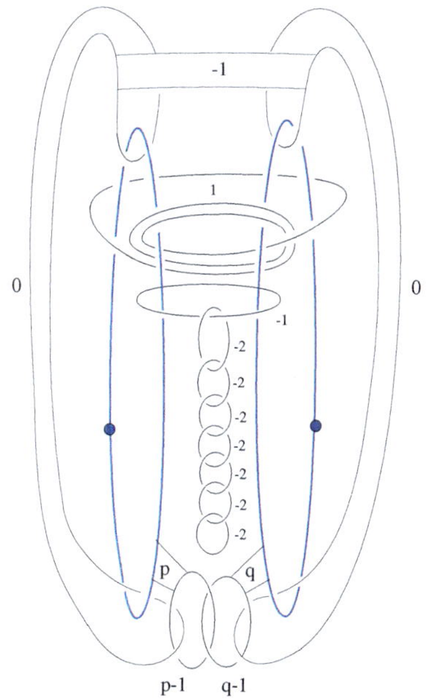

Figure 7.11: $E(1)_{p,q}$

Exercise 7.4. *By applying the above process to the cusp of* $E(n)$ *in Figure 12.32, draw a handlebody for* $E(n)_{p,q}$.

Remark 7.1. *In [A16] it was shown that* $E(1)_{2,3}$ *can be obtained from* $E(1)$ *by a "plug twisting" operation along the plug* $W_{1,2} \subset E(1)$ *(see Section 10.4); in particular* $E(1)_{2,3}$ *is obtained from* $E(1)$, *by twisting it along an imbedded copy* $\mathbb{RP}^2 \subset E(1)$.

7.3 PALFs

A special case of Lefschetz fibration, $\pi : X \to D^2$, is when the regular fiber is a smooth surface with a boundary, and the base D^2. We will call such a Lefschetz fibration a *positive Lefschetz fibration* or PLF for short, if every vanishing cycle is a homotopically essential curve on the fiber. Following [AO1], we will call a PLF a *positive allowable Lefschetz fibration* or PALF for short, if every vanishing cycle is a homologically essential curve on the fiber. We will view a PLF $\mathcal{F}$ as an extra structure on the underlying smooth manifold X, and express this by denoting $X = |\mathcal{F}|$ (similar to viewing triangulation as an additional structure on a manifold). The boundary $\partial\mathcal{F}$ of a PLF $\mathcal{F}$ defines an open book structure on the boundary of the underlying manifold $|\partial\mathcal{F}| = \partial X$ (Figure 7.12).

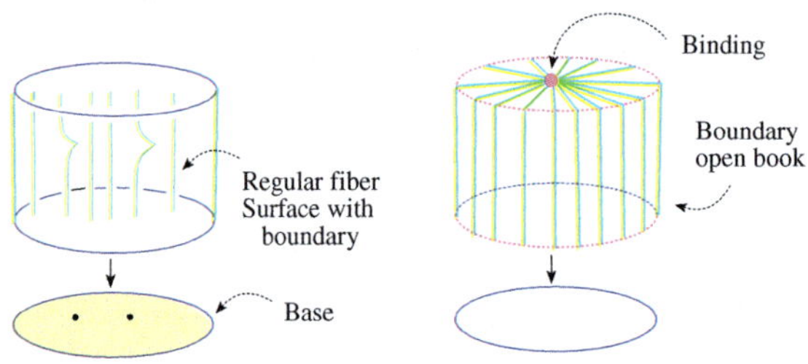

Figure 7.12: A PLF and its boundary open book

Definition 7.2. *An open book structure (or decomposition) of a 3-manifold Y is a surjective map, $\pi : Y \to D^2$, such that $\pi^{-1}(intD)$ is a disjoint union of solid tori and*

(a) π is a fibration over $S^1 = \partial D^2$.

(b) On each of the solid tori, π is the projection map $S^1 \times D^2 \to D^2$.
The centers of these solid tori are called binding.

So in the complement of the binding π is a fibration over S^1, with fibers surfaces with boundary. Alternatively, we can define an open book structure on Y just in terms of its pages and the monodromy (F, ϕ) of the fibration $\pi| : Y \to S^1$. Here F is an oriented surface with boundary and $\phi : F \to F$ is a diffeomorphism which is identity on ∂F. Then Y is the union the mapping torus of ϕ and solid tori glued along their boundaries in the obvious way. The monodromy of the open books arising from the boundary of PALFs is a composition of right handed Dehn twists (coming from the vanishing cycles).

Exercise 7.5. *Let $\pi : X \to S^2$ be a Lefschetz fibration with closed surface fibers and a section, then show that removing the tubular neighborhood of the section and a regular fiber as indicated in Figure 7.13, gives a PALF ([AO2]).*

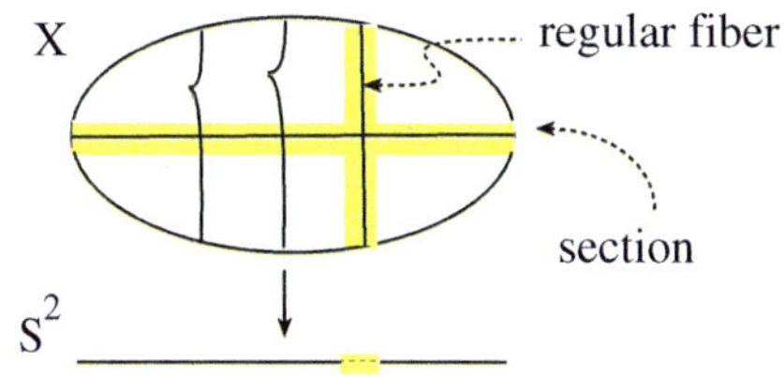

Figure 7.13

Remark 7.3. *This exercise gives so-called concave–convex decomposition of any Lefschetz fibration $\pi : X \to S^2$ with closed fibers and a section: The closed tubular neighborhood of the section union a regular fiber is the concave part, called concave LF; and the closure of its complement is the convex part, which is a PALF. Notice attaching 2-handles to the bindings of a PALF $\pi : Z \to D^2$ gives an LF over D^2 with closed fibers, and then by attaching 2-handles to LF (over D^2 with closed fibers) along the sections of $\pi| : \partial Z \to S^1$ gives the concave part of LF. Notice the boundary of a concave LF is also an open book (pages live in the part of the boundary which is close to the regular fiber).*

Just as in the case of Lefschetz fibrations, any PALF can be described by a collection of imbedded curves $\mathcal{C} = \{C_j\}$ (vanishing cycles) lying on a surface with boundary F, with framings one less then the framing induced from F. A basic example is the standard PALF of $B^4 = |\mathcal{F}_0|$, which consists of an annulus with the center circle C (Figure 7.14). We can either draw B^4 as a canceling pair of 1- and 2-handles, with F as a thin cylinder containing the vanishing cycle C, or draw B^4 as an empty set (after cancelation) with F as looking like a twisted annulus (right hand picture of Figure 7.14).

More generally, a PALF $\mathcal{F} = (F, \mathcal{C})$ can be enlarged to another PALF $\mathcal{F}' = (F', \mathcal{C}')$, where F' is the surface obtained from F by attaching a 1-handle to ∂F, and $\mathcal{C}'$ is the union of $\mathcal{C}$ with the new vanishing cycle C consisting of the core of the 1-handle union any imbedded arc in F connecting the ends of the core (Figure 7.15). The operation $\mathcal{F} \mapsto \mathcal{F}'$ is called *positive stabilization of the PALF* $\mathcal{F}$. It is clear that $|\mathcal{F}| = |\mathcal{F}'|$ since this is just the operation of introducing canceling 1- and 2-handles. In particular the operation $\partial\mathcal{F} \mapsto \partial\mathcal{F}'$ is called positive stabilization of the corresponding open book.

The second picture of Figure 7.15 is an abstract picture of the PALF $\mathcal{F}'$, the third picture should be thought of as $\mathcal{F}'$ being obtained from a PALF $\mathcal{F}_0$ on B^4 by introducing vanishing cycles to the pages of the induced open book $S^3 = |\partial\mathcal{F}_0|$. Stabilization has an effect of plumbing with the Hopf band.

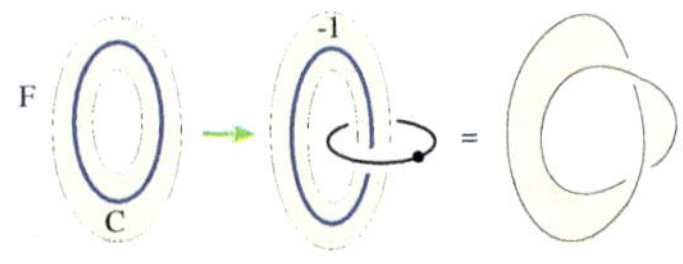

Figure 7.14: A PALF on B^4

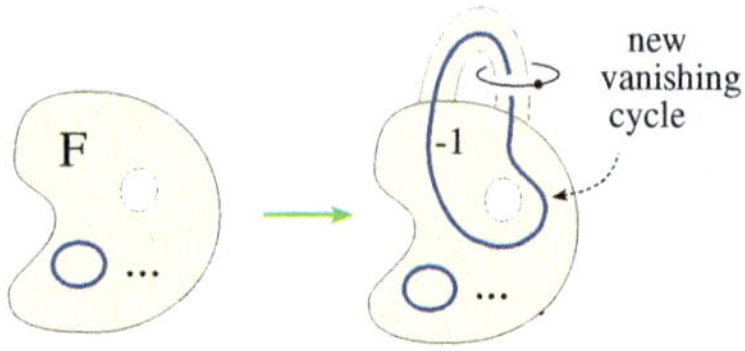

Figure 7.15: Positive stabilization

In some instances this process can be reversed as follows (cf. [AKa3]):

Exercise 7.6. *Let C be a non-separating curve on a surface $F \subset S^3$, such that the framing on C induced from this surface corresponds to -1 framing. Then F is isotopic rel C to the plumbing of another surface with a Hopf band whose core is C. (Hint: find a transverse arc intersecting C at one point with end points on ∂F, then do the indicated finger move of Figure 7.16 along C.)*

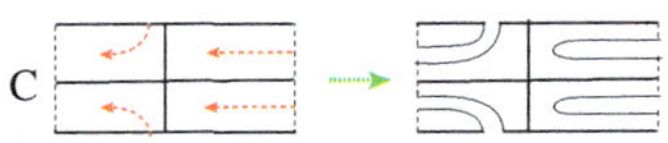

Figure 7.16

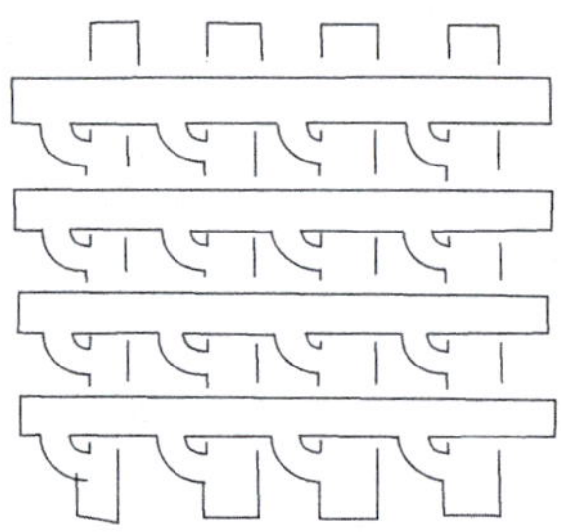

Figure 7.17: $F(p,q)$

The (p,q) torus knot $K(p,q)$ can be described as the boundary of the surface, $F(p,q)$ in S^3, obtained by connecting p horizontal squares and q vertical squares, underneath, by bands as shown in Figure 7.17 (figure drawn for $(p,q) = (4,4)$), [Ly]. A repeated application of Excercise 7.6 shows that $K(p,q)$ is obtained from S^3 by a sequence of plumbings by Hopf bands. So the Seifert surface $F(p,q)$ (fiber) of $K(p,q)$ represents a PALF of B^4. For future reference we will denote this PALF of the 4-ball by $\mathcal{F}_{p,q}$.

7.4 ALFs

Let $\pi : X \to \Sigma$ be a map from an oriented 4-manifold to a surface with $\Sigma = S^2$ or D^2. We say $p \in X$ is an *achiral Lefschetz singularity* of π, if we can find charts $(\mathbb{C}^2, 0) \hookrightarrow (X, p)$ and $\mathbb{C} \hookrightarrow \Sigma$, on which π is given by $(z, w) \mapsto z\bar{w}$. An *achiral Lefschetz fibration* (ALF in short) is a map, $\pi : X \to \Sigma$, which is a submersion in the complement of finitely many Lefschetz singularities $\mathcal{P}$, some of which are achiral; and π is injection on $\mathcal{P}$. So basically an ALF is a general version of a PALF, where we allow achiral singularities. Similar to the description of Lefschetz singularities in (7.2), any ALF is obtained from the trivial fibration $D^2 \times F \to D^2$ (F being the regular fiber) by attaching 2-handles $\{h_j\}$ to a collection of curves $\mathcal{C} = \{C_j\}$ in $p \times F \subset S^1 \times F$, with framing one less or one more than the page framing, corresponding to Lefschetz or achiral Lefschetz singularities of π, respectively. So any ALF is represented by such a pair $(F, \mathcal{C})$. We call any curve C_j with framing one more than the page framing a *defective handle.* A defective 2-handle h_j has the affect of introducing a left handed Dehn twist $\bar{\tau}_j : F \to F$ (i.e. the left handed version of Figure 5.1) to the monodromy of the boundary fibration over the circle, starting with $S^1 \times F \to S^1$. For example, Figure 7.18 describes an ALF structure of B^4. We call the operation of Figure 7.19 *negative stabilization*; which introduces the left handed Hopf band. Similar to PALF, the boundary of an ALF is an open book (F, ϕ), where the monodromy ϕ is the composition of is right and also left Dehn twists.

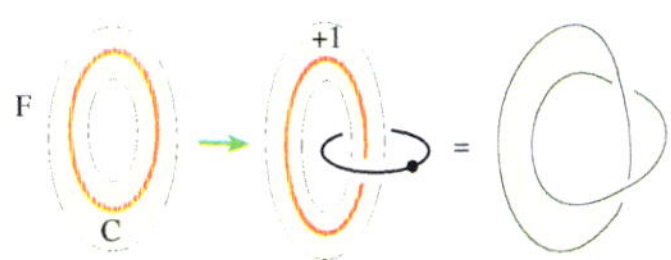

Figure 7.18: An ALF on B^4

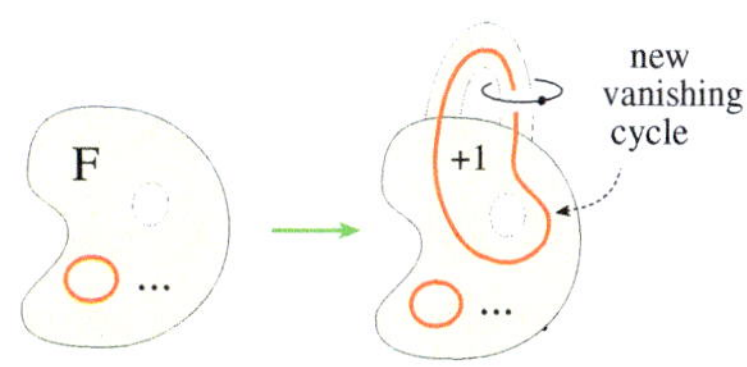

Figure 7.19: Negative stabilization

Proposition 7.4. *([Ha]) Every 4-manifold X without 3- and 4-handles, admits an ALF structure $\pi : X \to D^2$, with fibers having nonempty boundary.*

Proof. ([AO1]) We are given $X = X_\Lambda$, where $\Lambda = \{K_1^{r_1}, \ldots, K_n^{r_n}, C_1, \ldots, C_s\}$. We will first prove the claim when there are no 1-handles C_j present. First by an isotopy we can put the link $\mathcal{L} = \{K_1, \ldots, K_n\}$ in a *square bridge position* (why?).

This means that the link consists of vertical and horizontal line segments glued along their corners, and at the crossings the horizontal segments go over the vertical segments. Figure 7.22 demonstrates this for the trefoil knot. This implies that we can put the link $\mathcal{L}$ on the surface $F(p,q)$ as demonstrated in Figure 7.20. If the framings $\{k_1, k_2, \dots\}$ were one more, or one less then the page framings of the surface $F(p,q)$ we would be done, i.e. X would be the ALF obtained by attaching vanishing cycles to the PALF $\mathcal{F}_{p,q}$ of B^4. If not, by positively (or negatively) stabilizing the PALF, as shown in Figure 7.21, we can increase (or decrease) page framing (why?) and hence put ourselves in this case. These operations have the effect of attaching canceling pairs of 1- and 2-handles to X to create the ALF.

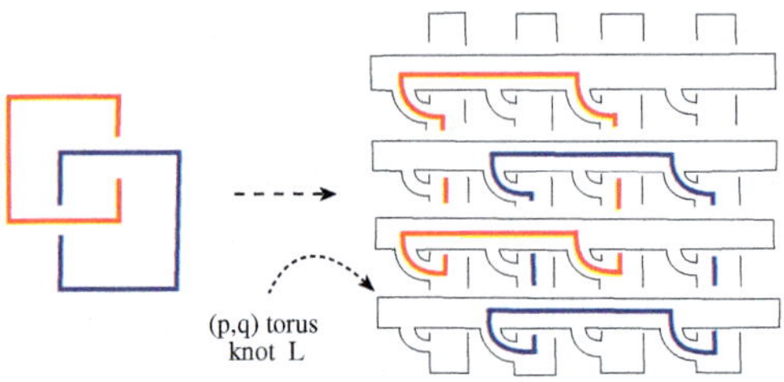

Figure 7.20

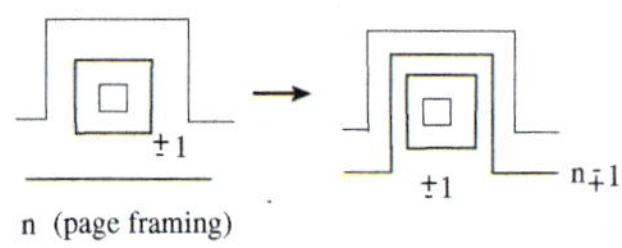

Figure 7.21

Now we do the case when there are 1-handles present (i.e. C_j's of Λ). First by treating each unknotted circle C with-a-dot as part of the framed link $\mathcal{L}$, by the same process we put C on the surface $F(p,q)$. Then by an isotopy we make C transverse to the page, so that it intersects each page of the open book $\partial\mathcal{F}_{p,q}$ once, as indicated in Figure 7.23 (see the last paragraph). Then as explained in Section 1.1, to create the 1-handle which C represents, we push the interior of the disk D, which C bounds, into B^4 then excise the tubular neighborhood of D from B^4. This has the effect of puncturing each fiber of the ALF given by $\mathcal{L}$ once. Clearly this makes X an ALF.

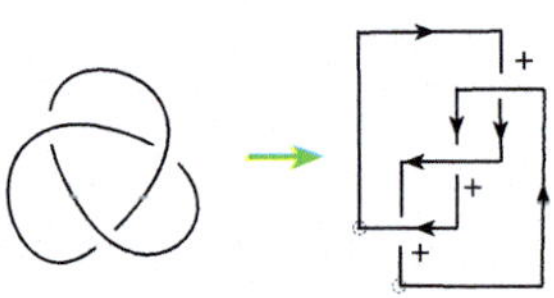

Figure 7.22: K

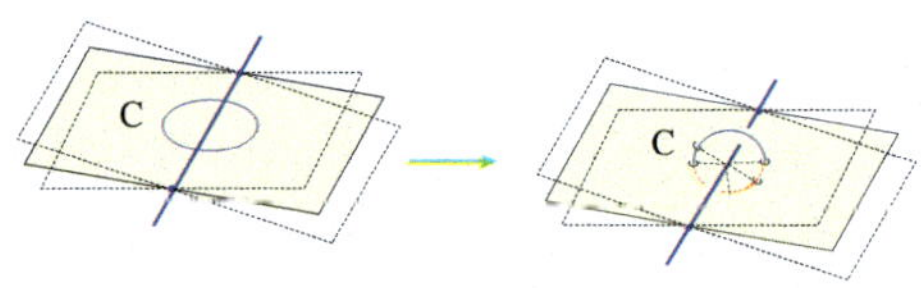

Figure 7.23

Finally, here is an explanation of how we make C transverse to all pages. C, being an unknot, when putting the link $\mathcal{L}$ in a square bridge position, we can make sure that C looks like Figure 7.24 on the surface $F(p,q)$. Hence in the handle notation of the PALF $\mathcal{F}_{p,q}$ (as in Figure 7.14) C will look like Figure 7.25; note that the second equivalent picture of this figure (after the indicated handle slides) makes it clear that C meets each fiber of $\mathcal{F}_{p,q}$ once, since it is isotopic to the small linking circle of $\partial F(p,q)$. □

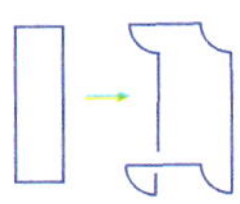

Figure 7.24: C

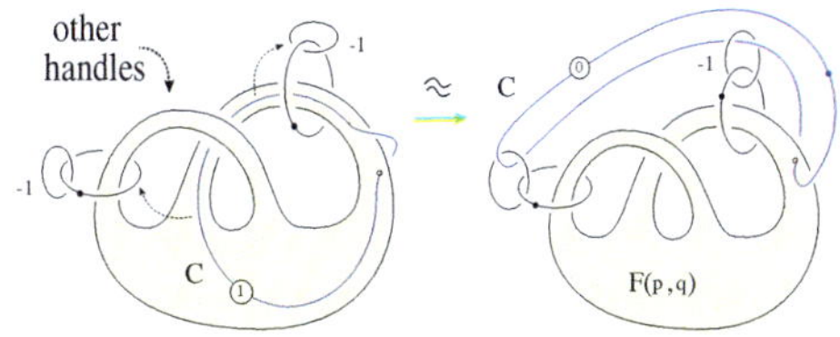

Figure 7.25: $\mathcal{F}_{p,q}$

Let $Y = \partial X_{\Lambda}$ with $\Lambda = \{K_1^{r_1}, \ldots, K_n^{r_n}, C_1, \ldots, C_s\}$. By replacing the circles with dots C_j (1-handles) with zero framed circles, we can assume that Y bounds a 4-manifold with no 1-handles. Hence Y is the boundary of the ALF obtained by stabilizing the PALF $\mathcal{F}_{p,q}$ on B^4, and then attaching vanishing cycles to it. We will call such an ALF an *extension of a PALF* on B^4. More generally, we will call an ALF $\mathcal{F}$ an *extension* of another ALF $\mathcal{F}'$, if it is obtained from $\mathcal{F}'$ by adding vanishing cycles.

Remark 7.5. *Since any diffeomorphism $\phi : F \to F$ can be written as a composition of (right or left) Dehn twists along imbedded curves of F, for any given open book (F, ϕ) on Y, there is an ALF $\mathcal{F}$ such that $\partial \mathcal{F}$ gives this open book. Furthermore we can assume that $\mathcal{F}$ is an extension of a PALF on B^4.*

To justify the last claim of this remark, we first compose the monodromy ϕ by a pair of right and left Dehn twists, along two parallel copies of each of the core 1-handle curves of F. This does not change the isotopy class of ϕ. Then by using the right handed monodromy curve (-1 curve) we cancel all the corresponding 1-handles of the ALF of (F, ϕ). We are left with a PALF on B^4 obtained from the trivial PALF of $B^4 = B^2 \times B^2$ by positive stabilizations, along with (right or left) vanishing cycles on it (Figure 7.26).

Exercise 7.7. *Given $X = X_{\Lambda}$ with $\Lambda = \{K_1^{r_1}, \ldots, K_n^{r_n}, C_1, \ldots, C_s\}$, put an ALF structure $\mathcal{F}$ on X which is an extension of an ALF structure $\mathcal{F}'$ on the 1-handles $\#_s(S^1 \times B^3)$. (Hint: first put the 2-handles in a square bridge position in $\#_s(S^1 \times B^3)$, and put them on the surface of Figure 7.27. Then construct the required ALF $\mathcal{F}'$ on $\#_s(S^1 \times B^3)$ by first stabilizing $\mathcal{F}_{p,q}$ in S^3, then taking away vanishing cycles).*

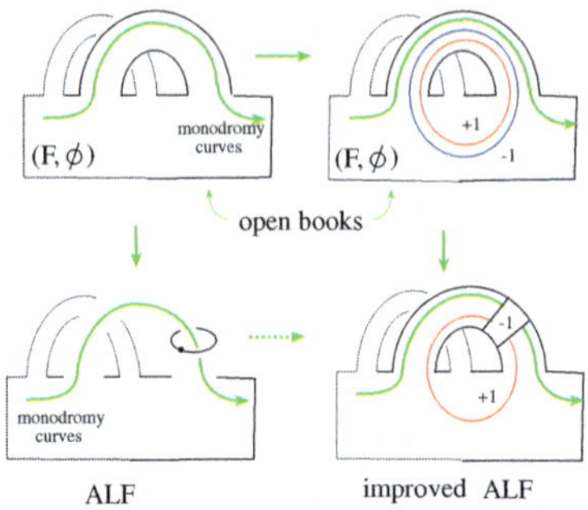

Figure 7.26

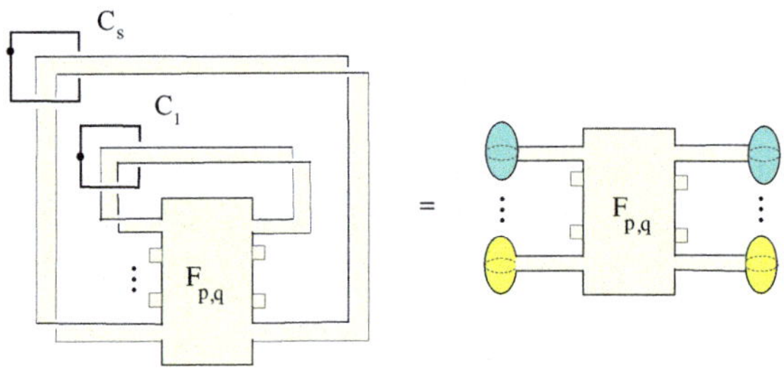

Figure 7.27

Exercise 7.8. *Let $K \subset R^3$ be an oriented knot in a square bridge position, sitting on the surface $F(p,q)$. Show that the framing of K induced from the surface $F(p,q)$ is given by the so-called Thurston–Benequin number:*

$$TB(K) = w(K) - c(K) \tag{7.3}$$

where $w(K)$ is the writhe (Exercise 1.1), and $c(K)$ is the number of south-west corners.

For example, the unknot of Figure 7.24 has $TB(C) = -1$, and the trefoil knot K of Figure 7.22 has $TB(K) = 3 - 2 = 1$. Also for a given oriented knot $K \subset R^3$ in square bridge position, we define a corner of K an *up-corner* (or a *down-corner*) if it is either the south-west or north-east corner, and the orientation arrow turns up (or turns down) as it traverses around this corner.

Definition 7.6. *The rotation number of an oriented knot K in square bridge position:*

$$r(K) = 1/2\,(\textit{number of down corners} - \textit{number of up corners}). \tag{7.4}$$

For example, for the knot K of Figure 7.28 we have $r(K) = 1/2(1-3) = -1$.

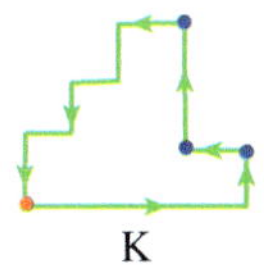

Figure 7.28: $r(K) = -1$

Definition 7.7. *(compare (8.7)) If an ALF $\mathcal{F}$ on X has no 1-handles, and it is in square bridge position (i.e. an extension of $\mathcal{F}_{p,q}$), we define its characteristic class by*

$$\mathbf{c} = \mathbf{c}(\mathcal{F}) = \sum_{j=1}^{r} r(K_j)K_j^* \in H^2(X;\mathbb{Z}) \tag{7.5}$$

where $K_j^* \in H^2(X)$ are the hom duals of the 2-dimensional homology classes represented by the framed knots K_j of X, i.e. $< K_j^*, K_i >= \delta_{i,j}$. If $\mathcal{F}$ has 1-handles, by turning its 1-handles (dotted circles) to zero-framed circles, while twisting everything going through them -1 times (cf. Figure 7.27), we transform $\mathcal{F}$ to an ALF $\mathcal{F}'$ without 1-handles, then we define $\mathbf{c}(\mathcal{F}) = \mathbf{c}(\mathcal{F}')$ (relate this to the process in Remark 7.5).

Exercise 7.9. *Show that the ALF modification in Figure 7.29 doesn't change the underlying smooth manifold X, while changing the ALF so that $r(K)$ changes by ± 2.*

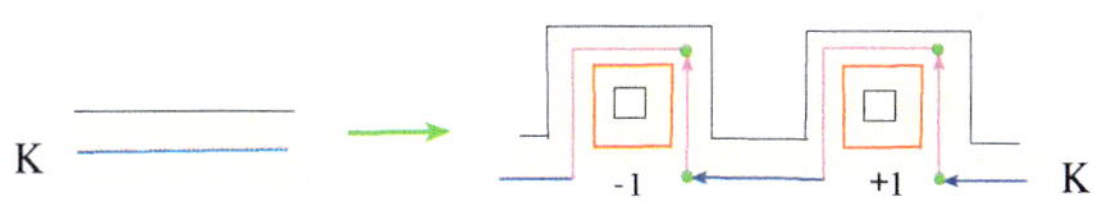

Figure 7.29

7.5 BLFs

A *Broken Lefschetz fibration* (BLF in short) is a singular version of LF, it is a Lefschetz fibration $X^4 \to S^2$ with closed surfaces as regular fibers, where we allow circle singularities away from the Lefschetz singularities (so-called fold singularities). This means that on neighborhoods of some imbedded circles in X we can choose coordinates such that map π looks like the map $S^1 \times B^3 \to \mathbb{R}^2$, given by $(t, x_1, x_2, x_3) \to (t, x_1^2 + x_2^2 - x_3^2)$, otherwise outside of these circles π is a Lefschetz fibration. Obviously these imbedded circles (along with LF singularities) are the singular points of π. Similarly we can define BLF over a disk $\pi : X \to D^2$, where the fold singularities are in the interior of X. So a BLF $\pi : X \to S^2$ is a LF with varying fiber genus, that is, there are imbedded circles $C \subset S^2$ where the genus of the fiber $\pi^{-1}(t)$ changes as t moves across C. The circles C are the images of the singular circles of π mentioned previously. BLFs first introduced in [ADK2]. In [GK] it was observed that by enlarging PALF handlebodies by round handle attachments we can construct BLF's; this helps to produce examples.

Definition 7.8. *Let X be a 4-manifold with boundary. A round 1-handle is a copy of $S^1 \times D^1 \times D^2$ attached to ∂X along $S^1 \times \{-1\} \times D^2 \smile S^1 \times \{1\} \times D^2$. A round 2-handle is a $S^1 \times D^2 \times D^1$ attached to ∂X along $S^1 \times S^1 \times D^1$ ([As]).*

Figure 3.10 gives a handlebody description of a round 1-handle attached to X along pair of circles $\{C_1, C_2\}$ in ∂X.

Exercise 7.10. *Show that any BLF $\pi : X \to D^2$ extends to another BLF after performing the following operations:*

(a) Attaching a round 1-handle along a pair of circles, which are sections of the boundary fibration $\pi| : \partial X \to S^1$. Also show that this operation increases the fiber genus, introducing a fold singularity along $S^1 \times 0 \times 0$.

(b) Attaching a round 2-handle along a (page framed) circle lying on a fiber of the fibration $\pi| : \partial X \to S^1$ (that is lying on a page of the open book). Show that this move decreases the fiber genus if the attaching circle is essential on the page, introducing a fold singularity along $S^1 \times 0 \times 0$.

Broken Lefschetz fibrations can be used to convert convex LFs to concave LFs by introducing broken singularity. The following are constructions from [GK], as presented in [AKa3] (also see [Ba1]).

Proposition 7.9. *Let F be a closed oriented surface, then $F \times D^2$ can be made a concave BLF over D^2.*

Proof. To turn the fibration $\pi : F \times D^2 \to D^2$ to a concave fibration, we need to attach a 2-handle along a section of $\pi| : F \times S^1 \to S^1$ (Remark 7.3), but we have to do this without changing the topology of $F \times D^2$. For this, we attach a canceling pair of 2- and 3-handles, and notice that this is equivalent to a (-1 framed) section 2-handle and a round handle, as shown in Figure 7.30. □

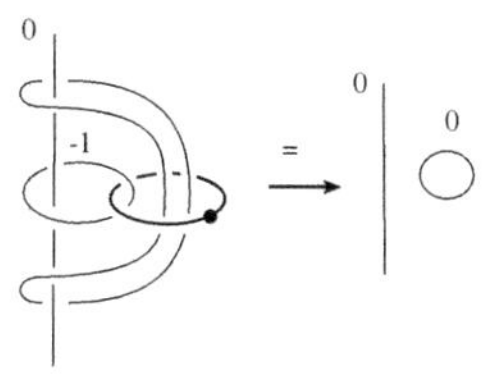

Figure 7.30

A *concave (or convex) BLF* is a concave (or convex) LF where we allow fold singularities. Introducing fold singularities allows us to perform some operations otherwise impossible. The following exercise is a good example.

Exercise 7.11. *Given a concave BLF X, show that we can modify the BLF structure on X so that the corresponding open book on ∂X gets positively stabilized, and also gets negatively stabilized. (Hint: First introduce a canceling 2/3- handle pair, then as in Figures 7.31 and 7.32 describe them with round handles. Check that -1 framed handles on the right side of both figures live on the page, and round handles are attached along the sections. Keep in mind that 2-handles can pass through the bindings, since they are not handles).*

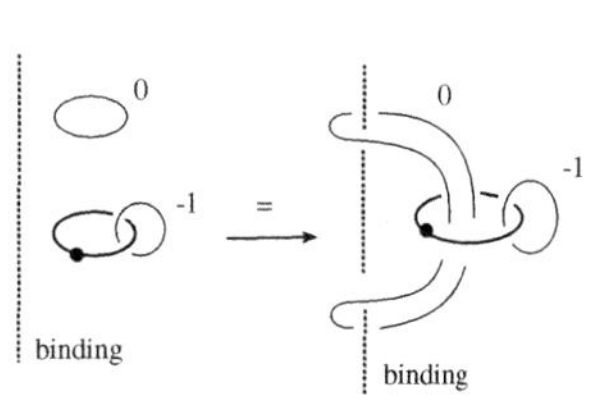

Figure 7.31: Positive stabilization

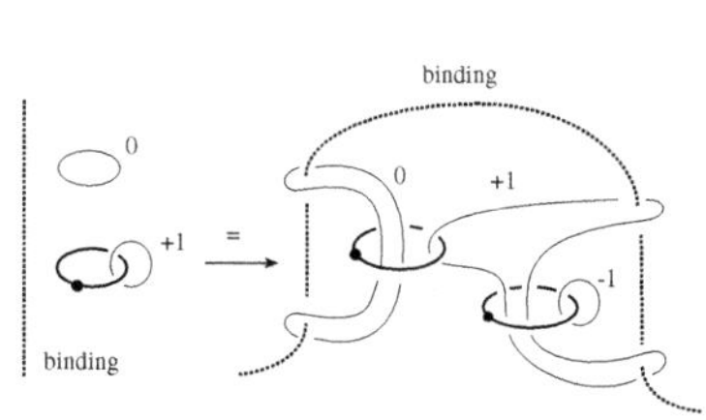

Figure 7.32: Negative stabilization

Exercise 7.12. *Let X be a closed smooth 4-manifold. Show that we can find a closed imbedded surface with trivial normal bundle $F \times D^2 \subset X$, such that its complement has no 3- and 4-handles. (Hint: As in Figure 7.33, introduce canceling 2/3- handle pairs to turn the 4-ball with 1-handles to $F \times D^2$; also recall Figure 2.3, then push the excess 2- and 3-handles to the other side as 1- and 2- handles).*

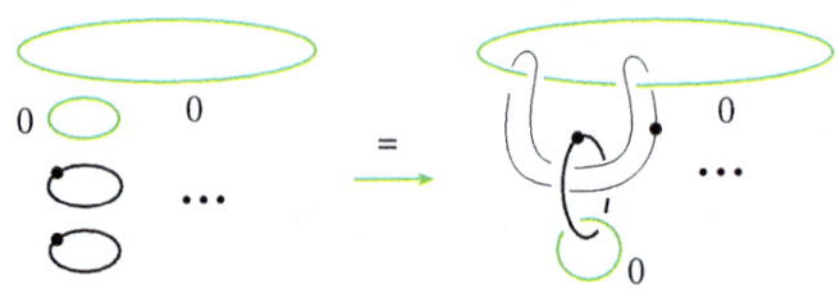

Figure 7.33

Chapter 8

Symplectic manifolds

A *symplectic structure* on a closed oriented smooth 4-manifold X^4 is a non-degenerate closed 2-form ω (non-degenerate means $\omega \wedge \omega > 0$ pointwise). It is known that any symplectic manifold (X, ω) admits a compatible almost complex structure J, which is a bundle map $J : TX \to TX$ satisfying the following (e.g. [MSa]):

(a) $J^2 = -id$.

(b) $\omega(Ju, Jv) = \omega(u, v)$.

(c) $\omega(u, Ju) > 0$, for all $u \neq 0$.

(c) implies that $g(u, v) := \omega(u, Jv)$ is a Riemannian metric on X. The set of compatible complex structures is contractible. In particular, all such almost complex structures are homotopic to each other. So we can talk about Chern classes, $c_i(X) = c_i(X, \omega) \in H^{2i}(X; \mathbb{Z})$ and splitting $\Lambda^{p,q}(T^*X_{\mathbb{C}})$, and hence the canonical line bundle $\mathbb{K} \to X$.

Definition 8.1. *The canonical line bundle of* (X, ω) *is defined to be*

$$\mathbb{K} = \Lambda^{2,0}(T^*X_{\mathbb{C}}) := \Lambda^2(T^{1,0}X) \to X.$$

Kähler surfaces (X, ω) are examples of symplectic 4-manifolds, they are complex surfaces $X \subset \mathbb{CP}^N$ (for some N), where ω is induced from the Kähler form of $\mathbb{CP}^N$.

Definition 8.2. *A symplectic* 4*-manifold is minimal if it does not contain a symplectic* 2*-sphere with self intersection* -1. *Similarly, a Kähler surface is minimal if it does not contain a complex* 2*-sphere with self intersection* -1.

Theorem 8.3. *([G4],[D5]) If a closed manifold X admits a Lefschetz fibration structure $\pi : X \to S^2$ such that the homology class of the regular fiber $[F] \neq 0$ in $H_2(X;\mathbb{Z})$, then it has a symplectic structure so that regular fibers are symplectic submanifolds. Conversely, given a symplectic manifold X, after blowing up finitely many times we can turn it to a Lefschetz fibration, $\pi : X \#_k \bar{\mathbb{CP}}^2 \to S^2$.*

A celebrated theorem of Taubes (Theorem 13.29) states that symplectic manifolds have nonzero Seiberg–Witten invariants (13.30); and the surfaces in manifolds with nonzero Seiberg–Witten invariants satisfy the adjunction inequality (Theorem 13.35).

Theorem 8.4. *([Ta]) Let (X,ω) be a closed symplectic 4-manifold with $b_2^+(X) > 1$, then $SW_X(\mathbb{K}) = \pm 1$.*

Theorem 8.5. *([KM], [MST], [OS2]) If X is a closed smooth 4-manifold with $b_2^+(X) > 1$ and $SW_X(L) \neq 0$ for some $L \to X$, and $\Sigma \subset X$ a closed oriented surface of genus $g(\Sigma) > 0$ with $[\Sigma] \neq 0$, and if either $\Sigma.\Sigma \geq 0$ or X is of simple type (Definition 13.15) then:*

$$2g(\Sigma) - 2 \geq \Sigma.\Sigma + |\langle c_1(L).\Sigma\rangle| \tag{8.1}$$

Proofs of Theorems 8.4 and 8.5 (when $\Sigma.\Sigma \geq 0$) will be given in Sections 13.14 and 13.17. For a discussion of the general case, including the $g = 0$ case, see [OSt] and [FS3].

Theorem 8.6. *([B], [MF], [FS3], [OS]) A minimal Kähler (or minimal symplectic) 4-manifold X with $b_2^+(X) > 1$, cannot contain a smooth 2-sphere $S \subset X$ with $S.S \geq -1$.*

8.1 Contact manifolds

A *contact structure* ξ on an oriented 3-manifold Y^3 is a totally non-integrable 2-plane field on its tangent bundle. Locally the 2-plane field ξ can be given by the kernel of a 1-form $\alpha \in \Omega^1(Y)$ (α can be taken to be global if ξ is co-orientable), and the totally non-integrable condition means that $\alpha \wedge d\alpha > 0$ pointwise (e.g. [Ge], [OSt]). We call a contact structure *overtwisted* if ξ can be made transversal to the boundary of an imbedded 2-disk in Y, otherwise it is called *tight*. For example, Figure 8.1 is a picture of the tight contact structure $(\mathbb{R}^3, ker(dz + xdy))$, given by the 2-plane field on Euclidean 3-space (along the lines parallel to x-axis contact planes rotating 180^0 uniformly from $-\infty$ to ∞).

Exercise 8.1. *Show that each tangent plane of the unit sphere $S^3 \subset \mathbb{C}^2$ contains a unique complex line, and this complex line field defines a contact structure on S^3, given by the kernel of the 1-form, $\alpha_0 = i/4 \sum_{J=1}^2 (z_j d\bar{z}_j - \bar{z}_j dz_j)$.*

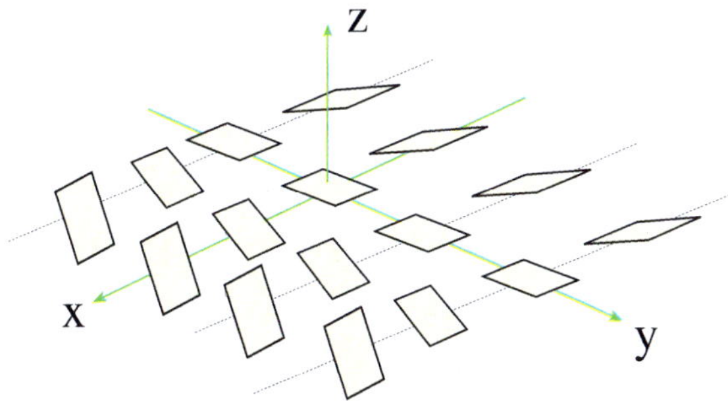

Figure 8.1: The tight contact structure of $\mathbb{R}^3$

We say that two contact structures ξ_1 and ξ_2 on Y^3 are isomorphic if $\xi_2 = f^*(\xi_1)$ for some diffeomorphism $f : Y \to Y$; if furthermore f is isotopic to identity then we say that ξ_1 and ξ_2 are isotopic contact structures.

Definition 8.7. *We say a contact structure $\xi = ker(\alpha)$ on Y^3 is carried (or supported) by the open book structure (F, ϕ) on Y, if $d\alpha$ is a positive area form on each page of the open book, and $\alpha > 0$ on the binding ∂F.*

It is known that every closed oriented 3-manifold Y^3 admits an open book structure, and every open book on Y^3 supports a contact structure; conversely, Giroux proved that every contact structure comes from (is supported by) an open book, and furthermore:

Theorem 8.8. *([Gi]) Two open book structures on an oriented 3-manifold Y^3 have common positive stabilization if and only if they support isotopic contact structures.*

Classifying contact structures ξ on Y^3 is hard to do, but classifying the homotopy classes of the underlying oriented 2-plane fields of ξ is easier. For example, when $H^2(Y;\mathbb{Z})$ has no 2-torsion, they are determined by two basic invariants: $c_1(\xi) \in H^2(Y;\mathbb{Z})$ and $d(\xi) \in \mathbb{Q}$ ([G2]). The first is the Chern class of the complex line bundle $\xi \to Y$, and when it is a torsion homology class, the second invariant is defined by

$$d(\xi) = \frac{1}{4}(c_1(X,J)^2 - 2\chi(X) - 3\sigma(X)) \tag{8.2}$$

where (X^4, J) is an almost complex manifold bounding (Y^3, ξ), extending ξ as complex lines in X i.e. $\xi = TY \cap J(TY)$, and $\chi(X)$ and $\sigma(X)$ are the Euler characteristic and the signature of X, respectively. To make $c_1(X,J)^2 \in \mathbb{Q}$ a well-defined quantity, we need $c_1(\xi) = c_1(X,J)|_Y$ to be torsion.

To compute these quantities we choose an open book (F,ϕ) supporting the contact structure ξ, then by using Remark 7.5 we find an ALF $\mathcal{F}$ on X which induces this open book on its boundary $Y = \partial X$, and given by a framed link in a square bridge position, $\mathcal{F} = \{K_1, \dots, K_r\}$, such that each component has framing $TB(K_j) \pm 1$. Then, even though X^4 may not be almost complex we can compute ([St]):

$$d(\xi) = \frac{1}{4}(\mathbf{c}^2 - 2\chi(X) - 3\sigma(X)) + q \tag{8.3}$$

where $\mathbf{c} \in H^2(X)$ is the characteristic homology class of the ALF $\mathcal{F}$ as defined in Definition 7.7, and q is the number of defective 2-handles, i.e. the number of K_j's with framing $TB(K_j) + 1$. Also we can compute $c_1(\xi)$ by restricting the homology class $c_1(\xi) := \mathbf{c}|_Y$. When $H^2(Y;\mathbb{Z})$ has no 2-torsion, the two invariants $c_1(\xi)$ and $d(\xi)$ determine the homotopy class of the 2-plane fields on Y when $c_1(\xi)$ is torsion (e.g.[OSt]).

Theorem 8.9. *([E3]) Two overtwisted contact structures on a 3-manifold have homotopic 2-plane fields if and only if they are isotopic.*

8.2 Stein manifolds

Let (X^4, J) be an almost complex manifold, then a map $f : X \to \mathbb{R}$ (Figure 8.2) is called a *J-convex function* if the symmetric 2-form

$$(u,v) \mapsto g_f(u,v) := -d(J^*df)(u, Jv)$$

defines a metric on X (e.g. [CE]). This implies that $\omega := -d(J^*df)$ is a symplectic form on X. If we call $\alpha = J^*df$, by solving the equation $i_u\omega = \alpha$ for u, we get a vector field u. Since $df(u) = \omega(u, Ju)$, it is easy to check that this vector field is *gradient-like* i.e. $u(f) \geq 0$, and also f is *ω-convex* with respect to u, that is, $\mathcal{L}_u\omega = \omega$. We can perturb f (outside of points) to a Morse function still satisfying these properties.

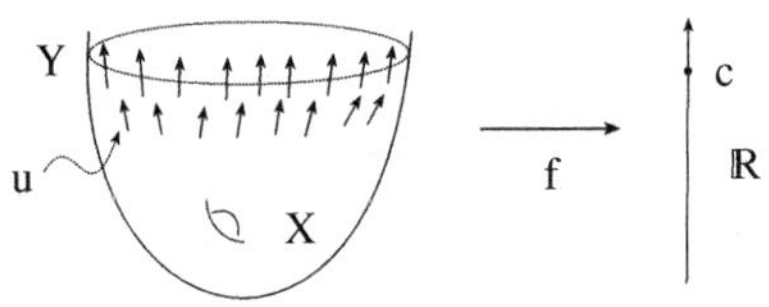

Figure 8.2

These properties imply that X^4 is collared at infinity $Y^3 \times [0, \infty)$, and u is an outward pointing vector field, and α induces a contact structure (Y, ξ) on Y, where $Y = f^{-1}(c)$ for some regular value c of f.

Grauert proved that all properly imbedded complex submanifolds $X \subset \mathbf{C}^N$ admit J-convex functions. We define any proper complex submanifold $X \subset \mathbf{C}^N$ to be a *Stein manifold*, and using this we define a *compact Stein manifold* to be a region $M \subset X$ cut out from X by $f \leq c$, where $f : X \to \mathbb{R}$ is a J-convex proper Morse function, and c is a regular value of f. Then $\omega = \frac{i}{2}\partial\bar{\partial}f$. We call the interior of a compact Stein manifold a *Stein domain*. The data (X, f, ω, u) we constructed from (X, J), consisting of a symplectic manifold, with an ω-convex proper Morse function with respect to a gradient-like vector field, is called a *Weinstein manifold*. This shows that Stein manifolds are Weinstein, the converse is also true but harder to prove [CE].

Definition 8.10. *We say that (M, ω) is a Stein filling of the contact manifold $(Y, \alpha) = \partial(M, \omega)$. The contact 2-plane field $\xi = ker(\alpha)$ can be identified with the restriction of the dual $\mathbb{K}^*$ of the canonical line bundle $\mathbb{K} \to M$ (see Definition 5.4).*

For example, the unit ball $B^4 \subset \mathbb{C}^2$ is a Stein manifold with the J-convex function $f(z_1, z_2) = |z_1|^2 + |z_2|^2$, and it induces the so-called "standard tight" contact structure (S^3, ξ), where the contact planes are the complex lines in $\mathbb{C}^2$ that are tangent to S^3. This is the contact structure of Example 8.1.

8.3 Eliashberg's characterization of Stein

Let (Y^3, ξ) be a contact 3-manifold. We call a curve $K \subset Y$ a *Legendrian knot* if it is tangent to the contact plane field ξ. Every knot in (Y, ξ) can be isotoped to a Legendrian position (knot). For example in $(\mathbb{R}^3, dz+xdy)$, the Legendrian curves appear as the knots with left handed crossings and cusps (southeast–northwest line going over the southwest–northeast line), as shown in the middle picture of Figure 8.3. Of course rotating the Legendrian knot counterclockwise 45° turns it to a square bridge position. By attaching a 1-handle to B^4, we can extend the Stein structure on B^4 to a Stein structure on $S^1 \times B^3$ (see Theorem 8.11). This extends the contact structure (S^3, ξ) to a contact structure on $S^1 \times S^2$, where the Legendrian knots look like curves with left handed crossings and cusps, that are placed between the attaching balls of the 1-handles, as in the last picture of Figure 8.3. By repeating this process we get a contact structure on $\#_k(S^1 \times S^2)$ with a similar description of the Legenderian knots inside. In [G2] the corresponding Stein manifold, $\#_k(S^1 \times B^3)$, is described as 1-handles piled over each other, attached along two parallel lines R_- and R_+, with the Legendrian knots in between.

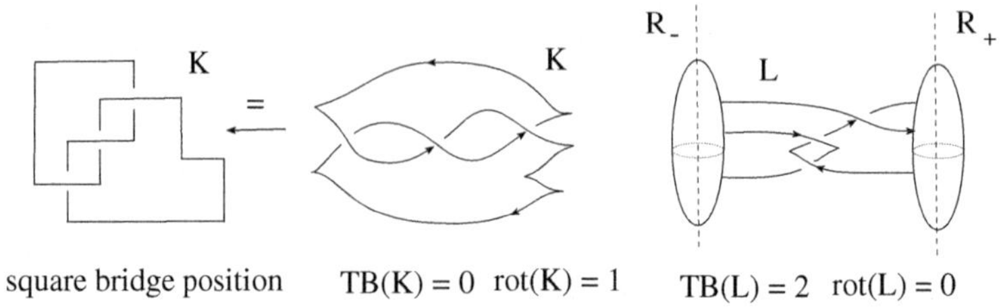

Figure 8.3

The framing of a Legendrian knot $K \subset (Y, \xi)$, which is induced by the contact planes, is called *Thurston–Bennequin framing*, denoted by $TB(K)$. In the case, $(Y, \xi) = (\mathbb{R}^3, dz + xdy)$, or more generally the induced contact structure on $\#_k(S^1 \times S^2)$, this framing can be computed by the integer:

$$TB(K) = w(K) - c(K) \tag{8.4}$$

where the first term is the writhe (Exercise 1.1) and the second term is the number of right (or left) cusps. Notice that by introducing zig-zags to K we can always decrease TB(K) as much as we want. Notice also this is the surface framing of the knot when it is rotated to square bridge position and put on the surface $F(p,q)$ (Exercise 7.8). The following is the 4-dimensional version of the Stein characterization theorem ([E2], [G2]).

Theorem 8.11. *([E2]) Let X^4 be a 4-manifold consisting of B^4 with 1 and 2-handles, i.e. $X = X_\Lambda$, where $\Lambda = \{K_1^{r_1}, \dots, K_n^{r_n}, C_1, \dots, C_s\}$. Then,*

(i) The standard Stein structure of B^4 extends to a Stein structure over the 1-handles $X_{\Lambda'}$, where $\Lambda' = \{C_1, .., C_s\}$.

(ii) If $\{K_1^{r_1}, \dots, K_n^{r_n}\}$ is a Legendrian framed link in $\#_{i=1}^s(S^1 \times S^2)$ with $r_j = TB(K_j) - 1$ for all j, then the Stein structure extends over X.

Furthermore, every Stein manifold admits such a handlebody structure.

Remark 8.12. *The above theorem is relative in the sense that we can replace B^4 with any Stein manifold X; then if we attach 1- handles and 2- handles with $TB - 1$ framing to X, then the Stein structure extends to the new handlebody.*

Exercise 8.2. *Show that the manifold $B_{p,q}$ of Figure 6.17 is Stein.*

Let (X, ω) be a Stein manifold and (Y, ξ) be the induced contact structure on its boundary $Y = \partial X$. Let $K \subset (Y, \xi)$ be an oriented Legendrian knot bounding an imbedded oriented surface F in X. We define the *rotation number* as the evaluation of the relative Chern class of the restriction $\mathbb{K}^* \to F$ of the canonical line bundle $\mathbb{K}^* \to X$

$$rot(K, F) = \langle c_1(\mathbb{K}^*, v), F \rangle \tag{8.5}$$

with respect to the tangent vector field v of K (the obstruction to extending v to a section of $\mathbb{K}^*$ over F). In the case, $X = B^4$, $rot(K, F)$ of an oriented Legendrian knot $K \subset S^3$ does not depend on the choice of the Seifert surface F it bounds in S^3, and it can be calculated as $r(K)$ of Definition 7.6 (after rotating K into square bridge position):

$$rot(K, F) = 1/2\ (\text{number of down cusps} - \text{number of up cusps}) \tag{8.6}$$

In [G2] the definition of the rotation number is extended by the same formula to the Legendrian knots $K \subset \#_s(S^1 \times S^2)$ even if they don't bound surfaces in $\#_s(S^1 \times B^3)$. This is possible since $c_1(\mathbb{K}^*) = 0$ on $\#_s(S^1 \times B^3)$ and hence similar interpretation as (8.5) is possible. Let $X = X_\Lambda$ with $\Lambda = \{K_1^{r_1}, \dots, K_n^{r_n}, C_1, \dots, C_s\}$ be the Stein manifold obtained by Theorem 8.11. As in Section 1.6, by viewing each $K_j \in C_2(X)$ as a 2-chain, we can then compute $\langle c_1(X), K_j \rangle = rot(K_j)$, hence we have (compare Definition 7.7)

$$c_1(\mathbb{K}^*) = c_1(X) = \sum_j rot(K_j) K_j^* \tag{8.7}$$

where $K_j^* \in C^2(M) = Hom(C_2(X); \mathbb{Z})$ are the hom duals of K_j, i.e. $\langle K_j^*, K_i \rangle = \delta_{ij}$.

8.4 Convex decomposition of 4-manifolds

The following theorem says that every closed smooth manifold is a union of two Stein manifolds glued along their common boundaries. It is a good application of the carving technique discussed in Section 1.1.

Theorem 8.13. *([AM3]) Given any decomposition of a closed smooth 4-manifold $M = N_1 \cup_\partial N_2$ by codimension zero submanifolds, such that each piece consists of 1- and 2-handles, after altering pieces by a homotopy inside M, we can get a similar decomposition $M = N_1' \cup N_2'$, where both pieces N_1' and N_2' are Stein manifolds, which are homotopy equivalent to N_1 and N_2, respectively.*

Proof. Theorem 8.11 (and remarks before) says that manifolds with 1- and 2-handles are Stein if the attaching framings of the 2-handles are sufficiently negative (say *admissible*), that is any 2-handle h has to be attached along a Legendrian knot K with framing less than $TB(K)$. Call $\Sigma = \partial N_1$. The idea is by local handle exchanges near h (but away from h) to alter the boundary $(\Sigma, K) \rightsquigarrow (\Sigma', K')$ by homotopy so that it results an increase of the Thurston–Bennequin framing:

$$TB(K) \rightsquigarrow TB(K') = TB(K) + 3$$

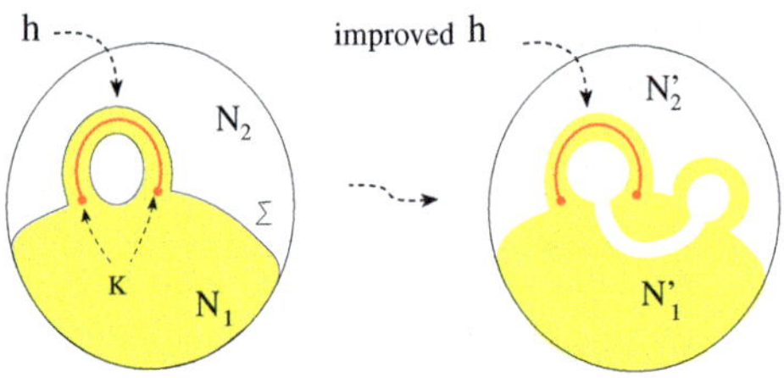

Figure 8.4

This is done by first carving N_1 (which means expanding N_2), then homotopically undoing this carving, as demonstrated by Figure 8.4. For example, as indicated in Figure 8.5, carving out a tubular neighborhood of a properly imbedded 2-disc **a** from the interior of N_1 increases $TB(K)$ by 3. Carving in the N_1 side corresponds to attaching a 2-handle **A** from the other N_2 side, which itself might be attached with a "bad" framing. To prevent this, we also attach a 2-handle **B** to N_1 linking **a**, which corresponds to carving a 2-disk from the N_2 side. This makes the framing of the 2-handle **A** in the N_2 side admissible, also **B** itself is admissible. So by carving a 2-disc **a** and attaching a 2-handle **B** we improved the framing of the attaching circle K of the 2-handle h to K', without changing other handles (we only move Σ by a homotopy). □

Exercise 8.3. *Show that by using Section 8.6, and by positive and negative stabilizing (Sections 7.3, and 7.4) we can arrange it so that:*

(a) The induced contact 2-plane fields in the middle level, $\Sigma := \partial N_1' = \partial N_2'$ are homotopic provided $H^2(\Sigma;\mathbb{Z})$ has no 2-torsion. (Hint: Equate the two homotopy invariants defined in Section 8.1.)

(b) By applying Theorems 8.8 and 8.9, we can make the induced open books in the middle level Σ match, up to orientation; and furthermore after vanishing cycle exchanges from one side to the other we can make both N_1' and N_2' PALF's ([Ba2]).

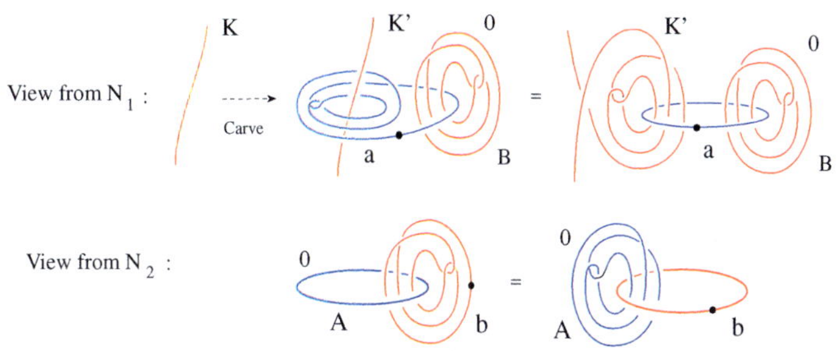

Figure 8.5: Making N_1 and N_2 Stein

Definition 8.14. *A scarfed 2-handle is a 2-handle attached along a Legendrian knot K, plus a small circle (scarf) trivially linking K positively several times (e.g. Figure 8.6).*

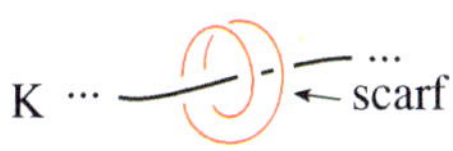

Figure 8.6

Exercise 8.4. *Show that carving a Stein 4-manifold X, along the scarf of the 2-handle in Figure 8.6 (as in the proof of Theorem 8.13) increases $TB(K)$ by 2.*

Exercise 8.5. *Using Exercise 8.4 prove that every ALF can be turned into a PALF after carving a scarf for each defective 2-handle. (Hint: Put all the scarfs on a page; this can be done by positive stabilizations.) The new pages are obtained by puncturing the old pages along scarfs. Twisting the scarf 1-handles repairs the defective framings.*

8.5 $M^4 = |BLF|$

In [GK] it was shown that any closed smooth 4-manifold admits a weaker version of BLF, which is a BLF $\pi : X \to S^2$ with achiral fibers, which they named *BALF*. In [Le] and [AKa3] two independent proofs that all closed 4-manifolds are BLF have been given; the first is a proof by singularity theory, and the second is a proof by handlebody. Also in [Ba3] another singularity theory proof was given, with a weaker conclusion, where on the circle singularities π is not required to be injective.

Theorem 8.15. *Every closed smooth 4-manifold admits a BLF structure.*

Proof. ([AKa3]) First, by Exercise 7.12, we can decompose X as a union of two codimension zero submanifolds along their common boundaries, $X = W \cup -(F \times D^2)$. Then by Propositions 7.4, and 7.9 we can make W an ALF, and $(F \times D^2)$ a concave BLF over D^2. Now we need to match the induced open books in the middle. Using Exercise 7.9 and 8.2, by positively and negatively stabilizing both sides, we can first match the two homotopy invariants (note that positively stabilizing one side corresponds to negatively stabilizing the other side), and then by Theorems 8.8 and 8.9 we can make the induced open books (induced from W and $F \times D^2$ sides) agree. We know how to stabilize ALF W, and by Exercise 7.11 we know how to stabilize the concave BLF $(F^2 \times D^2)$ side, and furthermore we know that negative stabilizations on the concave side can be gotten rid of by introducing more folds there. So we are left with a convex ALF and a concave BLF over D^2, with matching open books on their common boundary. Now, during all these processes we carry a scarf (Defintion 8.14) for each defective (i.e. $TB + 1$ framed) 2-handle on the W side. While moving things into a square bridge position (during the process of turning W into ALF), we treat scarfs as part of the framed link of W and make them lie on the page of the surface $F(p,q)$ as well. We then make them transverse to all the pages of the boundary open book (just as we treated circles with dots in the proof of Proposition 7.4), and carve W side along these scarfs, turning the ALF structure on W to a PALF structure (carving makes achiral 2-handles of W chiral, as in the proof of the Theorem 8.13). By Remark 7.3 this keeps the other side concave BLF. This is just the twisting diffeo across the 1-handle (dotted scarf), which is isotopic to identity, taking one open book to the other, as shown in Figure 8.7 and Figure 8.8 (left to right). □

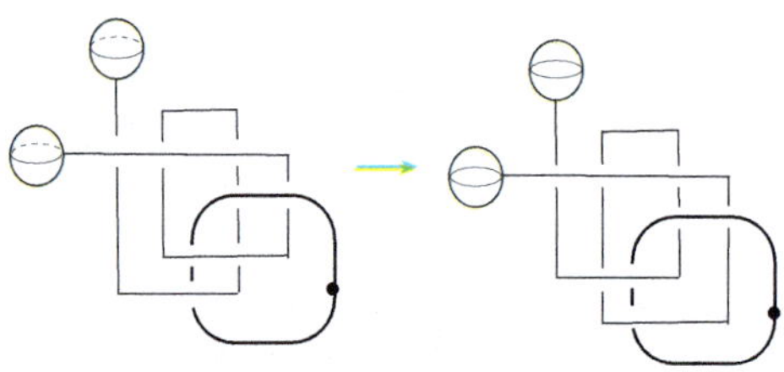

Figure 8.7: Twisting across 1-handle to turn ALF to PALF

Figure 8.8: Twisting across 1-handle described as an open book modification. The second diagram differs from the first by a Dehn twist along the curve α.

Remark 8.16. *(Making achiral singularity chiral by carving) The proof in [Le] starts with the result of [GK], and then by using singularity theory turns the achiral fibers to chiral fibers. The magic of carving can also handle this approach, e.g. as in Figure 8.9, first by carving make the achiral singularity chiral, then by attaching a round 2-handle to keep its topology unchanged (check that the zero-framed knot in the third picture can be made to lie on the page of the open book, then recall Exercise 7.10).*

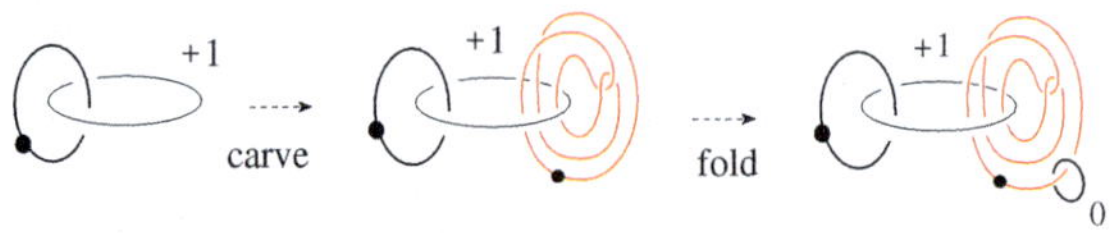

Figure 8.9

8.6 Stein = $|PALF|$

Here we prove that for every Stein manifold X, we can choose an underlying PALF stucture $X = |\mathcal{F}|$. Existence of this structure on Stein manifolds was first proven in [LoP], later in [AO1] an algorithmic topological proof along with its converse was given. More generally, consider the inclusion from (Section 7.3) $\{PALFs\} \subset \{PLFs\}$, then

Theorem 8.17. *There is a surjection* $\Psi : \{PLFs\} \longrightarrow \{\textit{Stein manifolds}\}$*, given by the map* $\mathcal{F} \mapsto |\mathcal{F}|$*. Morover, the restriction* Ψ *to* $\{PALF's\}$ *is a surjection.*

Proof. ([AO1]) The proof that a PALF $\mathcal{F}$ defines a Stein manifold $|\mathcal{F}|$ proceeds as follows: By Chapter 7 (and Section 7.3), a PALF $\mathcal{F}$ is obtained by starting with the trivial fibration $X_0 = F \times B^2 \to B^2$ and attaching a sequence of 2-handles to some curves $k_i \subset F$, $i = 1, 2...$ with framing one less than the page framing: $X_0 \rightsquigarrow X_1 \rightsquigarrow \ldots \rightsquigarrow X_n = \mathcal{F}$. We start with the standard Stein structure on $F \times B^2$ and assume that on the contact boundary F is a convex surface with the *dividing set* $c = \partial F$ ([Gi2], [To]). By the *Legendrian realization principle* "LRP" ([Ho]), there is an isotopy making the surface framings of $k_i \subset F$ agree with Thurston–Bennequin framings, then Theorem 8.11 applies.

In the case where $\mathcal{F}$ is only a PLF, the proof proceeds exactly the same with a small tweak. The only part of the proof we need to be careful is when applying LRP to k, because to apply it we need each $k \subset F$ to be non-isolating (i.e. the components of $F - k$ must touch the dividing curves). For this we choose any non-isolating curve $\alpha \subset F$ and isotope it to a Legendrian, then by the folding principle of Giroux, turn α to a pair of dividing curves ([Ho]); this makes k non-isolating, hence LRP applies.

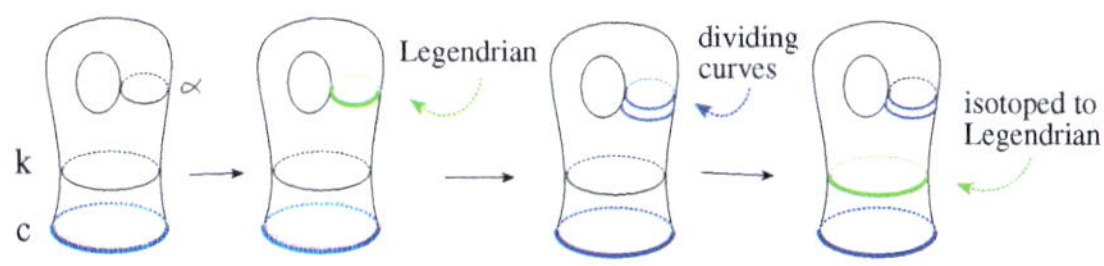

Figure 8.10

Conversely, the proof that a Stein manifold X is a PALF, can be obtained by an easier version of the proof of Proposition 7.4. When there are no 1-handles, by Theorem 8.11 X is a handlebody consisting of 2-handles, attached along a Legendrian framed link $\mathcal{L}$ in S^3, such that each of its components K is framed with $TB(K) - 1$ framing. Then by isotoping $\mathcal{L}$ to a square bridge position (by turning each component counterclockwise 45 degrees), we put the framed link $\mathcal{F}$ on the surface $F(p,q)$, as in Figure 7.20. Now by Exercise 7.8, the framings induced from $F(p,q)$ of each component K coincides with $TB(K)$. Hence $\mathcal{F}$ gives a PALF which is an extension of the PALF $\mathcal{F}_{p,q}$ on B^4. When there are 1-handles the proof is the same as the proof of Proposition 7.4 (by carving). In particular, this time we use Figure 7.27 instead of Figure 7.20. An improved version of this theorem was proven in [Ar]. □

Remark 8.18. *A Stein manifold can have many PALF structures. For example B^4 has a unique Stein structure, whereas it has infinitely many PALF structures induced from different fibered links (why?). Hence PALFs are finer structures than Stein structures.*

8.7 Imbedding Stein to symplectic via PALF

Stein manifolds compactify into closed symplectic manifolds, in fact more is true, they compactify into Kähler surfaces:

Theorem 8.19. *([LMa]) Every Stein manifold imbeds into a minimal Kähler surface as a Stein domain.*

Choosing an underlying PALF structure on a Stein manifold enables us to imbed it into a symplectic manifold in a concrete algorithmic way by the following theorem.

Theorem 8.20. *([AO2]) Any PALF, more generally PLF $\mathcal{F}$, determines a simply connected compactification of the corresponding Stein manifold $X = |\mathcal{F}|$ into a closed symplectic manifold.*

Proof. Given $X = |\mathcal{F}|$, by attaching a 2-handle to the binding of the open book on the boundary (Figure 8.11) we get a closed surface bundle over the 2-disk with monodromy consisting of the product of positive Dehn twists $\alpha_1.\alpha_2...\alpha_k$. Let F^* be the new closed fiber.

Next we extend this fibration by adding new monodromies, $\alpha_1, \alpha_2...\alpha_k \alpha_k^{-1}...\alpha_1^{-1}$ (i.e. attaching corresponding 2-handles) and capping off with $F^* \times B^2$ on the other side (the outside monodromy is identity). We do this after converting each negative Dehn twist α_i^{-1} in this expression, into products of positive Dehn twists by using the relation $(a_1 b_1 ... a_g b_g)^{4g+2} = 1$ of the standard Dehn twist generators of the genus g surface F^* (Figure 8.12). In particular this imbeds X into a Lefschetz fibration, $Z \to S^2$, with closed fibers, which is a symplectic manifold, by [GS]. □

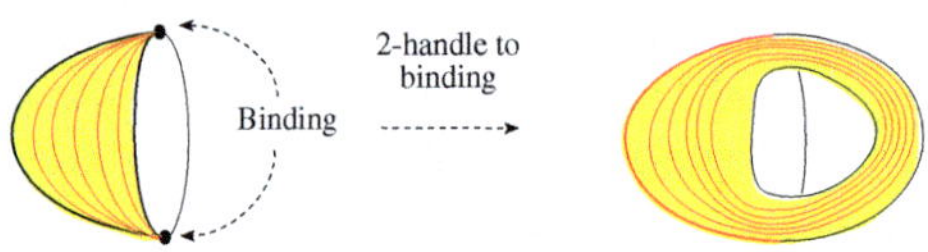

Figure 8.11

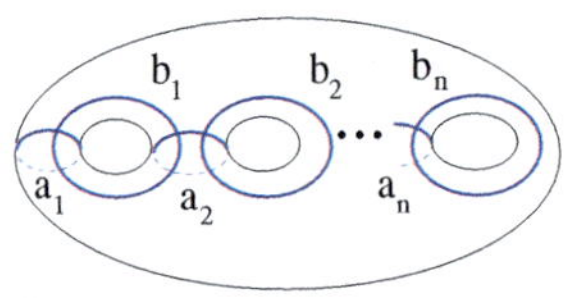

Figure 8.12

Exercise 8.6. *Show that we can make Z simply connected, and $b_2^+(Z) > 1$. (Hint: First expand X to a larger Stein manifold by introducing extra vanishing cycles to $\mathcal{F}$ so that X is simply connected and $b_2^+(X) > 1$, then continue with the compactification process.)*

After choosing a PALF $\mathcal{F}$ on X, compactifying X becomes an algorithmic process. Different PALFs on X could give different compactifications. Even in the case of the Stein ball B^4, with different PALFs on B^4 can give different symplectic compactifications.

Exercise 8.7. *Show that this process compactifies $B^4 = |unknot|$ to $S^2 \times S^2$, and compactifies $B^4 = |trefoil|$ into the K3 surface.*

8.8 Symplectic fillings

In [E1] and [Et1], Theorem 8.20 is extended to the symplectic category. Here we will mostly follow the approach of [Et1] and [Et2] (there is also a nice survey in [Oz1]).

Definition 8.21. *A symplectic manifold* (X,ω) *is said to be a symplectic filling of a contact manifold* (Y^3,ξ)*, if* $\partial X = Y$ *as oriented manifolds, and* $\omega|_\xi \neq 0$*. So if* ξ *is given by the kernel of a contact 1-form* α*, then* $\alpha\wedge\omega|_{TY}\neq 0$*.*

Definition 8.22. *We say* (X,ω) *is a strong convex (concave) symplectic filling of* (Y,ξ) *if there is a vector field* u *defined in a neighborhood of the boundary of* X*, transversally pointing out of (in to)* X *along* Y*, such that* $\xi = ker(\alpha)$*, where* $\alpha = i_u\omega$*.*

Definition 8.23. *A symplectization of a contact manifold* (Y,ξ) *with contact form* α *is the symplectic manifold* $Z = Y\times\mathbb{R}$ *with the symplectic form,* $\omega_Z = d(e^t\alpha)$*.*

Theorem 8.24. *([E1], [Et1]) Any symplectic filling* (X,ω) *of a contact manifold* (Y^3,ξ)*, admits a symplectic imbedding* $(X,\omega)\hookrightarrow(Z,\omega^*)$ *into a closed symplectic manifold* (Z,ω^*)*.*

Proposition 8.25. *([E1], [E4], [OO]) If* (X,ω) *is a symplectic filling of a contact manifold* (Y,ξ)*, which is a rational homology sphere. Then the symplectic form* ω *on* X *can be modified in the neighborhood of* ∂X *to a new symplectic form* $\tilde{\omega}$*, so that* $(X,\tilde{\omega})$ *becomes a strong symplectic filling of* (Y,ξ)*.*

Proof. (from [Ge]): Let $N = [0,1]\times Y$ be a collar of X with $\{1\}\times Y = \partial X$. Since $H^2(N) = 0 \Rightarrow \omega = d\eta$ for some $\eta\in\Omega^1(N)$. Pick $\alpha\in\Omega^1(Y)$ with $\xi = ker(\alpha)$. By assumption, $\alpha\wedge\omega|_{TY} > 0$. For smooth functions $f:[0,1]\to[0,1]$, $g:[0,1]\to\mathbb{R}_{\geq 0}$, define

$$\tilde{\omega} = d(f\eta) + d(g\alpha) = f'dt\wedge\eta + f\omega + g'dt\wedge\alpha + gd\alpha$$

We claim, by choosing f, g appropriately we can make $\tilde{\omega}$ a symplectic form on $[0,1]\times Y$, which agrees with ω in a neighborhood of $\{0\}\times Y$ and strongly fills $\{1\}\times Y$. Clearly this claim implies the result. The proof of this claim is contained in the following exercise. □

Exercise 8.8. *First observe that the assumption of Proposition 8.25 implies that all the 4-forms* $\omega\wedge\omega$*,* $dt\wedge\alpha\wedge\omega$*,* $dt\wedge\alpha\wedge d\alpha$ *are positive volume forms of* X*. Show that:*

(a) By choosing f *to be a decreasing function which is 1 near 0, and 0 near 1; and by choosing* g *to be an increasing function which is 0 near 0, we can arrange* $\tilde{\omega}\wedge\tilde{\omega}$ *to be a positive volume form. (Hint: choose* g *so that* g' *very large compare to* f *and* g*.)*

(b) By taking $s = \log g(t)$, check that near the boundary we have $\tilde{\omega} = d(e^s\alpha)$ and

$$\mathcal{L}_{\partial/\partial s} d(e^s\alpha) = d(e^s\alpha)$$

.

Proposition 8.26. *([EH], [Ga]) Any contact manifold (Y^3, ξ) has a concave filling.*

Proof. In the case where (Y^3, ξ) has a Stein filling (X, ω) this follows from [LMa], which says that (X, ω) symplectically imbeds into a compact Kähler minimal surface of general type, which is a closed symplectic manifold. In the case where (Y^3, ξ) does not have a Stein filling, we pick an open book (F, ϕ) representation of (Y, ξ), and express $\phi : F \to F$ as a product of positive and negative Dehn twists, $\phi = \alpha_1^{\pm}\alpha_2^{\pm}\dots a_k^{\pm}$. For each negative Dehn twist α^- along a curve $\gamma \subset F$, we compose ϕ with another positive Dehn twist along a parallel copy of γ. Enhancing ϕ by this processes cancels the effects of all negative Dehn twists, and this process results in a Stein cobordism from (Y, ξ) to another contact manifold (Y', ξ') represented by (F, ϕ'), where ϕ' is a product of positive Dehn twists, hence (Y', ξ') is Stein fillable, and so by the previous case it bounds a concave filling. □

Proof. (of Theorem 8.24) ([Et1]) The proof starts similarly to that of Theorem 8.20 (except here we won't cap off the boundary of the surface F with a 2-handle). We pick an open book (F, ϕ) supporting (Y, ξ), and express $\phi : F \to F$ as a product of positive and negative Dehn twists, $\phi = \alpha_1^{\pm}\alpha_2^{\pm}\dots a_k^{\pm}$. The mapping class group of F is known to be generated by positive Dehn twists along any loops of F and negative Dehn twists along the boundary parallel loops. So we can write $\phi^{-1} = \phi_0 \circ \phi_1^{-1}$, a composition of positive Dehn twists ϕ_0, and boundary parallel negative Dehn twists, ϕ_1^{-1} (we can put ϕ_1 at the end since boundary parallel Dehn twists commute with other Dehn twists). Therefore, by composing ϕ by positive Dehn twists we end up with a monodromy ϕ_1 of F consisting of some number of (say n) boundary parallel positive Dehn twists. This means that we have constructed a Stein cobordism from (F, ϕ) to (F, ϕ_1), then add to (F, ϕ_1) further positive Dehn twists along the cores of the 1-handles of F, as shown in Figure 8.13, to get (F, ϕ_2), which clearly gives an open book on the 3-manifold obtained by doing $1/n$ surgery to a connected sums of trefoil knots (see Exercise 5.4).

Hence by attaching 2-handles to X we have constructed a cobordism from (Y, ξ) to some contact manifold (Y', ξ'), which is a rational homology sphere, i.e. we have enlarged the symplectic manifold (X, ω) to a larger symplectic manifold (X', ω') with boundary rational homology sphere (Y', ξ'). So by Proposition 8.25 we can deform this to a strong filling of (Y', ξ'). Also by Proposition 8.26 we can find a concave filling of (Y', ξ') on the other side. Then again by Proposition 8.25 we can deform that filling to a strong concave filling (X'', ω''). Application of Exercise 8.9 then finishes the proof. □

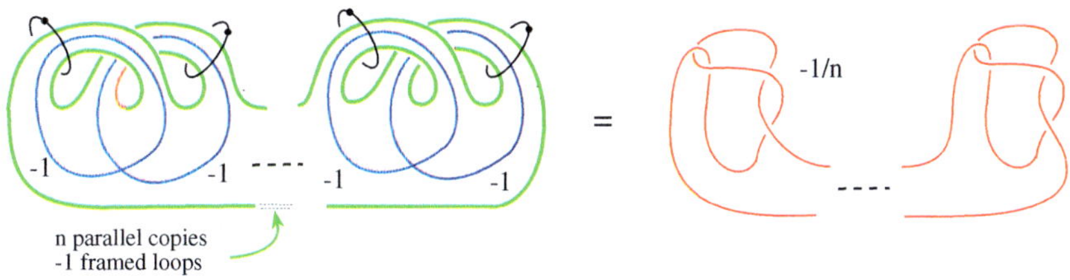

Figure 8.13

Exercise 8.9. *Prove that a strong convex filling (X', ω') and a strong concave filling (X'', ω'') of the same contact manifold (Y^3, ξ) can be glued together along their boundaries to form a closed symplectic manifold. (Hint: The induced contact forms α' and α'' from each side must be related by $\alpha' = g\alpha''$ for some $g : Y \to (0, \infty)$, which we can write as $g = e^f$). Show that $\alpha'' = F^*(i_v \omega_Z)$, where (Z, ω_Z) is the symplectization of (Y, α') (Definition 8.23), $v = \frac{\partial}{\partial t}$ and $F(x) = (x, f(x))$ is the graph (Figure 8.14)).*

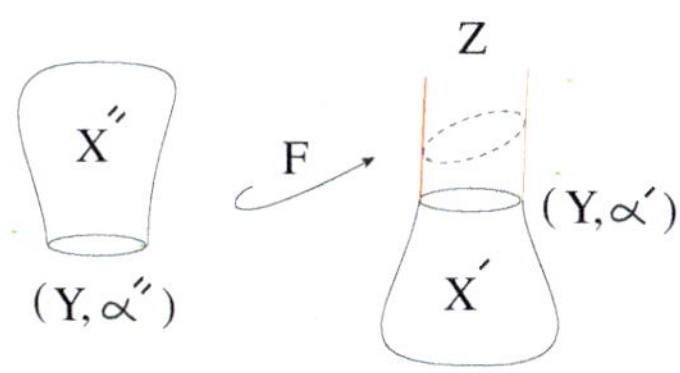

Figure 8.14

Chapter 9

Exotic 4-manifolds

9.1 Constructing small exotic manifolds

Theorem 9.1. *([AM1]) Let $F^2 \subset X^4$ be a properly imbedded surface in a compact Stein manifold, such that $K = \partial F \subset \partial X$ is Legendrian with respect to the induced contact structure. Let n denote the framing of K induced from trivialization of the normal bundle of F (recall definitions of $TB(K)$ and $rot(K)$ from Section 8.3), then:*

$$-\chi(F) \geq (TB(K) - n) + |rot(K)| \tag{9.1}$$

Proof. Attach a 2-handle to X along K with $TB(K) - 1$ framing to get a new Stein manifold $Z = X \smile h^2$ (Remark 8.12). The knot K has two framings, one is the contact framing $TB(K)$ induced from the contact structure of ∂X, and the other is the framing n coming from F. Then the closed surface, $\Sigma = F \cup_\partial D \subset Z$ (where D is the core of the 2-handle h^2), has self intersection $\Sigma.\Sigma = TB(K) - n - 1$, and by (8.7) $rot(K) = \langle c_1(Z), [\Sigma] \rangle$. Then the result follows from the following proposition. □

Proposition 9.2. *Let Z^4 be a compact Stein manifold, and $\Sigma \subset Z$ be a smoothly imbedded closed oriented surface of genus g, representing a nontrivial homology class $[\Sigma] \neq 0$, then the following inequality holds:*

$$2g - 2 \geq [\Sigma]^2 + |\langle c_1(Z), [\Sigma] \rangle| \tag{9.2}$$

Proof. First by attaching 2-handles with $TB - 1$ framing (as in Exercise 8.6) we enlarge Z to have the property $b_2^+(Z) > 1$. Then by [LMa] we imbed Z into a minimal Kähler manifold as a Stein domain. The result follows from Theorems 8.4, 8.5 and 8.6. For the $g = 0$ case, see discussion in [OSt] and [AY7]. □

Theorem 9.3. *([A11]) Let W be the contractible manifold given in Figure 9.1, and let $f : \partial W \to \partial W$ be the involution induced from the obvious involution of the symmetric link in S^3, with $f(\gamma) = \gamma'$, where γ, γ' are the circles in ∂W, as shown in the figure. Then $f : \partial W \to \partial W$ does not extend to a diffeomorphism $F : W \to W$ (but it does extend to a homeomorphism).*

Proof. ([AM1]) From Theorem 8.11, we see that W is Stein (for this we use the description of W given in the second picture of Figure 9.1). Since γ' is slice and $f(\gamma') = \gamma$, if f extended to a diffeomorphism $F : W \to W$ then γ would also be slice in W. But this violates the inequality of Theorem 9.1 (here $F = D^2$, $n = 0$ and $TB(\gamma) = 0$). The fact that f extends to a homeomorphism of W follows from the Freedman theorem [F]. □

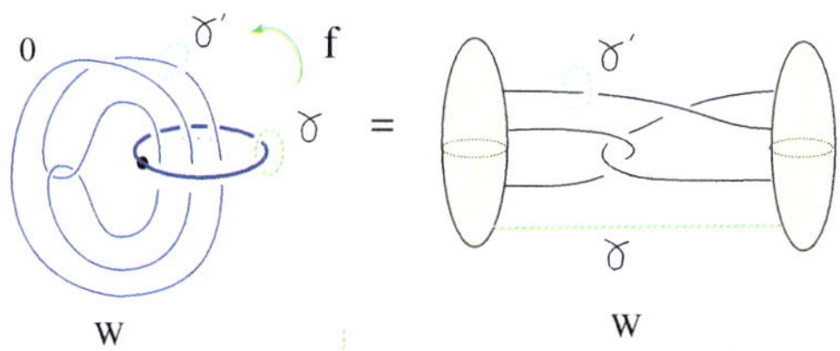

Figure 9.1

Remark 9.4. *If $F : W \to W$ is a homeomorphism extending f, it pulls back the smooth structure of W to a smooth structure which is an exotic copy of W rel boundary. Also note that the contact structure ξ on ∂W (induced from the Stein structure of W) and the pull-back contact structure $f^*(\xi)$ are isomorphic but not isotopic to each other [AM2].*

Such contractible manifolds with involution on their boundaries (W, f) are called *corks*; they will be studied formally in Chapter 10. Recently it was shown that corks can be used to construct *absolutely exotic manifolds* (i.e. no condition on their boundaries).

Theorem 9.5. *[AR] There are compact contractible smooth 4-manifolds V and V' with diffeomorphic boundaries, such that they are homeomorphic but not diffeomorphic to each other. Similarly, there are absolutely exotic smooth manifolds which are homotopy equivalent to S^1.*

This theorem was proved by constructing an invertible cobordism X^4 from $\Sigma = \partial W$ to a homology sphere N^3, with the property that any self-diffeomorphism of N extends to X in such a way that it is isotopic to identity on Σ. Then we let $V = W \cup_\Sigma X$ and $V' = W \cup_f X$. Here invertible cobordism means that there is another cobordism $\bar{X}$ from N to Σ, such that $X \cup_N \bar{X} = \Sigma \times [0,1]$ (show that this implies Theorem 9.5).

Theorem 9.6. *([A12]) Let Q_1 and Q_2 be the manifolds obtained by attaching 2-handles to B^4 along the knots K_1 and K_2 with -1 framing respectively, as shown in Figure 9.2. Then Q_1 and Q_2 are homeomorphic but not diffeomorphic to each other; even their interiors are not diffeomorphic to each other.*

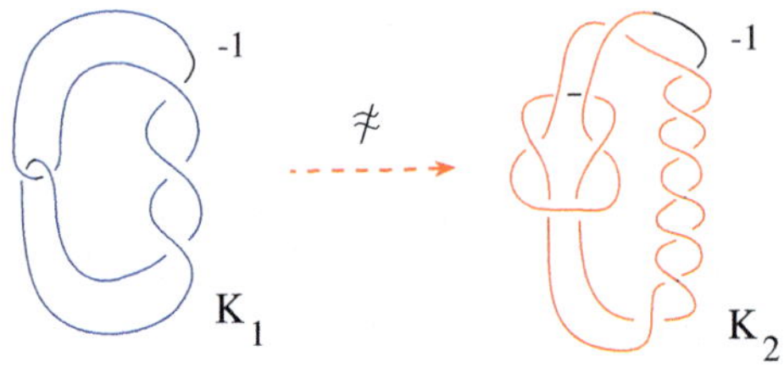

Figure 9.2

Proof. ([AM1]) In Figure 9.3 we obtain Q_1 by canceling the -1 framed 2-handle with the 1-handle; and obtain Q_2 by first sliding the 2-handle over the small -1 framed 2-handle then used it to cancel the 1-handle. So we have the identification, $\partial Q_1 \approx \partial Q_2$, and the manifolds Q_1 and Q_2 are obtained from each other by removing the copy of W and regluing with the involution f of Theorem 9.3 (cork twisting). So they are homotopy equivalent to each other, and by [F] they are homeomorphic to each other.

By Theorem 8.11, Q_1 is a Stein manifold. By [LMa] Q_1 is a domain in a minimal Kähler surface X with $b_2^+(X) > 1$. If Q_1 and Q_2 were diffeomorphic to each other the generator of $H_2(Q_1;\mathbb{Z})$ would be represented by a smoothly imbedded 2-sphere with -1 self intersection (since K_2 is a slice knot, the generator of $H_2(Q_2;\mathbb{Z})$ is represented by a smoothly imbedded sphere). This violates Theorem 8.6. □

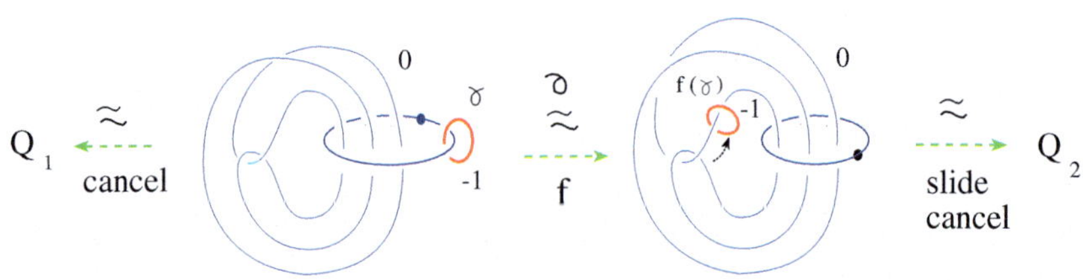

Figure 9.3

In [Y], Theorem 9.6 is generalized to the case of 2-handles attached to B^4 with any framings, for example in the case of 0-framings we have:

Theorem 9.7. *([Y]) Let P_1 and P_2 be the manifolds obtained by attaching 2-handles to B^4 along the knots L_1 and L_2 with 0-framing respectively, as shown in Figure 9.4. Then P_1 and P_2 are homeomorphic but not diffeomorphic to each other. Furthermore both manifolds are Stein manifolds.*

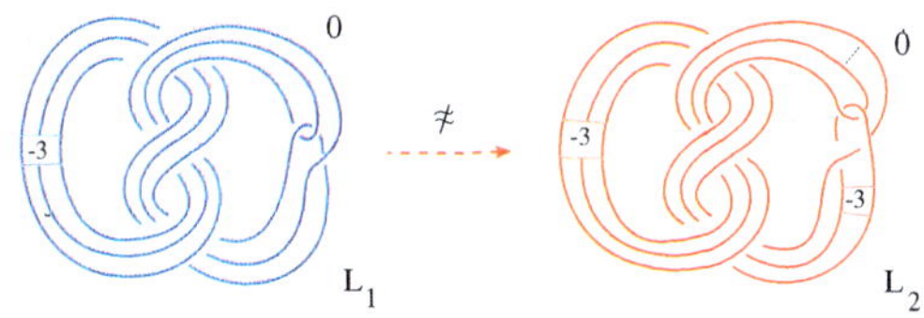

Figure 9.4

Proof. (a sketch) The proof proceeds similarly to that for Theorem 9.6, with a small tweak, which gives the interesting map g. Namely we start with a cork twist, as before.

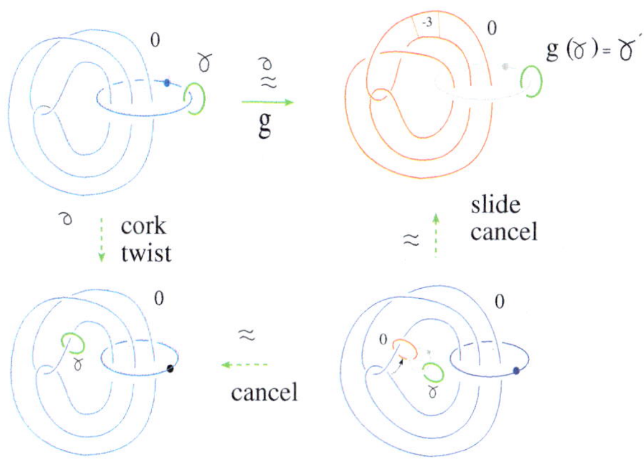

Figure 9.5

In this case we separate the linking loop γ from the 2-handle with the help of a canceling 1/2-handle pair of Figure 9.5. Now the 2-handle can slide over the new 2-handle of the canceling pair, be simplified and then cancel the 1-handle of the cork.

The end result is a useful diffeomorphism $g : \partial W \xrightarrow{\approx} \partial W$ taking the loop γ to γ' in the top horizontal picture of Figure 9.5 where the target W is drawn in a nonstandard way. Let K and K' denote the loops γ and γ' with the Trefoil knots tied onto them. Then attach 2-handles h_K and $h_{K'}$ to the source and target copies of W, along K and K'. Call these manifolds $P_1 = W + h_K$ and $P_2 = W + h_{K'}$. Then the diffeomorphism $g : (\partial W, K) \to (\partial W, K')$ first extends to a homotopy equivalence, $G : P_1 \to P_2$ (here use [Bo]), which by Freedman's theorem can be assumed to be a homeomorphism. Now check that after canceling the 2-handles h_K and $h_{K'}$ with the 1-handles of W we get the manifolds of Figure 9.4. □

Exercise 9.1. *(a) Show that a generator of $H_2(P_2)$ is represented by T^2. (Hint: Do the slice move to L_2 along the dotted line of Figure 9.4.) (b) By imitating the proofs of Theorems 9.6 and 10.10 show that P_1 cannot be diffeomorphic to P_2. (Hint: First by considering a Legendrian projection of L_1 show that P_1 is Stein, then imbed into a symplectic manifold and apply the adjunction inequality to show that any surface representing a generator of $H_2(P_1)$ must have genus ≥ 2.)*

9.2 Iterated 0-Whitehead doubles are non-slice

The inequality of Theorem 9.2 can obstruct the sliceness of some knots as follows. Let $K \subset S^3$ be a Legendrian knot, and let K_0' be a 0-framing push off of K, then:

Proposition 9.8. *([AM1]) We can move K_0' to a Legendrian representative K_0 by an isotopy which fixes K, such that $TB(K_0) = -|TB(K)|$.*

Proof. Figure 9.6 shows 0-framing push offs of K, when $w(K) \leq 0$ and $w(K) \geq 0$, respectively. Clearly in the first case of $w(K) \leq 0$ we have $w(K_0) = w(K)$ and $c(K_0) = c(K)$, hence $TB(K_0) = TB(K)$ (see Section 8.3 to recall definitions).

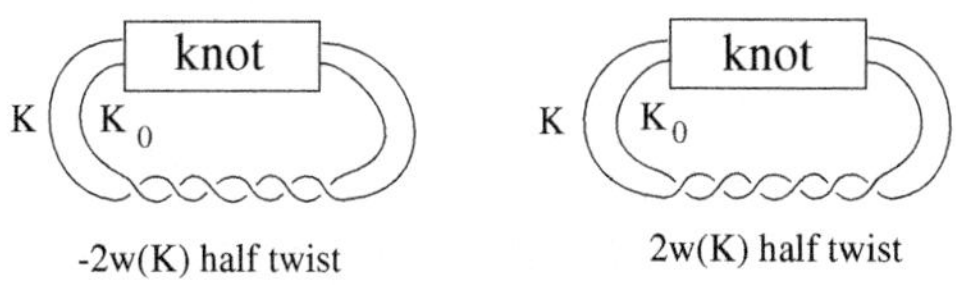

Figure 9.6: $w(K) \leq 0$ and $w(K) \geq 0$ cases

When $w(K) > 0$ and $TB(K) \leq 0$, the $c(K) \geq w(K)$, hence there are enough (right and left) cusps to accommodate a $2w(K)$ half twist between two strands, as shown in Figure 9.7. Hence $TB(K_0) = TB(K)$.

Figure 9.7: The case $w(K) > 0$ and $TB(K) \le 0$

If $w(K) > 0$, $TB(K) > 0$, then $w(K) > c(K)$. Place $2c(K)$ half twist near the cusps, and leave the remaining $2w(K) - 2c(K)$ half twist with zig-zags (Figure 9.8) to make it Legendrian. So $TB(K_0) = TB(K) - [2w(K) - 2c(K)] = TB(K) - 2TB(K) = -TB(K)$. □

Figure 9.8: The case $w(K) > 0$ and $TB(K) > 0$

The $n = 1$ case of the following theorem was first proved by the author (unpublished) via branched covers (see Exercise 11.4), and the general case was proven by Rudolph [R]. The following is a strengthening of this from [AM1], which is an easy consequence of Theorem 9.1.

Theorem 9.9. *If $K \subset S^3$ is a Legendrian knot with $TB(K) \ge 0$, then all iterated positive 0-Whithead doubles $Wh_n(K)$ are not slice. In fact if $Q_n^r(K)$ is the manifold obtained by attaching a 2-handle to B^4 along $Wh_n(K)$ with framing $-1 \le r \le 0$, then there is no smoothly imbedded 2-sphere in $Q_n^r(K)$ representing the generator of $H_2(Q_n^r(K); \mathbb{Z})$.*

Proof. Since $TB(K) \ge 0$, $TB(K_0) = -TB(K)$ (Proposition 9.8). The first Whitehead double $Wh(K)$ is obtained by connecting K and K_0 by a left handed cusp (Figure 9.9).

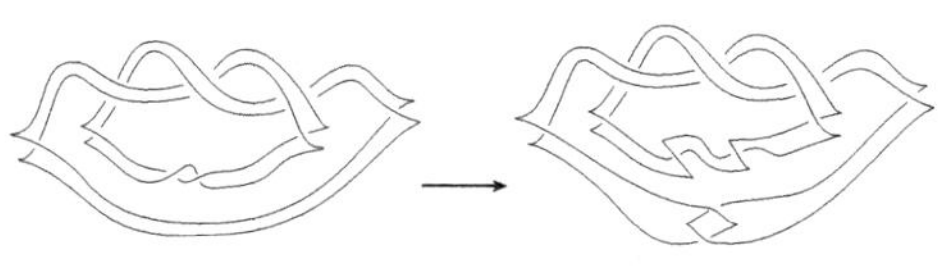

Figure 9.9

Forming the left handed cusp contributes +1 to TB, hence

$$TB(Wh(K)) = TB(K) + TB(K_0) + 1 = TB(K) - TB(K) + 1 = 1$$

By iterating this process we get $TB(Wh_n(K)) = 1$. But Theorem 9.1 (applied to the case $X = B^4$, hence $n = 0$) says that any slice knot L must have $TB(L) \leq -1$.

For the second part, observe that since $TB(Wh_n(K)) - 1 = 0$, the manifold $Q_n^r(K)$ is Stein when $r \leq 0$, hence it can be imbedded into a minimal Kähler surface S. Then by Theorem 8.6 we can not have a smoothly imbedded 2-sphere $\Sigma \subset Q_n^r(K) \subset S$ representing $H_2(Q_n^r(K);\mathbb{Z})$, since $\Sigma.\Sigma = r \geq -1$. (Why can we assume $b_2^+(S) > 1$?) □

Remark 9.10. *The* 0-*Whitehead double* $Wh(K)$ *of any knot* K *may not be slice in* B^4, *as we have seen above, but it is slice in some contractible manifold, when it is imbedded on its boundary [Lit]. Let* W *be the contractible manifold obtained blowing down* B^4 *along the obvious ribbon* D_- *which* $K\# - K$ *bounds (see Section 6.2), and let* $Wh(K) \subset W$ *be the imbedding as shown in the first picture of Figure 9.10. Then the moving pictures of Figure 9.10 demonstrates how* $Wh(K)$ *bounds an imbedded disk in* W. *More specifically, the ribbon move performed in the second picture results in two disjoint circles in* W *which we can cap by disjoint disks. The two disjoint circles first appear as linked Hopf circles in the fourth picture, but they get separated in the final picture after blowing down* D_-.

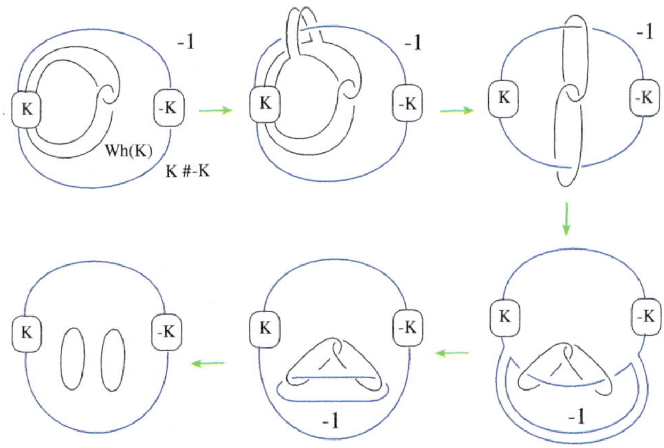

Figure 9.10

Remark 9.11. *In [F] Freedman proved that any knot* $K \subset S^3$ *with Alexander polynomial* $\Delta_K(t) = 1$ *is topologically slice (i.e. bounds a locally flat topologically imbedded* $D^2 \subset B^4$*). For example, the knots* $Wh_n(K)$ *above, where* K *is the trefoil and* $n \geq 1$, *are examples of knots which are topologically slice but not (smoothly) slice.*

9.3 A solution of a conjecture of Zeeman

Every knot in $K \subset S^3$ bounds a PL disk in B^4, which is the disk obtained by simply coning K from the center of B^4. In [Z] Zeeman conjectured that the analogue of this should not hold for contractible manifolds, more specifically he conjectured the knot $\gamma \subset \partial(W)$ of Figure 9.1 should not bound a PL disk in W. This conjecture was settled in [A2] as follows. We already know by Theorem 9.3 that the knot γ cannot bound a smooth disk in W, but after moving it by involution f, the image $\gamma' = f(\gamma)$ bounds a smooth disk in $D_2 \subset W$. But now if γ bounds a PL disk in W, by pushing the singularities together we can assume that it bounds a PL disk $D_1 \subset W$ with one singularity, around which it looks like a cone on a knot K in a small 4-ball $B_1 \subset int(W)$. We know that $-W \smile_f W = S^4$, hence the knot K bounds a smooth disk in the outside 4-ball $B_2 := S^4 - int(B_1)$ namely $D_3 = (D_1 - B_1) \cup D_2$. By replacing the ball B_1 with a copy B_2' of the outside ball B_2 (i.e. inverting outside with inside) we see that γ bounds a smooth disk in W, namely $(D_1 - B_1) \cup D_3'$ where D_3' is a copy of D_3; a contradiction!

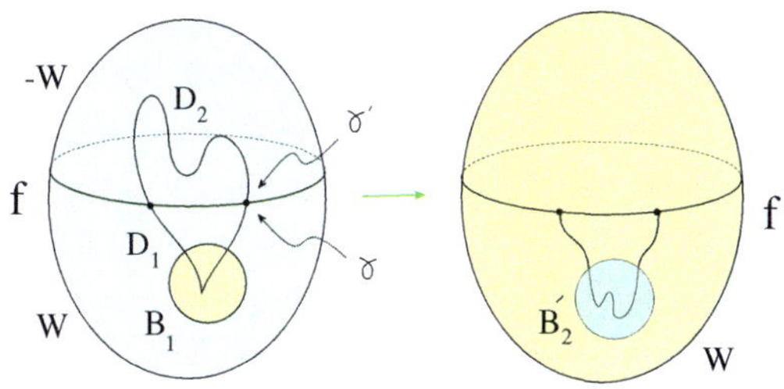

Figure 9.11: Inverting 4-balls

9.4 An exotic $\mathbb{R}^4$

From Remark 9.11, an exotic copy of $\mathbb{R}^4$ can be constructed by the following steps. Let $K \subset S^3$ be a knot which is topologically slice but not smoothly slice (Remark 9.11). Then the 4-manifold K^0 obtained by attaching a 2-handle to B^4 along K by 0-framing, can not smoothly imbed into $\mathbb{R}^4$, but it does topologically imbed $K^0 \subset \mathbb{R}^4$ (verify why?); we can then put a smooth structure $\mathbb{R}^4_\Sigma$ on this topological copy of $\mathbb{R}^4$ by simply extending the smooth structure of K^0 to its complement $\mathbb{R}^4 - K^0$ by using Theorem 1.4 (and by using the fact that 3-manifolds have a unique smooth structures). Then $\mathbb{R}^4_\Sigma$ cannot be the standard $\mathbb{R}^4$ by the initial remark. By more sophisticated use of adjunction inequality (minimal genus function) many more open exotic manifolds can be constructed [G5].

9.5 An exotic non-orientable closed manifold

An exotic copy of $\mathbb{RP}^4$ was constructed in [CS]. There is also an exotic copy Q^4 of the non-orientable manifold $M^4 := S^3 \tilde{\times} S^1 \# S^2 \times S^2$ [A15]. An interesting feature of Q^4 is that it can be obtained from M by a Gluck twisting (Section 6.1). We construct Q by identifying the two boundary components of a compact manifold homotopy equivalent to $S^3 \times [0,1] \# S^2 \times S^2$. Each boundary component of this manifold is the homology sphere $\Sigma(2,3,7)$ (Section 2.4). We construct an upside down handlebody of Q starting with $\Sigma(2,3,7) \times [0,1]$, which is Figure 9.12 (see Section 6.5), then attaching an orientation reversing 1-handle (Section 1.5), a pair of 2-handles and a 3- and a 4-handles, as in Figure 9.13 (3- and 4-handles are not drawn since they are attached in a unique way).

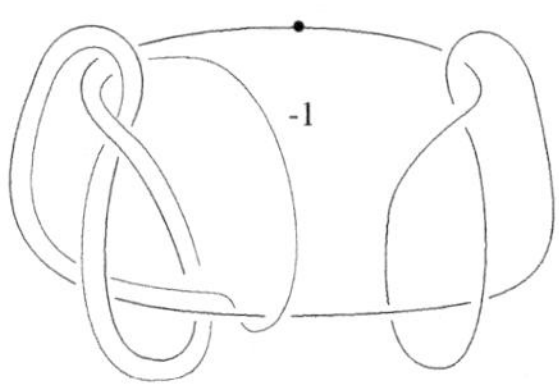

Figure 9.12: $\Sigma(2,3,7) \times [0,1]$

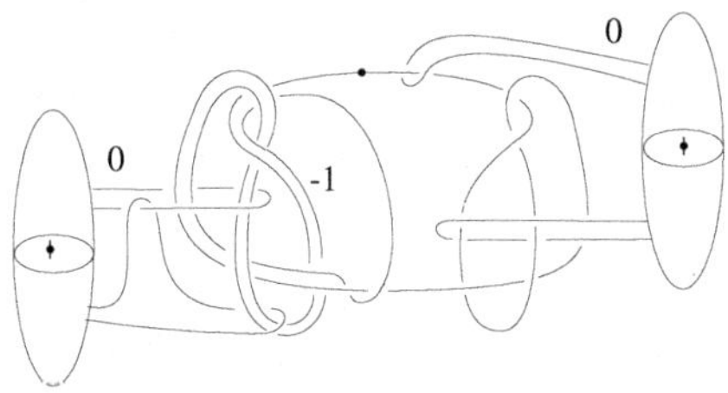

Figure 9.13: Q

Exercise 9.2. *By isotoping the attaching circles across the orientation reversing* 1-*handle, identify the manifold in Figure 9.13 with Figure 9.14 then check that its boundary is* $S^2 \tilde{\times} S^1$. *(Hint: By circle-with-dot and* 0-*framed circle exchanges, turn the two linking* 0-*framed circles in the middle, canceling* 1- *and* 2-*handle pairs, then cancel them).*

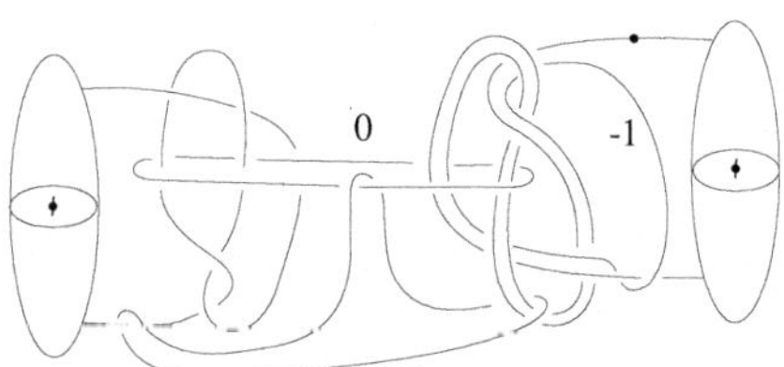

Figure 9.14: Q

Proposition 9.12. *Q is an exotic copy of M.*

Proof. By composing with the collapse map, any homotopy equivalence $Q \to M$ gives rise to a degree 1 normal map, $f : Q \to S^3 \tilde{\times} S^1$ (normal means f pulls back the stable normal bundle to the stable normal bundle [Br]). Let $\mathcal{S}(M)$ be the set of equivalence classes of pairs (Q^4, f), where Q^4 is a closed smooth 4-manifold and $f : Q \to S^3 \tilde{\times} S^1$ is a degree 1 normal map. We define the equivalence $(Q_1, f_1) \sim (Q_2, f_2)$ as a compact smooth cobordism Z^5 between Q_1 and Q_2, and a normal map $F : Z \to S^3 \tilde{\times} S^1$ which restricts to $f_1 \sqcup f_2$ on the boundary. Define $\lambda : \mathcal{S}(M) \to \mathbb{Z}$ (mod 32) as follows. For $(Q, f) \in \mathcal{S}(M)$, make f transverse to S^3 and let Σ be the framed 3-manifold $f^{-1}(S^3)$

$$\lambda(Q, f) = 2\mu(\Sigma) - \sigma(W) \pmod{32}$$

where $W = Q - \Sigma$. This is a well-defined invariant of (Q, f); this is because if we cut open Z along a transverse inverse image $N = F^{-1}(S^3)$, we get an oriented 5-manifold $H = Z - N$ with boundary $\partial H = -W_1 \cup N \cup W_2 \cup N$ (Figure 9.15) for some spin 4-manifold N (why?) with $\partial N = \Sigma_1 \cup -\Sigma_2$, and hence the signature of its boundary is zero, and by definition $\mu(\Sigma_1) - \mu(\Sigma_2) = \sigma(N)$ (mod 16) (Section 5.5).

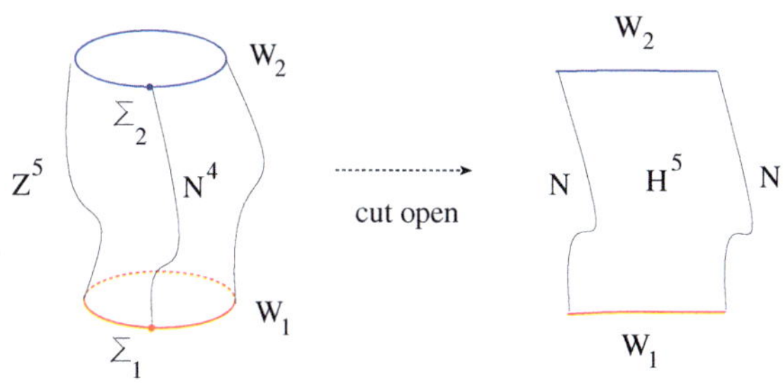

Figure 9.15

Now we can compute $\lambda(Q, f)$ for the specific manifold we constructed above:

$$\lambda(Q, f) = 2\mu(\Sigma(2,3,7)) - \sigma(S^3 \times [0,1] \# S^2 \times S^2) = 2\mu(\Sigma(2,3,7)) = -16 \pmod{32}$$

We claim Q cannot even be h-cobordant to $S^3 \tilde{\times} S^1 \# S^2 \times S^2$, because if it is, by attaching a 3-handle to this cobordism we can extend f across the cobordism and get $(Q, f) \sim (S^3 \tilde{\times} S^1, g)$, where $g : S^3 \tilde{\times} S^1 \to S^3 \tilde{\times} S^1$ is a homotopy equivalence. Therefore, $\lambda(Q, f) = \lambda(S^3 \tilde{\times} S^1, g) = 0$ (why?). Hence Q cannot be standard. □.

Exercise 9.3. *Show that Q is obtained from $S^3 \tilde{\times} S^1 \# S^2 \times S^2$ by a Gluck twisting across an imbedded 2-sphere. (Hint: Observe that the -1 framed circle of Figure 9.14 is isotopic to the trivially linked -1 framed circle of Figure 9.16 (see Section 6.2).)*

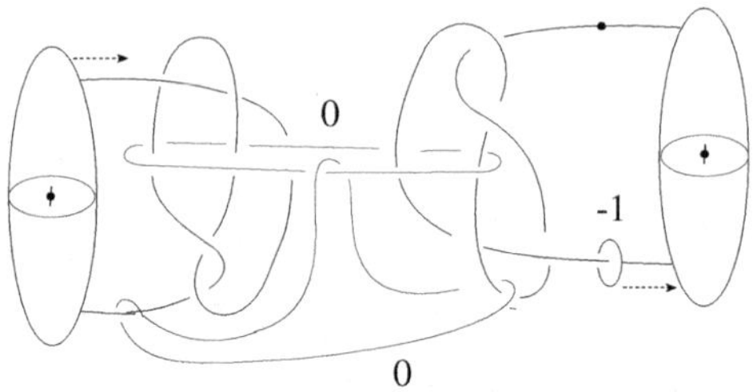

Figure 9.16

Exercise 9.4. *Show that Q is also obtained from $(S^3 \tilde{\times} S^1) \# \mathbb{CP}^2 \# \bar{\mathbb{CP}}^2$ by a Gluck twisting across an imbedded 2-sphere. (Hint: Slide -1 framed circle of Figure 9.16 across the orientation reversing 1-handle, then slide it over the 0-framed 2-handle, and then slide the other 0-framed handle over it to obtain Figure 9.17.)*

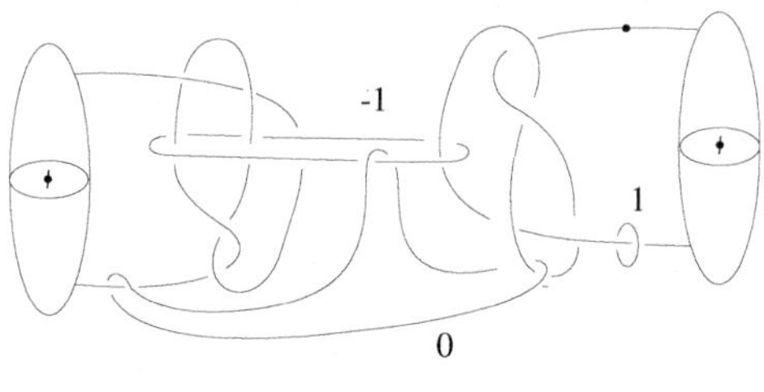

Figure 9.17

Chapter 10

Cork decomposition

In higher dimensions, 1-handles of simply connected manifolds can be eliminated (this is the first step of the proof of the h-cobordism theorem). In dimension 4 it is not clear if this can be done, in fact when the manifold has boundary it cannot be done, as in the case of the Mazur manifold (Figure 1.4) (see also Example 7.5 of [AY2]). Nevertheless, in dimension 4 there is a useful weaker analogue of this (e.g. [Ma] and [GS]). That is, we can decompose $X^4 = W \smile_\partial N$ as a union of contractible manifold W and a manifold N without 1-handles glued at the top of W along its boundary, and furthermore by applying Theorem 8.13 we can make each piece Stein.

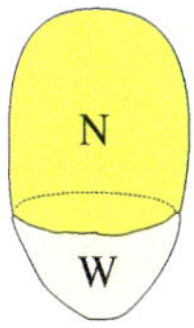

Figure 10.1: X^4

Proposition 10.1. *Let X be a smooth 1-connected 4-manifold given as a handlebody with $X^{(2)} = X_\Lambda$ where $\Lambda = \{K_1^{r_1}, \dots, K_n^{r_n}, C_1, \dots, C_p\}$, where $X^{(i)}$ denotes the subhandlebody of X consisting of handles of index $\leq i$. Then by introducing pairs of canceling 2- and 3-handles, we can obtain a new enlarged handlebody for X with $X^{(2)} = X_{\Lambda'}$, where*

$$\Lambda' = \{K_1^{r_1}, \dots, K_n^{r_n}, L_1^{s_1}, \dots, L_p^{s_p}, C_1, \dots, C_p\}$$

such that the attaching circle of each 2-handle L_j is homotopic in $X^{(1)}$ to the core of the 1-handle C_j. So this gives the decomposition of $X = W \cup N$, as in Figure 10.1, such that $W = W_{\Lambda''}$ with $\Lambda'' = \{L_1^{s_1}, \dots, L_p^{s_p}, C_1, \dots, C_p\}$, and $N = N_{\Lambda'''}$ with $\Lambda''' = \{K_1^{r_1}, \dots, K_n^{r_n}\}$.

Proof. Let γ be the core circle of a 1-handle C (the linking circle of the circle with dot). If γ bounded a disk in $\partial X^{(2)}$, by attaching a canceling pair of 2- and 3-handles, along γ, we could create a canceling 2-handle for C. But we don't know this. To prove the proposition we only have to create a 2-handle in X whose attaching circle L is homotopic to γ in $X^{(1)}$. Since X is simply connected we can find an immersed cylinder H in $X^{(2)}$ connecting γ to an unknotted trivial circle $c \subset \partial X^{(2)}$. By general position, H misses the cores of the 1-handles of X, but might meet 2-handles of X along points. By piping these points to γ, as shown in Figure 10.2, we obtain a circle $\gamma' \subset \partial X^{(2)}$, which is homotopic to c in in the complements of the cores of 1- and 2-handles (and also homotopic to γ in $X^{(1)}$). Hence this homotopy H can be pushed into the 3-manifold $\partial X^{(2)}$, where it can be viewed as moving pictures starting at γ', occasionally self crossing and ending at c. So modifying γ' by connected summing with the small dual circles of itself (when self crossings occur), we obtain L which is isotopic to c, and still homotopic to γ in $X^{(1)}$. So after attaching a canceling 2/3-handle pair to c, we see that L bounds an imbedded disk, which we can viewed as the attaching frame circle of a 2-handle. By repeating this process all the 1-handles we get the required frame link $\{L_1^{s_1}, \dots, L_p^{s_p}\}$. □

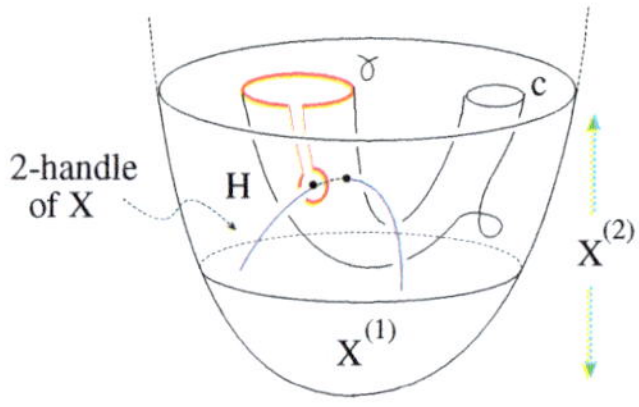

Figure 10.2

Remark 10.2. *When $dim(X) \geq 5$ we can get a stronger conclusion with less work. In the beginning of the proof, we can make the cylinder H miss all the 2-handles by general position, and imbed it into $\partial X^{(2)}$, which we can use to construct the 2-handle canceling the 1-handle. By this process we can cancel all the 1-handles of X.*

10.1 Corks

Definition 10.3. *A cork is a pair (W, f), where W is a compact contractible Stein manifold, and $f : \partial W \to \partial W$ is an involution, which extends to a self-homemorphism of W, but it does not extend to a self-diffeomorphism of W. We say (W, f) is a cork of M, if $W \subset M$ and cutting W out of M and re-gluing with f produces an exotic copy M' of M (a smooth manifold homeomorphic but not diffeomorphic to M). This means that we have a decomposition as in Figure 10.3 (where $N = M - int(W)$):*

$$M = N \cup_{id} W \ , \quad M' = N \cup_f W \tag{10.1}$$

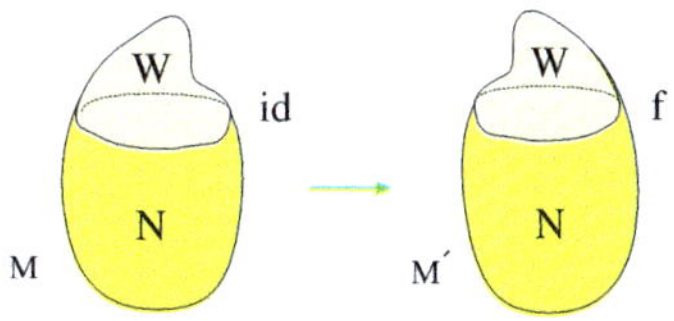

Figure 10.3: Cork decomposition

A Cork of 4-manifold first appeared in [A11] (Theorem 9.3). More generally in [Ma] and [CFHS], it was shown that any exotic copy M' of a closed simply connected 4-manifold M differs from its original copy by a cork (this was also discussed in [K2]). In [AM3], by using the convex decomposition Theorem 8.13, it was shown that in the cork decompositions (10.1), each of the W and N pieces can be made into Stein manifolds. Since then, the Stein condition has become a part of the definition of cork. So we have:

Theorem 10.4. *Let M' be an exotic copy of a closed smooth simply connected 4-manifold M. Then M and M' decompose as in (10.1), where W is a compact contractible, and both W and N are Stein manifolds.*

The operation $M \mapsto M'$ is called a *cork twisting.* Note that a cork is a fake copy of itself relative to its boundary. There are some natural infinite families of corks: W_n, $\bar{W}_n$, $n \geq 1$ as shown in Figure 10.4 ([AM3], [AY2]), where their involutions f are induced by the *zero and dot exchanges* on their underlying symmetric links. W_1 is a cork of $E(2)\#\bar{\mathbb{CP}}^2$ ([A11]), and $\bar{W}_1$ is a cork of $E(1)_{2,3}$ ([A7]). We call W_n's *Mazur corks*, since W_1 is a Mazur manifold ([Maz1]). In [AM3] $\bar{W}_n{}'s$ are called *positrons.*

Exercise 10.1. *Prove that W_n and $\bar{W}_n$ are corks. (Hint: By attaching -1 framed 2-handles to γ or $f(\gamma)$, where γ is the linking circle of the 1-handle, obtain two manifolds Q_1 and Q_2, such that Q_1 is Stein. Then by compactifying Q_1 to a Kähler surface conclude that, if $Q_1 \approx Q_2$ we get violation of Theorem 8.6.)*

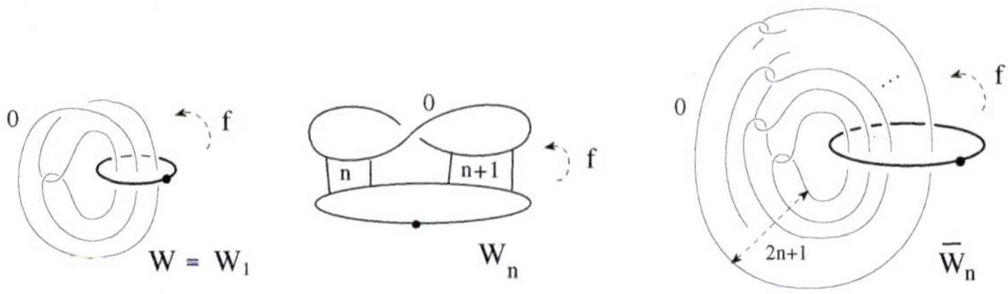

Figure 10.4: Variety of corks

Next we outline R. Matveyev's proof of the general *cork theorem*; here we will present a weaker version where two different contractible manifolds W are allowed on two sides of the decomposition Definition 10.3.

Proof. ([Ma]) Let $f : M \to M'$ be a homotopy equivalence between closed smooth simply connected 4-manifolds. This implies that there is a 5-dimensional h-cobordism Z^5 between M and M' ([Wa1]). We can assume that Z has no 1- and 4-handles (Remark 10.2). So M' is obtained from M by attaching by 2- and 3-handles. Upside down 3-handles are 2-handles. Let N be the middle level between 2- and 3-handles (Figure 10.5).

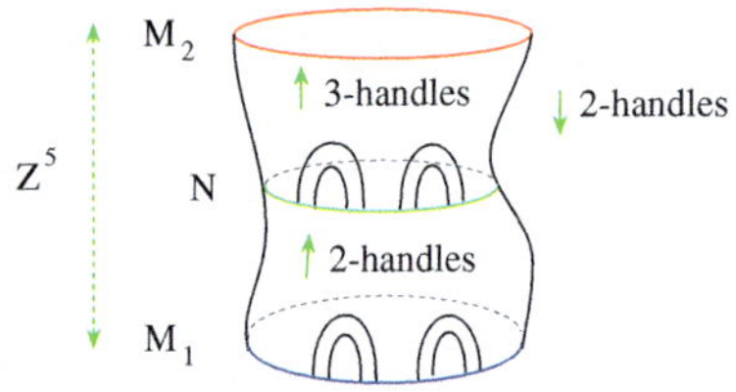

Figure 10.5

In the middle level we have following diffeomorphisms. This is because surgering a circle in a 1-connected 4-manifold has the effect of connected summing $S^2 \times S^2$ (why?).

$$\phi_1 : M \overset{k}{\#} (S^2 \times S^2) \to N$$

$$\phi_2 : M' \overset{k}{\#} (S^2 \times S^2) \to N$$

After introducing canceling 2- and 3-handle pairs we can assume

$$(\phi_2^{-1} \circ \phi_1)_* \mid_{H_2(M)} = f_*$$

and assume that the 2-spheres are mapped to homotopic copies of the corresponding 2-spheres by $\phi_2^{-1}\circ\phi_1$ in the connected sum decomposition. For simplicity assume $k=1$. So we have two homotopic imbeddings, $c_i : S_1^2 \vee S_2^2 \hookrightarrow N$, $i=1,2$, giving rise to an algebraically dual imbedded 2-spheres $\{S,P\}$ in N, where $S=c_1(S_1^2)$ and $P=c_2(S_2^2)$, with the property that surgering N along S gives M, and surgering N along P gives M'. Take the tubular neighborhood of the union of these two spheres

$$V_0 = N(S\cup P)$$

Now we need to expand V_0 a little so that surgering S and P inside would give the two contractible manifolds, W_1 and W_2, to be used in (10.1). If S and P intersect at more than one point, we can take a pair of oppositely signed intersection points, and take the corresponding Whitney circle λ. Clearly λ is null homotopic in the complements of either S or P (because surgering N along them gives simply connected manifolds M and M'). Let D and E be two immersed disks which λ bounds in the complements of S and P, respectively. Then as homology we have

$$[D\cup -E]\in H_2(M)\oplus\langle S,P\rangle$$

So we can decompose $[D\cup -E]=a+\alpha[S]+\beta[P]$. After, possibly by connected summing D (and E) by immersed 2-spheres lying in the complement of S (and in the complement of P), we can assume that D and E are homotopic to each other relative to the boundary λ. We can put this homotopy $\varphi_t : B^2\to N$ in general position so that the middle level $K=\varphi_{1/2}(B^2)$ might meet both S and P, but D and E are obtained from K by a series of Whitney moves. This means that there are imbedded Whitney disks $X=\{X_i\}$ between K and S, and $Y=\{Y_j\}$ between K and P, with interiors disjoint from everything else (but their boundaries might meet in K). Whitney moves across them produces the disks D and E. Now take the union of V_0 with the tubular neighborhood of $K\cup X\cup Y$

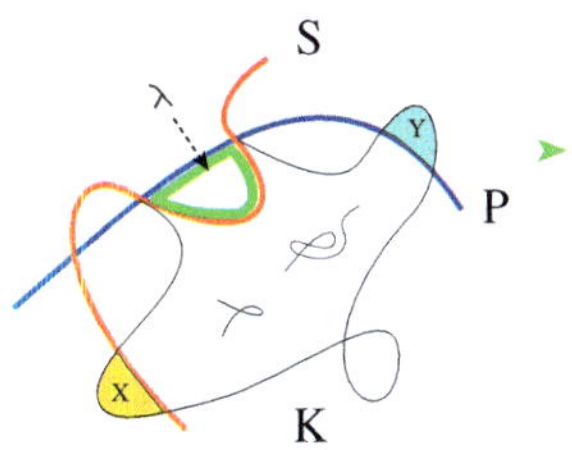

Figure 10.6

$$V_1 = V_0 \cup N(K \cup X \cup Y).$$

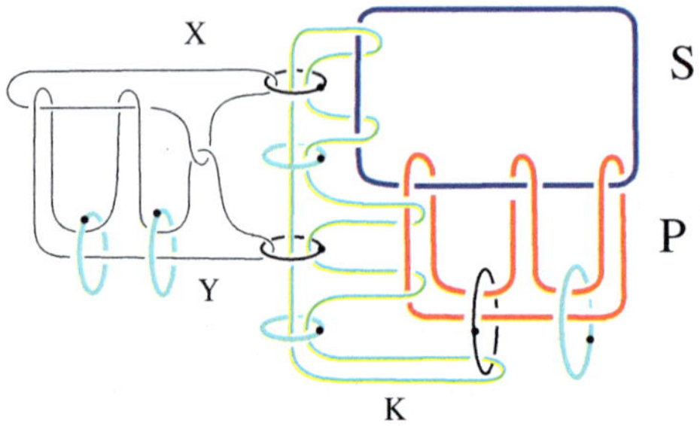

Figure 10.7: V_1

We have the following properties (the first two follow from the construction):

(a) $\pi_1(V_1)$ is a free group.

(b) $H_2(V_1) = \mathbb{Z} \oplus \mathbb{Z}$ generated by $\{S, P\}$.

(c) S, P have geometric duals $S^{\perp}, P^{\perp} \subset V_1$
(immersed spheres intersecting S and P at one point).

To prove (c), by using disks X we push K to a new disk K', whose interior is disjoint from S. Then by using K' we do the immersed Whitney trick to P eliminate the pair of points of opposite sign in $S \cap P$. By continuing this way we obtain $S^{\perp}$. Similarly we construct $P^{\perp}$.

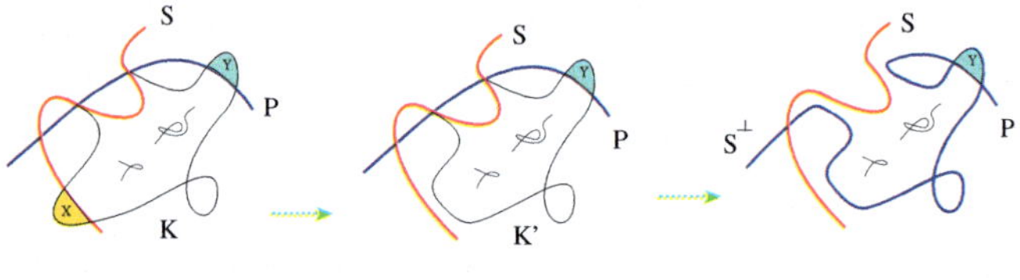

Figure 10.8: Creating a geometric dual $S^{\perp}$ to S

Finally, by first representing the free generators of $\pi_1(V_1)$ by 1-handles, and then enlarging V_1 by attaching the 2-handles algebraically canceling each one of these 1-handles (provided by Proposition 10.1), we obtain a simply connected manifold V_2 satisfying (b) and (c). Then surgering S (or P) from V_2 we get the required W_1 and W_2 □

Example 10.1. *(Relating a rational blowdown to a cork twist [AY3]) If there is an imbedding of plumbings $C_p \subset D_p \subset X$ of Figure 10.9, and $X \mapsto X_p$ is the rational blowdown of X along C_p (Section 6.6), then the cork twisting X along the cork W_p is equivalent to $X_p \# (p-1)\mathbb{CP}^2$. To see this, in Figure 10.10 we decompose $D_p \approx \bar{D}_p + h^2$, where h is a 2-handle, then in Figure 10.11 show this equivalence for $\bar{D}_p$ (identify Figure 10.10 with D_p, and verify the top diffeomorphism of Figure 10.11, and recall Figure 6.19).*

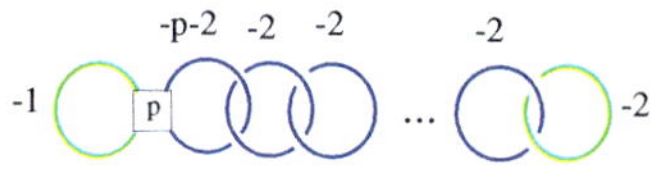

Figure 10.9: $C_p \subset D_p$

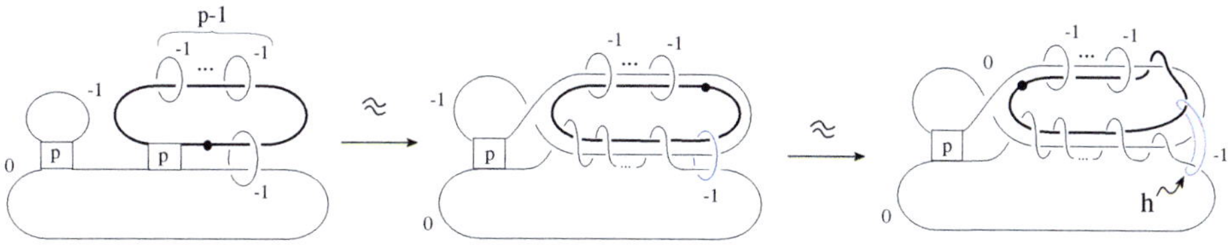

Figure 10.10: $D_p \approx \bar{D}_p + h^2$

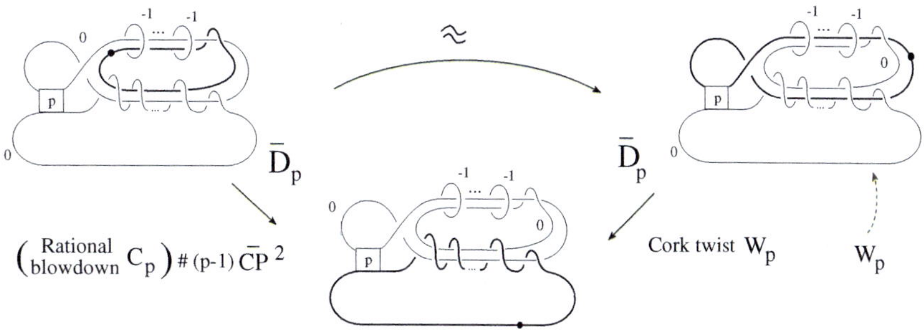

Figure 10.11

10.2 Anticorks

When the involution f of a cork (W, f) fixes the boundary of a properly imbedded disk $D \subset W$ up to isotopy, $f(\partial D)$ bounds a disk in W as well (isotopy in the collar union D), hence we can extend f across the tubular neighborhood of $N(D)$ of D by the carving process of Section 2.5. This results in a manifold $Q = W - N(D)$ homology equivalent to $B^3 \times S^1$, and an involution on its boundary, $\tau : \partial Q \to \partial Q$, which doesn't extend to Q as a diffeomorphism (otherwise f would extend to a self-diffeomorphism of W). Clearly when τ takes a null-homotopic loop $\delta \subset Q$ in Q to an essential loop in Q, it can not extend to Q by homotopical reason. Of course the interesting case is when τ extends to a self-homeomorphism of Q, in which case we can conclude that τ gives an exotic structure to Q relative to its boundary (just as in the cork case). This happens when $\pi_1(Q) = \mathbb{Z}$ by [F]. For example, when we apply this process to the Mazur cork we get the manifold Q, shown in Figure 10.12 (cf. [A2]). This leads us to the following definition.

Definition 10.5. *An anticork is a pair (Q, τ), where Q is a compact manifold which is homotopy equivalent to S^1, and an involution $\tau : \partial Q \to \partial Q$, which extends to a self-homeomorphism of Q, but it does not extend to a self-diffeomorphism of Q.*

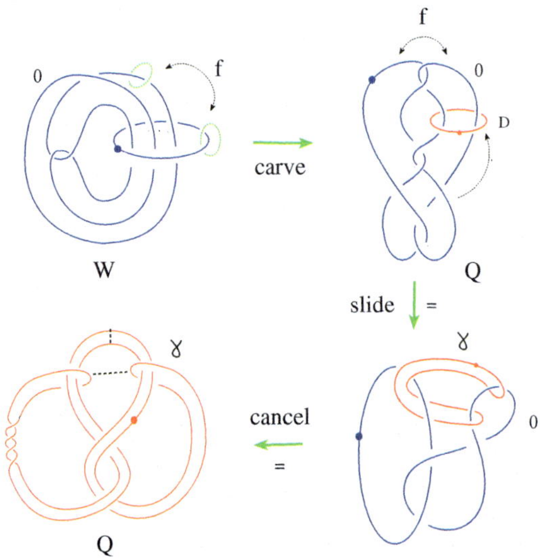

Figure 10.12

The reason for calling (Q, τ) an anticork, is that when it comes from a cork (W, f) as above, twisting Q by the involution τ undoes the effect of twisting W by f. Notice that the loop $\gamma = \partial D$ bounds two different disks in B^4 with the same complement Q (where the identity map between their boundaries does not extend to a diffeomorphism inside); they are described by the two different ribbon moves, indicated in the last picture of Figure 10.12. These two disks correspond to the obvious disks which γ bounds in the next to last picture of Figure 10.12, before and after zero and dot exchanges in the figure.

Remark 10.6. *Let M be the 4-manifold obtained by attaching a 2-handle to B^4 along the ribbon knot γ of Figure 10.12, with $+1$ framing. Clearly M has two imbedded 2-spheres S_i, $i = 1, 2$, of self intersection $+1$ representing $H_2(M) \cong \mathbb{Z}$, corresponding to the two different 2-disks which γ bounds in B^4. Blowing down either S_1 or S_2 turns M into the (positron) cork $\bar{W}_1$ of Figure 10.4, and the two different blowing down processes turns the identity map $\partial M \to \partial M$ to the cork involution $f : \partial \bar{W}_1 \to \partial \bar{W}_1$, i.e. the maps in Figure 10.13 commute (this can be seen by blowing down γ of Figure 10.12 by using the two different disks). In particular since $M = \bar{W}_1 \# \mathbb{CP}^2$, this shows that the blowing up $\mathbb{CP}^2$ operation undoes the cork twisting operation of the cork $(\bar{W}_1, f)$ (cf. [A20]).*

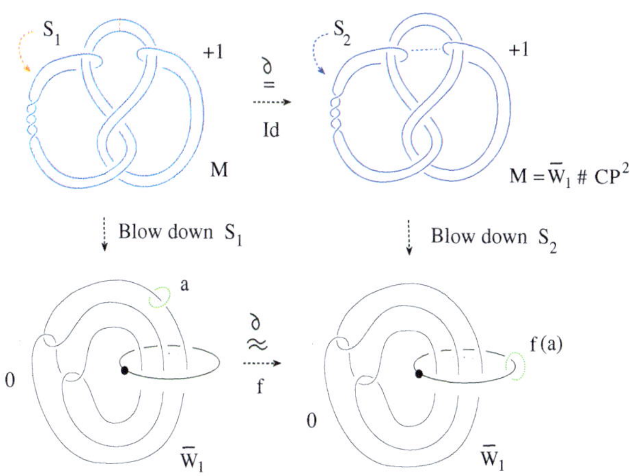

Figure 10.13

Exercise 10.2. *Explain why 2-spheres $S_1, S_2 \subset M$ above, are not isotopic to each other in M, but they become isotopic in $M \# S^2 \times S^2$. (Hint: Next to last picture of Figure 10.12.)*

10.3 Knotting corks

Even though corks are contractible objects, any one cork in a 4-manifold $f : W \hookrightarrow X^4$ should be thought of as analogous to a knot in $\mathbb{R}^3$. The imbedding of W can be knotted with different isotopy types, detected by the different exotic smooth structures it induces on X. The question of whether all different exotic smooth structures of X can be detected by a fixed cork is an intriguing one.

Theorem 10.7 ([AY3]). *There exists a closed smooth 4-manifold X^4, and infinitely many non-isotopic imbeddings $f_k : W \hookrightarrow X$ of a cork W, such that by cork twisting X along $f_k(W)$ for different $k = 1, 2, \ldots$ gives infinitely many different exotic copies of X.*

Proof. ([AY6]) Let S be the Stein manifold in Figure 10.14 (consider as a PALF). Compactify S to a symplectic manifold $\tilde{S}$ (Theorem 8.20). Notice that S contains the cork $W := W_1$ and the cusp C, imbedded disjointly. Then commutativity of the diagram in Figure 10.15, along with the facts (a), (b) and (c), gives the proof.

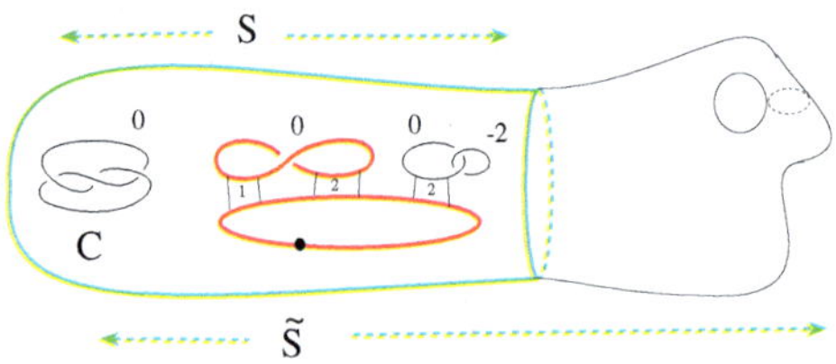

Figure 10.14

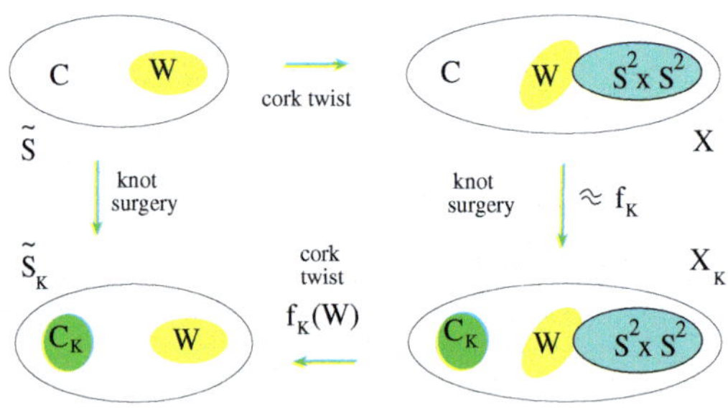

Figure 10.15

(a) Knot surgery of $\tilde{S} \mapsto \tilde{S}_K$ along C, by using knots K with different Alexander polynomials, produces different exotic manifolds (Section 6.5). This is indicated by the left vertical arrow.

(b) Cork twisting $\tilde{S}$ along W_1 splits an $S^2 \times S^2$ (top horizontal arrow).

(c) Knot surgery operation gets undone when connected summing with $S^2 \times S^2$, hence we have a diffeomorphism, $f_K : X \to X_K$ ([A14], [Au]).

$f_K : X \to X_K$ being a diffeomorphism, we can view $f_K(W) \subset X$ as an image of the fix cork under different imbedding $W \hookrightarrow X$, which cork twisting gives $\tilde{S}_K$. □

10.4 Plugs

Similar to corks, there are different behaving codimension zero submanifolds appearing naturally, which are also responsible for exoticness of 4-manifolds; they first appeared in [AY2], where they are named *plugs*. Plugs generalize the *Gluck twisting* operation.

Definition 10.8. *A plug is a pair* (W, f)*, where* W *is a compact Stein manifold, and* $f : \partial W \to \partial W$ *is an involution satisfying the conditions: (a)* f *does not extend to a self-homemorphism of* W*; (b)* W *imbeds into some smooth manifold* $W \subset M$ *such that the operation* $M = N \cup_{id} W \mapsto M' = N \cup_f W$ *produces an exotic copy* M' *of* M.

Manifold $W_{m,n}$ of Figure 10.16, with $m \geq 1, n \geq 2$, is an example of a plug, where f is the involution induced by 180° rotation.

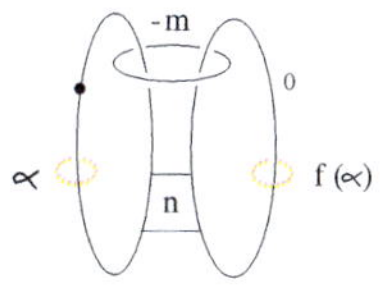

Figure 10.16: $W_{m,n}$

If the involution $f : \partial W_{m.n} \to \partial W_{m,n}$ extended to a homeomorphism inside, we would get homeomorphic manifolds, $W^1_{m,n}$ and $W^2_{m.n}$, obtained by attaching 2-handles to α and $f(\alpha)$ with -1 framings, respectively. But $W^2_{m,n}$ and $W^1_{m,n}$ have non-isomorphic intersection forms (show that one of them does not have an element of square -1):

$$\begin{pmatrix} -2n - mn^2 & 1 \\ 1 & -1 \end{pmatrix}, \begin{pmatrix} -2n - mn^2 & -1 - mn \\ -1 - mn & -1 - m \end{pmatrix},$$

Exercise 10.3. *Verify that* $(W_{m.n}, f)$ *satisfies the second condition of being plug (Hint: Use the hint of Exercise 10.1.)*

Exercise 10.4. *By canceling the* 1*-handle and the* $-m$ *framed* 2*-handle, show that* $W_{m,n}$ *can be obtained from* B^4 *by attaching a* 2*-handle to a knot with* $-2n - n^2m^2$ *framing.*

The degenerate case of plug twisting $(W_{1,0}, f)$ corresponds to the *Gluck twisting operation* (Exercise 6.1). Recall that there is an example of an exotic non-orientable manifold which is obtained from the standard one by the Gluck operation (Exercises 9.2). Corks and plugs can also be used to construct exotic Stein manifold pairs as follows:

Theorem 10.9. *([AY2]) The manifold pairs* $\{Q_1, Q_2\}$ *and* $\{M_1, M_2\}$ *of Figures 10.17 and 10.18 are simply connected Stein manifolds which are exotic copies of each other.*

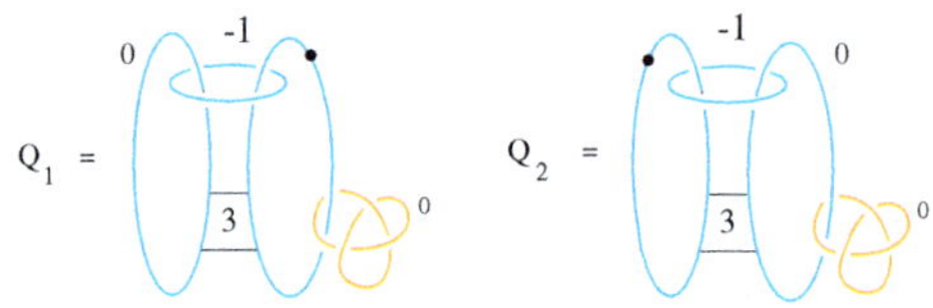

Figure 10.17: Q_1 and Q_2

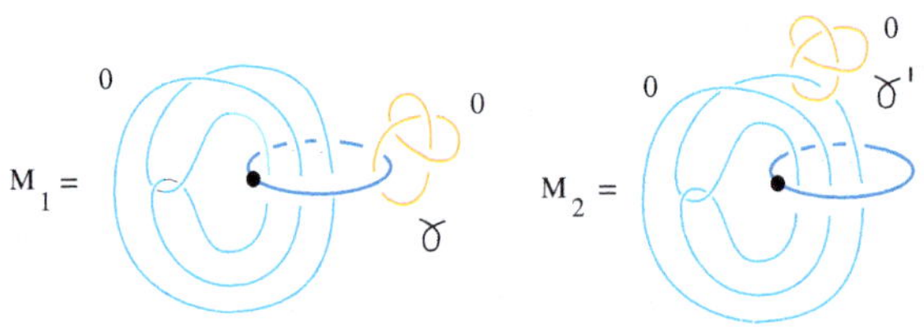

Figure 10.18: M_1 and M_2

Proof. Let us show $\{M_1, M_2\}$ are exotic Stein pairs (the proof of the other case is similar). By Theorem 8.11 both M_1 and M_2 are Stein manifolds. Then we compactify M_1 into a symplectic manifold S (Theorem 8.20). From the Legendrian handle picture of M_1 in Figure 10.19, we can compute $\mathbb{K}_S.\gamma = rot(\gamma) = \pm 2$ (Section 8.3). Then if γ is represented by a surface Σ of genus g in S, by the adjunction formula, $2g - 2 \geq \Sigma.\Sigma + |\mathbb{K}_S.\Sigma| = 2$.

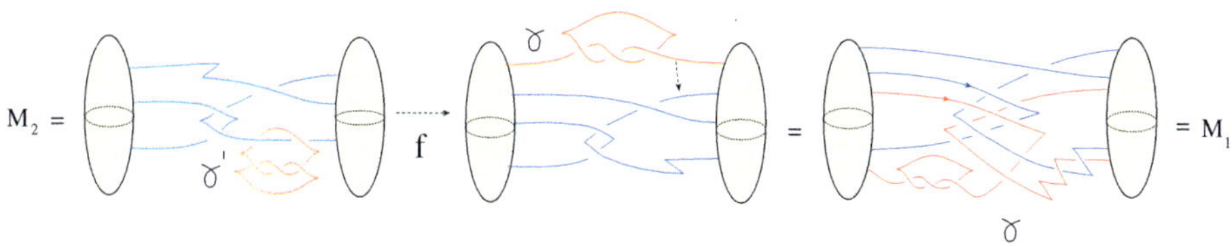

Figure 10.19

Hence $g \geq 2$. But on the other hand, we see from the picture of M_2 that the homology class γ' is respresented by a torus. So M_1 and M_2 cannot be diffeomorphic. □.

Notice from the pictures that, there are transformations, $M_1 \rightsquigarrow M_2$, and $Q_1 \rightsquigarrow Q_2$, which are obtained by twisting along the cork W, and twisting along the plug $W_{1,3}$, respectively. It is easy to check that the boundary diffeomorphisms extends to a homotopy equivalence inside, since they have isomorphic intersection forms (e.g. [Bo]). Hence by Freedman's theorem they are homeomorphic. For similar constructions see [AY4].

The existence of infinitely many simply connected Stein manifolds which are exotic copies of each other was established in [AEMS], where they produced such manifolds as fillings of a fixed contact manifold. Here we prove the same result in a more concrete way, by drawing small 4-dimensional handlebodies (of Betti number 2) with this property.

Theorem 10.10. *([AY5]) There are infinitely many simply connected Stein manifolds (with Betti number 2), which are exotic copies of each other; furthermore they are Stein fillings of the same contact 3-manifold.*

Proof. Let X_p be the 1-connected Stein manifold, drawn in Figure 10.20 (the second picture is the same manifold drawn as a Stein handlebody).

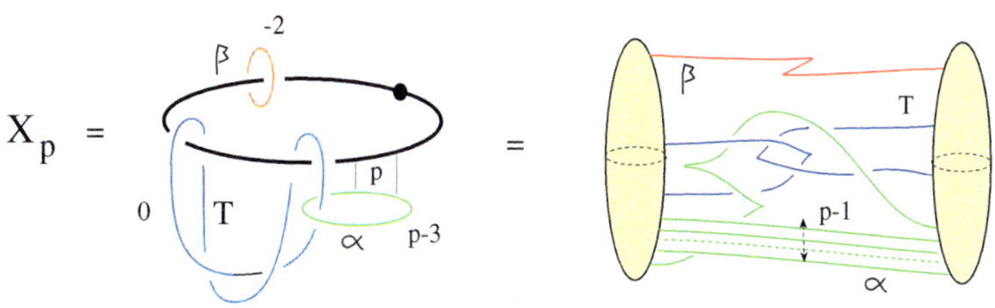

Figure 10.20

The boundaries of all X_p are diffeomorphic to each other; we can see this by replacing the dotted circle with a 0-framed circle, and blowing up a -1 circle on it and sliding this -1 circle over the 0-framed handle and blowing it down (this describes a diffeomorphism $\partial X_p \approx \partial X_{p+1}$). Note that $X_p \mapsto X_{p+1}$ is just the Luttinger surgery operation performed along the obvious torus T inside X_p, which is discussed in Section 6.4. Luttinger surgery preserves symplectic structure inducing isomorphic contact structures on the boundaries of X_p and X_{p+1} (cf. [EP]). Now it follows that all X_p are homotopy equivalent to each other, and hence by [F] are homemorphic to each other. Next we will show that for infinitely many different values of p these manifolds are exotic copies of each other.

First notice that ∂X_p is a homology sphere. Compactify X_p into a closed symplectic manifold S (Theorem 8.20). Let α, β and T be the homology classes of S with self-intersections $p-3$, -2 and 0, respectively. By computing rotational numbers from the picture, we get $K_S.(\alpha - p\beta) = 1 \pm p$ (depending upon how we draw the zig-zag). Also $K_S.T = 0$ (this follows from the adjunction inequality, since T is represented by a self-intersection zero torus). Then by taking p odd we can assume that the self-intersection number $(\alpha - p\beta)^2 = p - 3 - 2p^2$ is some even integer, $2k$. Then classes T and $\gamma = \alpha - p\beta - (k+1)T$ define a basis for $H_2(X_p;\mathbb{Z})$ with intersection matrix, $(\gamma.\gamma = -2)$,

$$\begin{pmatrix} 0 & 1 \\ 1 & -2 \end{pmatrix}$$

By Exercise 10.5 all self homemorphisms of X_p must preserve the homology class γ. Also $K_S.\gamma = 1 \pm p$, so if γ is represented by a surface Σ of genus g, then $2g-2 \geq -2+|1\pm p|$. Hence $g \mapsto \infty$ as $p \mapsto \infty$. So as $p \mapsto \infty$ infinitely many X_p's are not mutually diffeomorphic. □

Exercise 10.5. *Show that if $v \in H_2(X_p;\mathbb{Z})$ and $v.v = -2$, then we must have $v = \pm\gamma$*

Remark 10.11. *([AY5]) The contact structure on ∂X_p given by the Luttinger surgery might be different from the one induced from the Stein structure of X_p. Also, instead of using the property of the Luttinger surgery, alternatively we can use the following lemma to conclude that among the infinite family of contact manifolds ∂X_p (induced from the Stein structures of X_p) there is one which has infinitely many distinct Stein fillings.*

Lemma 10.12. *(Wendl). Any closed connected oriented 3-manifold has at most finitely many different strongly fillable contact structures up to isomorphisms.*

Proof. By Theorem 0.6 in [CGH] for any non-negative integer, every closed connected 3-manifold has at most finitely many contactomorphism classes of tight contact structures with Giroux torsion equal to that integer. Then the result from Corollary 3 in [Ga2] which says that a contact structure with Giroux torsion > 0 is not strongly fillable. □

Exercise 10.6. *([A13]) By using the description of $E(1)_{p,q}$ in Figure 7.11, show that $E(1)_{p,q}\#5\bar{\mathbb{CP}}^2$ contains the Stein submanifold $M_{p,q}$ of Figure 10.22 inducing the decomposition $E(1)_{p,q} = M_{p,q} \smile_\partial N$ for some codimension zero submanifold $N \subset E(1)_{p,q}$ (Hint: Identify Figure 10.22 with the second picture of Figure 10.21.)*

Remark 10.13. *([A13]) It turns out that the infinite family of exotic manifolds $E(1)_{p,2}$ with p odd, differ from each other by codimension zero Stein submanifolds Z_p with the same property of the manifolds of Theorem 10.10, i.e. as $p \to \infty$ they become distinct exotic Stein manifolds filling the same contact 3-manifold on their boundary.*

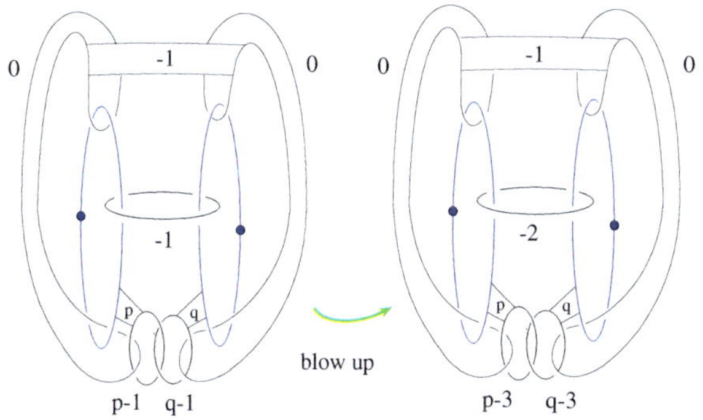

Figure 10.21

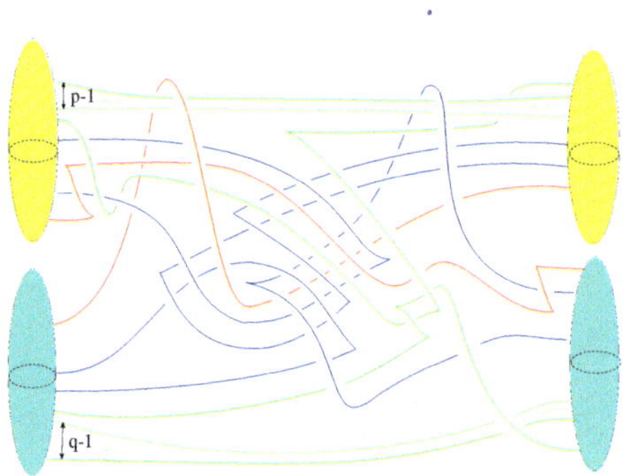

Figure 10.22: $M_{p,q}$

Chapter 11

Covering spaces

11.1 Handlebody of coverings

A quick way to draw a handlebody of a covering space of a 4-manifold is to first draw the part of the covering space lying over its 1-handles, which is a wedge of thickened circles $\#_k(S^1 \times B^3)$, then extend the covering space over its 2-handles. By uniqueness, extending over the 3-handles is automatic. 1-handles of a 4-manifold X^4 give generators $x_1, x_2, \ldots, x_n$, and the 2-handles $h_1, .., h_k$ give the relations of the fundamental group, $\pi_1(X) = \langle x_1, x_2, \ldots, x_n | h_1(x_1, \ldots, x_n) = 1, \ldots, h_k(x_1, \ldots, x_n) = 1\rangle$. Let $\pi_1(X) \to G$ be an epimorphism mapping $x_1, x_2, \ldots, x_k$ to the generators of G $(k \leq n)$. To draw the covering corresponding to the subgroup given by the kernel of this map, we first draw the *Cayley graph* K_G of the group G, which is a 1-complex with vertices consisting of elements of G, and any two vertices a and b are connected by a 1-simplex whenever $a = xb$ for some generator $x^{\pm 1} \in G$. Then thicken this 1-complex to get the 1-handles of the covering space, then naturally extend it over them to covering of the 2-handles. The other 1-handles $x_{k+1}, \ldots, x_n$ just lift trivially to $|G|$ copies of 1-handles above. For example Figure 11.1 describes the 4-fold cyclic covering $\pi : \tilde{X} \to X$ of the fishtail X.

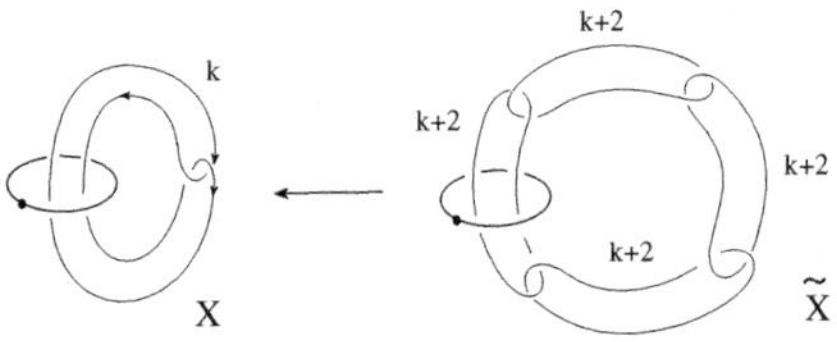

Figure 11.1

Figure 11.3 describes the 60-fold covering space $\pi : \tilde{W} \to W$ corresponding to the symmetric group $A_5 = \langle x, y \mid x^5 = y^3 = (xy)^2 = 1 \rangle$. Here the handlebody of W consists of one 0-handle, two 1-handles x, y and one 2-handle attached along the contractible -1 framed knot K, as shown in Figure 11.3. Hence $\pi_1(W)$ is the free group $\langle x, y \rangle$ in two generators with the obvious epimorphism $\pi_1(W) = \langle x, y \rangle \to A_5$. Figure 11.2 is the Cayley graph of A_5, which describes the covering space of the wedge of two circles. By thickening this, we first get a 60-fold covering space of $\#2(S^1 \times B^3)$, then by lifting the attaching circle K of the 2-handle to the 60 framed circles, as shown in Figure 11.3, we get the handlebody of the covering space $\pi : \tilde{W} \to W$ ([A18]).

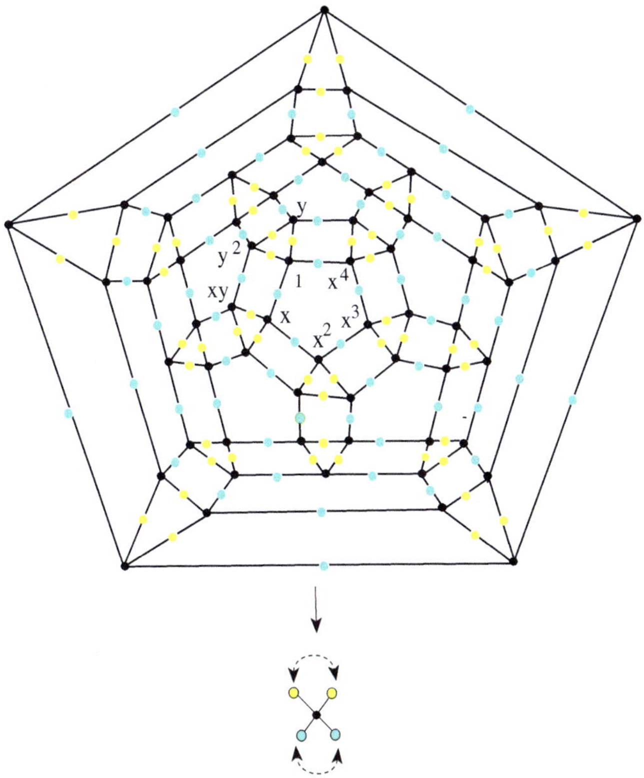

Figure 11.2

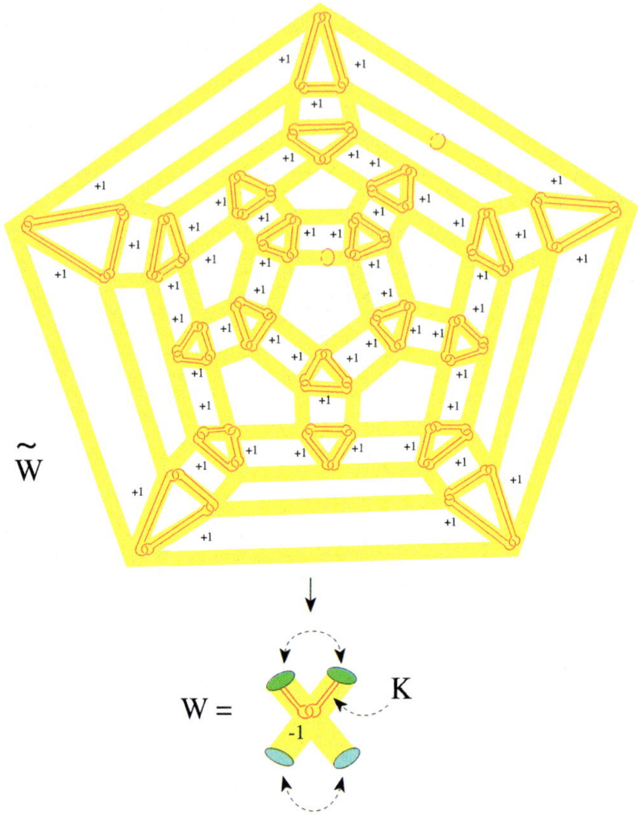

Figure 11.3

Exercise 11.1. *In the examples in Figure 11.1 and Figure 11.3, by lifting the normal vector fields of the attaching circles of the 2-handles of X and W to $\tilde{X}$ and $\tilde{W}$, verify that the framings on the lifted 2-handles are as indicated in the figures.*

11.2 Handlebody of branched coverings

Let $F \subset B^4$, a properly imbedded surface obtained by pushing the interior of an imbedded surface $F \subset S^3$ into B^4 by a small isotopy. We want to draw a handlebody picture of the cyclic branched covering, $\pi : \tilde{B}^4 \to B^4$, branched along F ([AK3]). First of all, when $F = D^2$ the n-fold cyclic branched covering of B^4 is B^4. We can construct this by first cutting B^4 open along the trace of this small isotopy of D^2, creating a ball with a small codimension zero "lip" $A \cup A'$ on ∂B^4, then by cyclically gluing n-copies of balls pairs $(B_i^4, A_i \cup A_i'), i = 1, \ldots, n$ with identification $A_i' \leftrightarrow A_{i+1}$, $B^4 = \underset{A_i' = A_{i+1}}{\sqcup} B_i^4$ (where $A_{n+1}' = A_1$).

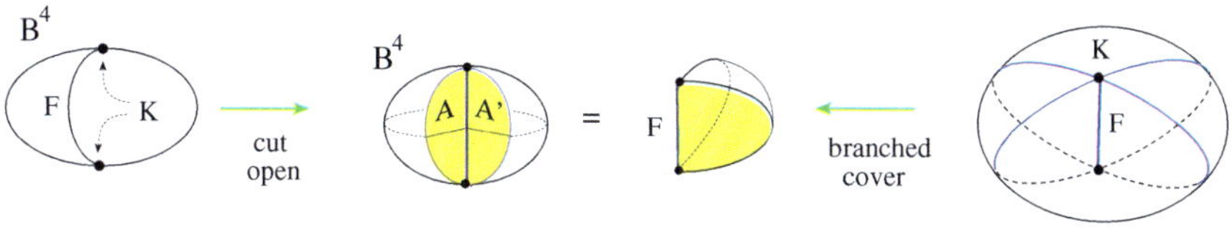

Figure 11.4

This process easily generalizes to the case $D^2 \subset M^4 = B^4 + 2 - handles$, provided the attaching circles of the 2-handles of M link the circle ∂D^2 *zero* (mod n) times. In this case, to form the n-fold cyclic branched covering, $\pi : \tilde{M} \to M$, we first take the branched covering B^4 along D, then simply lift n copies of each handle (i.e. a framed knot) up above as a framed knot in the most natural way with appropriate framing. Figure 11.5 demonstrates this when $M = \bar{\mathbb{CP}}^2 - B^4$.

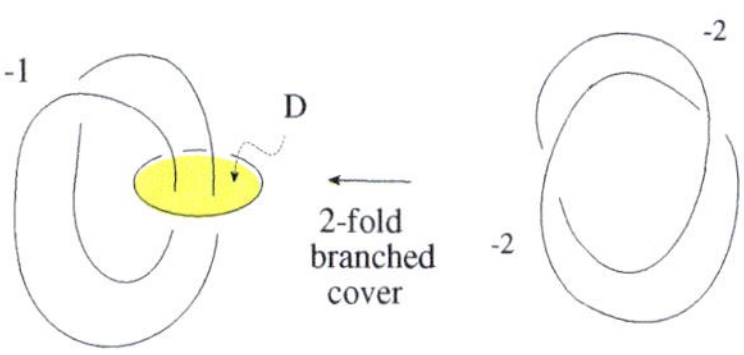

Figure 11.5

Exercise 11.2. *In Figure 11.5, by lifting the normal vector field of the* -1 *framed attaching circle of the* 2*-handle, show that the lifted* 2*-handles are attached by* -2 *framings.*

One useful application of this is, if $K \subset S^3$ is a knot and we are only interested in finding the 3-manifold which is the cyclic branched covering of S^3 along K, we simply blow up the interior of B^4 to $M = B^4 \# \pm k\mathbb{CP}^2$ for some k, where the knot $K \subset \partial M$ looks like an unknot $K = \partial D$ for some properly imbedded $D^2 \subset B^4$, and the handles of M link ∂D zero times, we then apply this technique (e.g. as demonstrated in Figure 11.6).

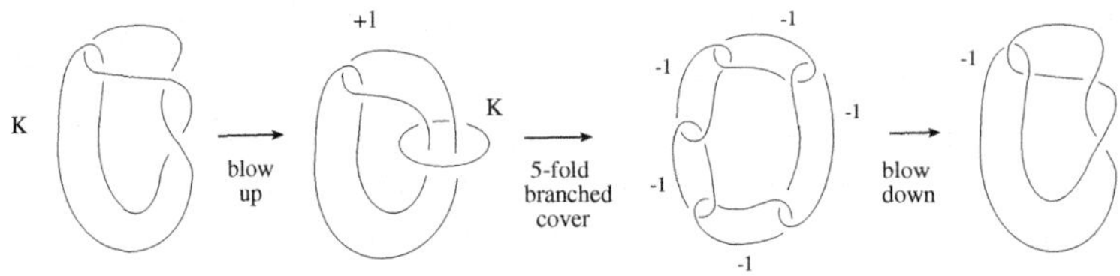

Figure 11.6: $\Sigma(2,3,5)$ as the 5-fold branched cover of the trefoil knot K

Exercise 11.3. *By the steps in Figure 11.7 show that the homology sphere* $\Sigma(2,3,11)$ *(Excercise 2.6) is the* 2*-fold branched covering of the knot* K *in the figure.* $\Sigma(2,3,11)$ *is also the* 2*-fold branched covering space of the* $(3,11)$*-torus knot (Section 12.1).*

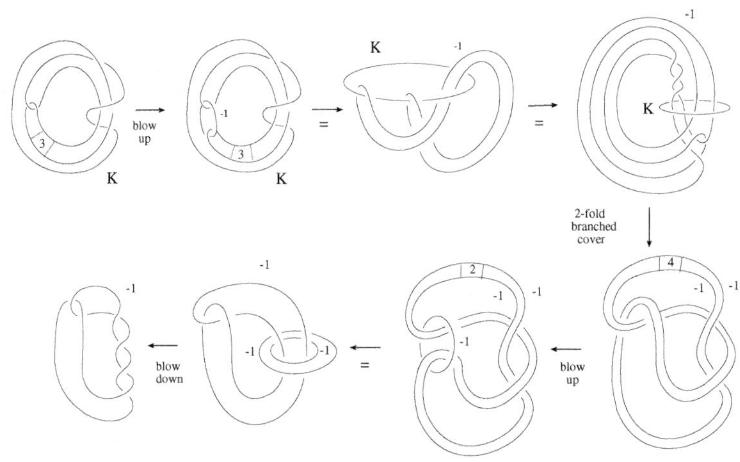

Figure 11.7

Exercise 11.4. *By steps of Figure 11.8 and blowing up the last figure once, show that 2-fold branched covering of the 0-Whitehead double $Wh(K)$ of the trefoil knot (Section 9.2) bounds a compact smooth manifold X^4 with intersection form $E_8 \oplus \langle -1 \rangle \oplus \langle -1 \rangle$ (Exercise 2.6). This shows $Wh(K)$ is not slice, if it were the branched covering of B^4 branched along the slice disk would give a rational ball; together with X you get a closed 4-manifold realizing a nondiagonizable definite intersection form, violating Theorem 13.25.*

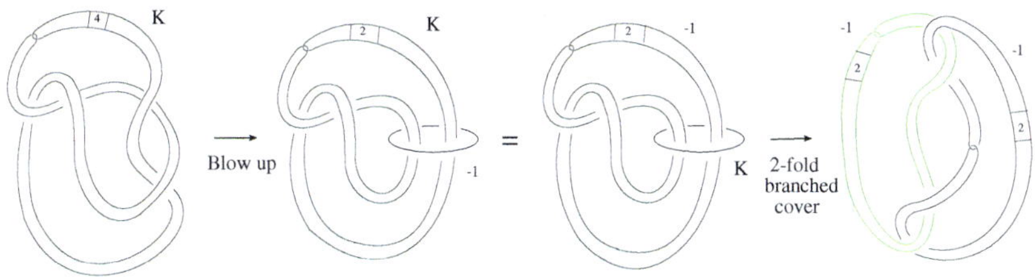

Figure 11.8

Now we can construct the n-fold cyclic branched covering, $\pi : \tilde{M}^4 \to B^4$, branched along a properly imbedded connected surface $F \subset B^4$ (isotoped from S^3). As described in Figure 11.4, $\tilde{M}$ is obtained by cutting B^4 along F then gluing n copies of the resulting "balls with lips" along their lips $\cong F \times I$. By writing F as a 2-disk with handles $D^2 \cup h$ we obtain a ball with 1-handles $F \times I = A \cup a$. We do the gluings in two steps: First glue along the balls, then extend it to the identification of the 1-handles. By cutting B^4 open along F, we create lips which are the two halves of a ball with 1-handles $(A \cup a) \cup (A' \cup a')$. Then as before, we glue 4-balls with lips $(B^4_i, (A_i \cup a_i) \cup (A'_i \cup a'_i)), i = 1, ..n$, along the balls to get $B^4 = \bigsqcup_{A'_i = A_{i+1}} B^4_i$, then identify $a'_i \leftrightarrow a_{i+1}$ by attaching $n-1$ 2-handles to B^4.

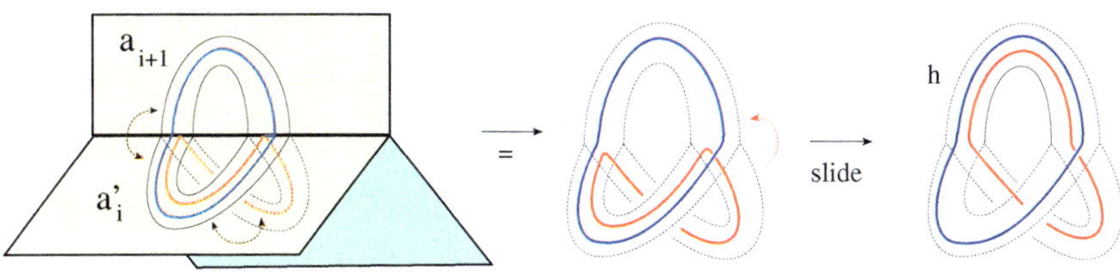

Figure 11.9

This process is demonstrated in Figure 11.9. By sliding the 2-handles over each other we can make all the half arcs K_i' of the attaching circles $K_i = K_i' \cup_\partial K_i''$, $i = 1, 2, \ldots, n-1$ of the 2-handles go over one copy of h of the surface F, the other halves K_i'' are just doubling of K_i' s where they pile over each other as if lying on different pages of a book. Framings of K_i are twice the page framing induced from F (e.g. Figure 11.10).

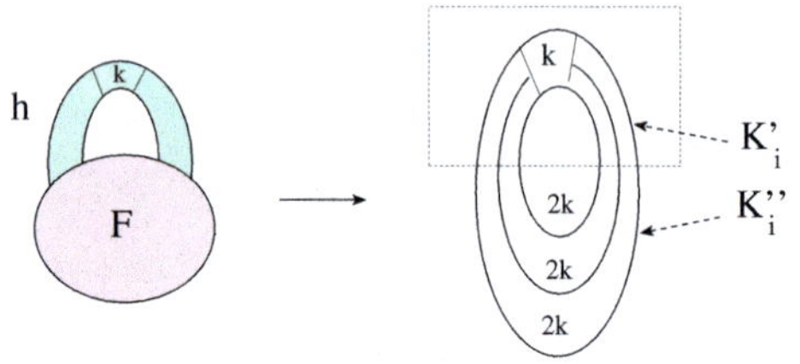

Figure 11.10

For example, the right and left pictures of Figure 11.11 give the handlebodies of the 2- and 3-fold branched coverings of B^4 branched along the surface F, shown in the middle, respectively (also compare to Figure 11.14). Also the case $F \subset M^4 = B^4 + handles$, where the attaching circles of the 2-handles of M link the circle ∂F *zero* (mod n), proceeds as before, that is we lift the handles as the most natural way.

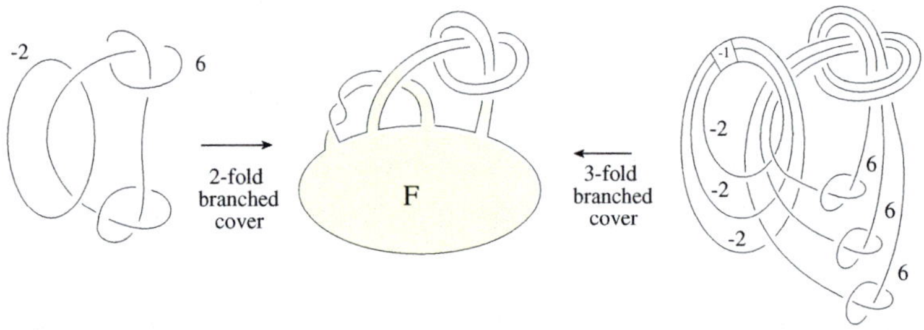

Figure 11.11

Next we discuss the case of branched covering B^4 along disconnected surface F. As before, we first deal with the simple case of when the surface is disjoint union disks $F = \sqcup_{i=1}^{n} D_j$. Surprisingly, even in this case there are a variety of very different branched coverings. To see these, as before we cut B^4 along the disks of F creating n lips on B^4

(Figure 11.12). We codify this picture of B^4 by denoting it with a star-shaped graph, where the outer edges are indexed by integers $k_1, \ldots, k_n$; we call this graph a $(k_1, \ldots, k_n)$ *switch* or just simply a *switch*.

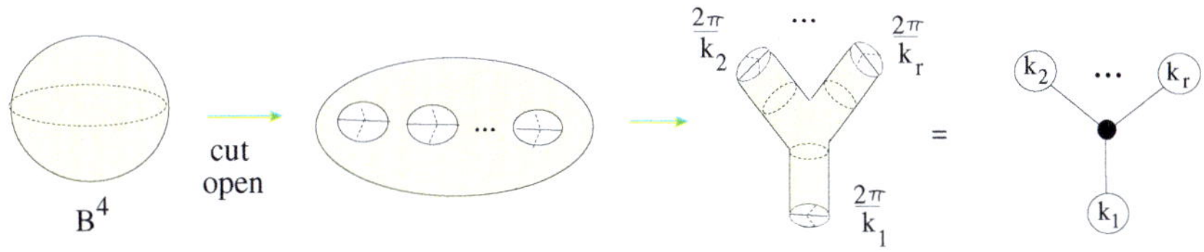

Figure 11.12

To construct various (regular or irregular) branch coverings of B^4 branched along $\sqcup_{i=1}^{n} D_j$, we glue many copies of these switches (i.e. balls) along their outer vertices, with the rule that k copies of index k vertices of switches meet at a vertex (index 1 switches are unmatched). Imagine an index k vertex of a switch as a lip of a 4-ball with $2\pi/k$ solid angle, so that k of them fit nicely together creating a k-fold branching around that disk. So this process constructs a branched covering, $\pi : \#_k S^1 \times B^3 \to B^4$, as indicated in Figure 11.13 (to emphasize various $S^1 \times B^3$'s we put circles with dots in the figure).

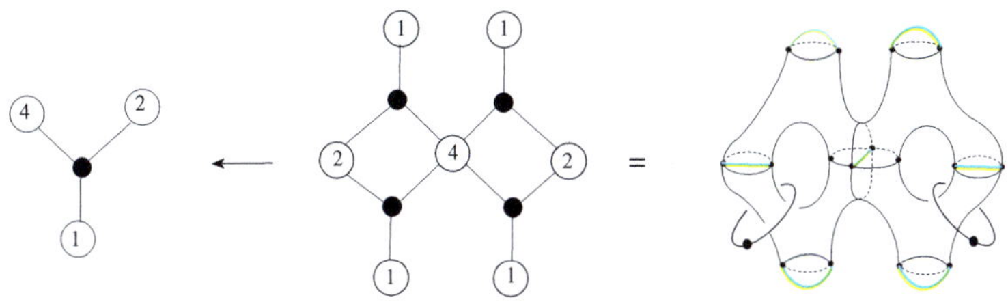

Figure 11.13

Now, again, by the gluing process of Figure 11.9, we can extend the handlebody of branched covers, $\#_k S^1 \times B^3 \to B^4$ of B^4, branched along $\sqcup_j D_j$ to the branched covers branched along the (disconnected) surfaces, $F \subset B^4$, which are obtained from $\sqcup_j D_j$ by attaching 1-handles (bands). In fact as before, we can go one step further and extend this to the case of disconnected surfaces, $F \subset M^4 = B^4 + handles$, provided the attaching circles of the 2-handles of M link the $\partial(\sqcup_j D_j)$ *zero* (mod n) times.

Exercise 11.5. *By first taking the branched covering of B^4 branched over the disjoint union of two disks $D = D_1 \sqcup D_2$, and then extending this over the surface F consisting of D with two 1-handles, show that the picture on the right in Figure 11.14 gives the handlebody of the 3-fold cyclic branched cover of B^4 branched over F.*

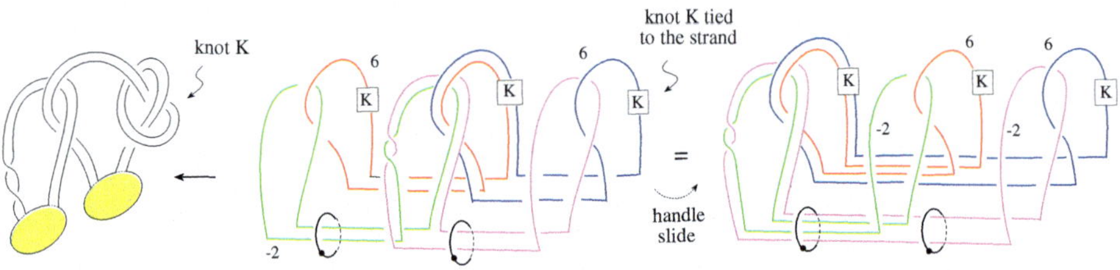

Figure 11.14: 3-fold cyclic branched cover

Exercise 11.6. *By first taking the branched covering of B^4 along $\sqcup_{j=1}^{3} D_j$ by the switches indicated in Figure 11.15, and extending over the handles of the surface $F = D +$ bands, and extending over the -2 framed 2-handle, show that figure on the right gives the irregular 3-fold covering space of the manifold on the left branched along the surface F.*

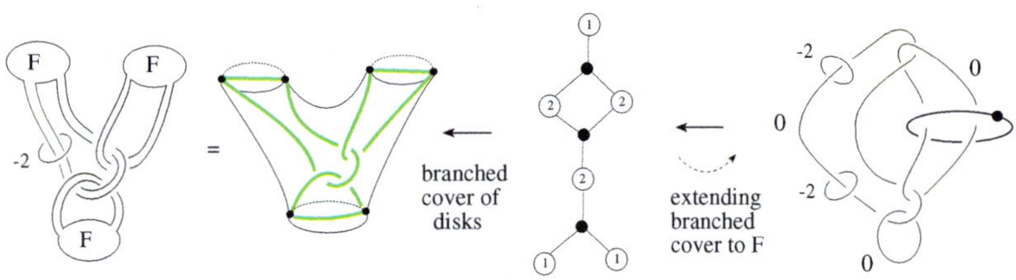

Figure 11.15: 3-fold irregular branched cover

11.3 Branched covers along ribbon surfaces

In the previous section we described the handlebodies of various branched covering spaces of 4-manifolds $\pi : \tilde{M} \to M$ branched along properly imbedded surface $F \subset M$, which is an isotopic copy of a surface lying on the boundary of the 0-handle $F \subset \partial B^4 \subset M$. Now we generalize this to the case of when the surface $F \subset M$ is an isotopic copy of a ribbon-immersed surface $f : F \looparrowright \partial B^4$. This case can be dealt with similarly to the imbedded case ([Mo]), because a ribbon surface $F = \sqcup D_j + \sum h_i$ is just a collection of disks $D = \sqcup D_j$ connected with ribbon 1-handles $\{h_i\}$, rather then imbedded 1-handles. Hence, once we put the ribbons into *standard position*, as shown in the second picture of Figure 11.16, we extend the branched covering space of B^4 branched over D to F by the same way as before (third picture of Figure 11.16).

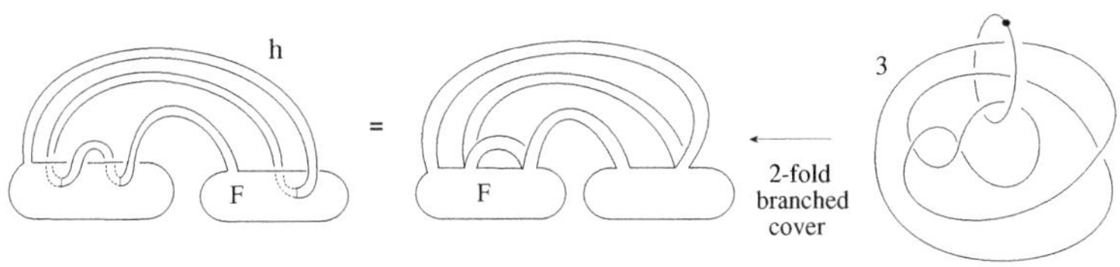

Figure 11.16: Mazur manifold as the 2-fold branched cover of B^4

Now it is clear from Figure 11.16 that this process can be reversed, if the 2-handles of the 4-manifold exhibit some symmetry in the following sense:

Definition 11.1. *A link $\{K_1, \ldots, K_k\}$ in $\#_k S^1 \times S^2$ is called strongly invertible if the "standard" involution of $\#_k S^1 \times S^2$ (induced by 180° rotation) induces an involution on each of the knots K_j with two fixed points (Figure 11.17).*

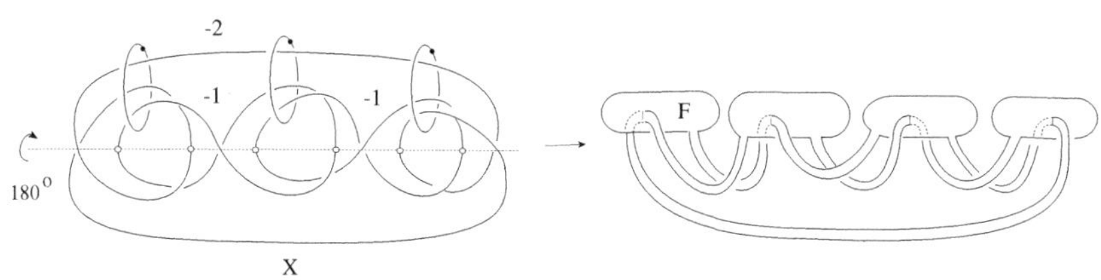

Figure 11.17: X as a 2-fold branched cover over B^4

Theorem 11.2. *([Mo]) If a compact smooth 4-manifold has only handles of index ≤ 2, and its 2-handles are attached along a strongly invertible link in $\#_k S^1 \times S^2$, then M is a 2-fold branched covering space of B^4 branched along a properly imbedded surface $F \subset B^4$.*

Corollary 11.3. *If the 2-handles of any closed smooth 4-manifold M are attached along a strongly invertible link in $\#_k S^1 \times S^2$, then M is a 2-fold branched covering space of S^4 branched along a closed smooth 2-manifold $F \subset S^4$.*

Proof. Let $M^{(2)}$ be the sub-handlebody of M consisting of handles of index ≤ 2; we have $M - int(M^{(2)}) = \#_k S^1 \times B^3$. By Theorem 11.2 we have a branched covering, $\pi : M^{(2)} \to B^4$, branched along properly imbedded $F \subset B^4$, and by [KT] the induced branched covering on the boundary (by restriction) $\pi_| : \#_k S^1 \times S^2 \to S^3$ has to be standard with branch set $S^1 \sqcup \ldots \sqcup S^1$. We can then extend $\pi_|$ trivially to another branched covering map $\pi' : \#_k S^1 \times B^3 \to B^4$. By putting π and π' together we get the desired branched covering $M \to S^4$. □

Exercise 11.7. *Show that there is a 2-fold branched covering, $T^3 \to S^3$, branched along the link L of Figure 11.18 (with branching index 2 for each component).*

Figure 11.18: L

We can easily draw handlebodies of the k-fold cyclic branched coverings of plumbed 4-manifolds, where the branching spheres have Euler class divisible by k and they are separated from each other by other 2-spheres, as in the example of Figure 11.19.

Exercise 11.8. *Generalize the above process of taking a 2-fold branched covering of B^4 along a ribbon disk (of Figure 11.16) to taking a k-fold cyclic branched covering of B^4 along a ribbon disk $D \subset B^4$. (Hint: imitate the Seifert surface case.)*

Exercise 11.9. *Show that the k-fold cyclic branched covering of the manifold M of Figure 11.19 branched along 2-spheres $A \sqcup B \sqcup C \sqcup D$ is given by the plumbing $\tilde{M}$, where the small case letters indicate Euler classes of the corresponding spheres (Hint: Compute how branched covering changes the Euler class of each 2-sphere. While branched covering along A, B, C, D, the spheres e, f, g get branched covered along two points.)*

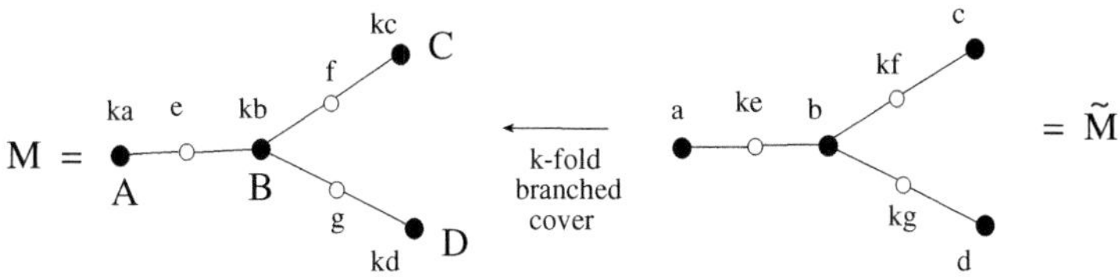

Figure 11.19

Chapter 12

Complex surfaces

A rich source of 4-manifolds are complex surfaces (2-dimensional complex algebraic sets). Next we will construct handlebodies of complex surfaces in $\mathbb{C}^3$ and $\mathbb{CP}^3$ ([AK3]).

12.1 Milnor fibers of isolated singularities

Let $F : \mathbb{C}^3 \to \mathbb{C}$ be a complex polynomial map with an isolated singularity, that is, $dF(z)$ is nonzero on $U - \{0\}$, where U is some open subset of $\mathbb{C}^3$ containing the origin. Let $z_0 \in intF(U)$ be a regular value; then the compact smooth manifold with boundary $M_F = F^{-1}(w_0) \cap B^6$, where $B^6 \subset U$ is a ball centered at the origin, is called the *Milnor fiber* of F. When $F(x, y, z) = f(x, y) + z^d$, the manifold M_F is the d-fold cyclic branched covering of the unit ball B^4 branched over the Milnor fiber M_f of $f : \mathbb{C}^2 \to \mathbb{C}$ (Figure 12.1).

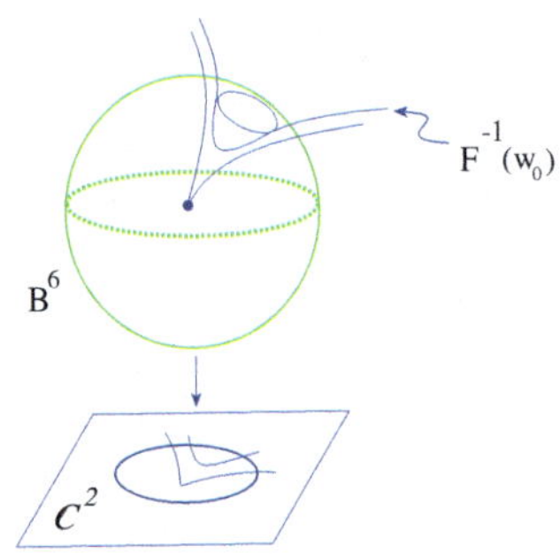

Figure 12.1

An interesting special case is when $F(x,y,z) = x^a + y^b + z^c$. Denote the Milnor fiber by $M(a,b,c) = M_F$. So $M(a,b,c)$ is a c-fold cyclic branched covering space of B^4 branched along the Milnor fiber $M(a,b)$ of $x^a + y^b$. The boundary $\partial M(a,b,c) := \Sigma(a,b,c)$ is called *Brieskorn manifold.* The Milnor fiber $M(a,b)$ is the Seifert surface of the (a,b) torus link whose interior is pushed into B^4 ([M3] p. 53). Hence $M(a,b)$ is the fiber (a,b) torus link obtained by by stacking a plates connected by b half twisted bands (Figure 12.2). Then, by isotopies we can turn $M(a,b)$ into a 2-disk with 1-handles as in Figure 12.3.

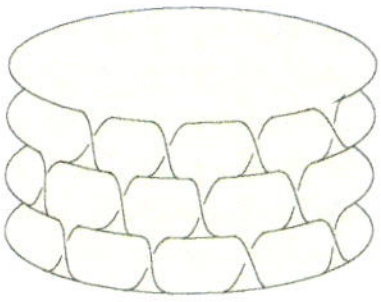

Figure 12.2: $M(4,4)$

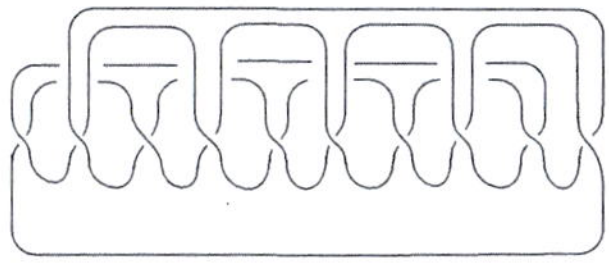

Figure 12.3: $M(3,5)$

Figure 12.4 is $M(2,3,5)$, a 5-fold branched cover of B^4 branched along $M(2,3)$.

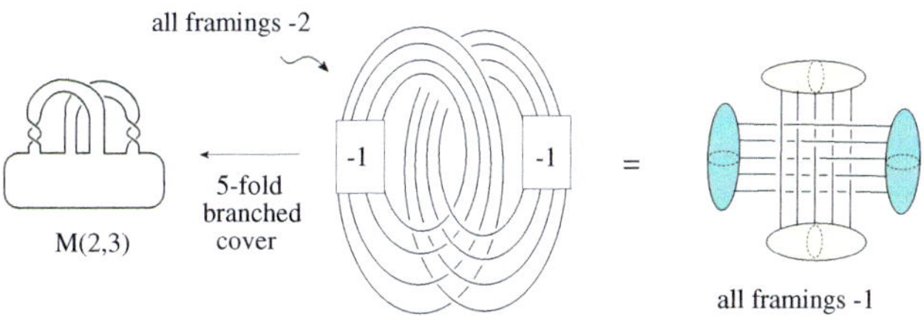

Figure 12.4: $M(2,3,k)$, when $k = 5$

We can identify $M(2,3,5)$ with E_8 by taking the 2-fold branched covering of B^4 branched along $M(3,5)$ (Figure 12.3), and by doing the handle slides of Figure 12.5.

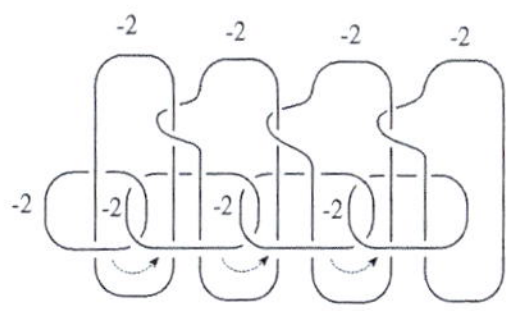

Figure 12.5: $M(2,3,5) = E(1)$

Proposition 12.1. *$M(2,3,k)$ is given by Figure 12.6 when $k = 2m+1$, and given by Figure 12.7 when $k = 2m$.*

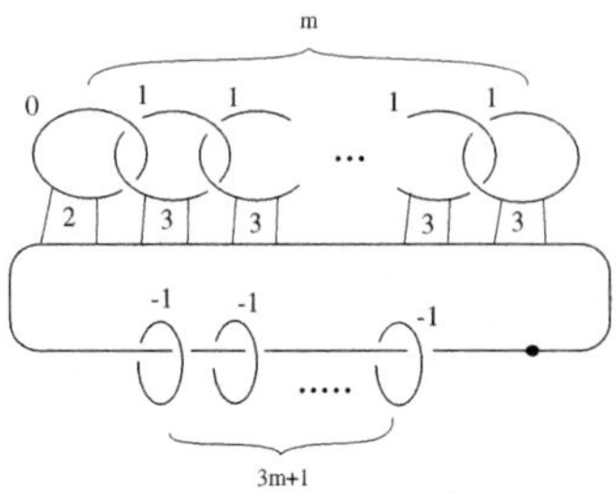

Figure 12.6

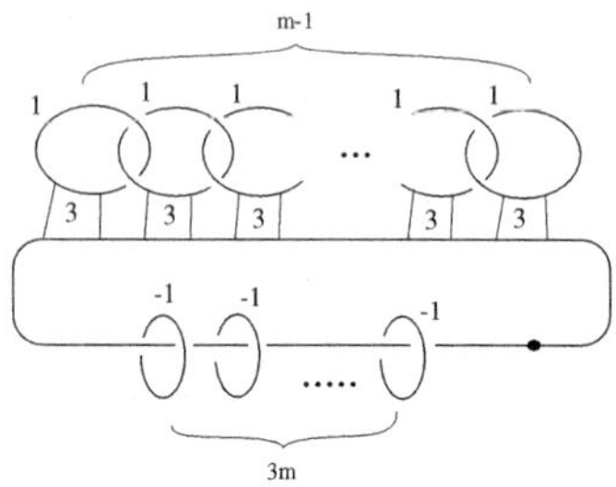

Figure 12.7

Proof. Let us do the case $k = 5$ (general case is similar). The first picture of Figure 12.8 is the 5-fold branched covering of $M(2,3)$ (check), subsequent steps establish the proof. □

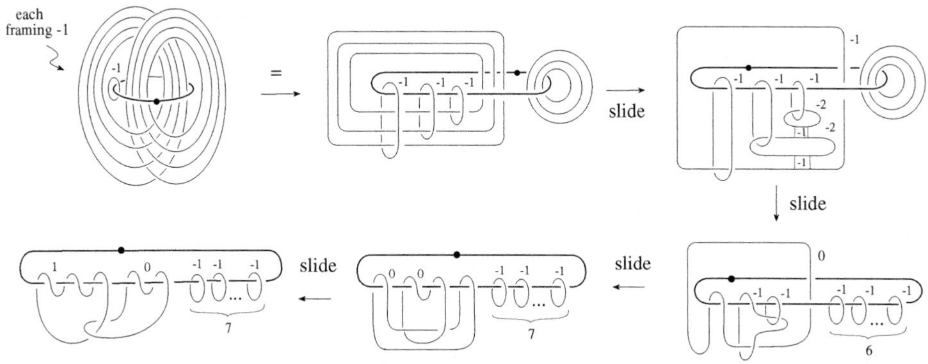

Figure 12.8

Exercise 12.1. *By using the description given in Proposition 12.1, identify $M(2,3,5)$ with E_8. (Hint: Figure 12.9.)*

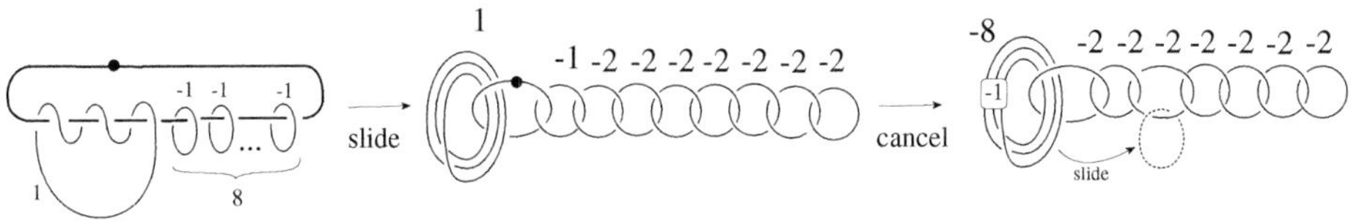

Figure 12.9

12.2 Hypersurfaces that are branched covers of $\mathbb{CP}^2$

A degree d hypersurface $V_d \subset \mathbb{CP}^n$ is defined by the locus of any degree d homogeneous polynomial $f(z_0, z_1, \ldots, z_n) = 0$. We say V_d is nonsingular, if $df(z)$ is nonzero on V_d. Let $\{m_j(x)\}_{j=0}^{N}$ be the collection of degree d monomials on $\mathbb{C}^{n+1}$, hence any V_d is the locus of $\sum a_j m_j(x) = 0$ for some $a = [a_1, \ldots, a_N]$. All degree d nonsingular hypersurfaces in $\mathbb{CP}^n$ are diffeomorphic to each other, because they are inverse images of the regular values of π (the singular values has codimension 2 since it is a complex algebraic set),

$$Z = \{(x, a) \mid \sum a_j m_j(x) = 0\} \subset \mathbb{CP}^n \times \mathbb{CP}^N \xrightarrow{\pi} \mathbb{CP}^N$$

When $n \geq 3$, V_d are simply connected, because we can describe V_d as a hyperplane section $\lambda(\mathbb{CP}^n) \cap H$ of the imbedding, given by $\lambda : \mathbb{CP}^n \to \mathbb{CP}^N$, $\lambda(x) = [m_0(x), \ldots, m_N(x)]$, and by Lefschetz hyperplane theorem (e.g. [M1]), $\pi_r(\lambda(\mathbb{CP}^n), \lambda(\mathbb{CP}^n) \cap H) = 0$ for $k \leq n-1$. V_4 is also known as the *Kummer surface*, and can be identified with $E(2)$ (Section 7.1).

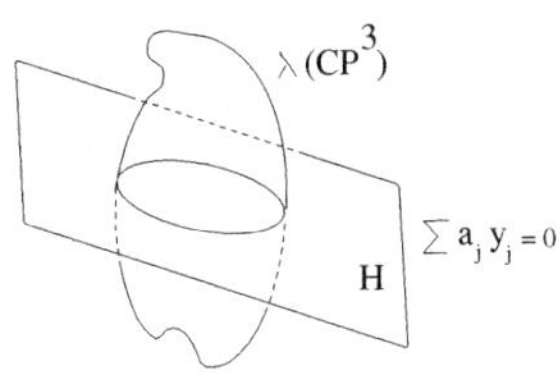

Figure 12.10

By taking $f(z) = z_0^d + z_1^d \ldots + z_n^d$, we see that the restriction of the projection gives a branched covering map $\pi : V_d \to \mathbb{CP}^{n-1}$, branched along a nonsingular degree d hypersurface in $\mathbb{CP}^{n-1}$. Since $V_d \subset \mathbb{CP}^3$ is simply connected, its homotopy type is determined by its intersection form $I(V_d)$ [M2]. In turn the intersection form is determined by three data: signature $\sigma(V_d)$, second Betti number $b_2(V_d)$ and the parity of $I(V_d)$ (i.e. the knowledge of whether the intersection form is even or odd).

Proposition 12.2. *$V_d \subset \mathbb{CP}^3$ is spin if and only if d is even and*

- $b_2(V_d) = d^3 - 4d^2 + 6d - 2$
- $\sigma(V_d) = -(d^2 - 4)d/3$
- *If d is odd $I(V_d) = \lambda_d\langle 1\rangle \oplus \mu_d\langle -1\rangle$, where*

  $\lambda_d = (d^3 - 6d^2 + 11d - 3)/3,$
  $\mu_d = (2d^3 - 6d^2 + 7d - 3)/3,$

- *If d is even $I(V_d) = m_d(E_8) \oplus l_d\begin{pmatrix} 0 & 1 \\ 1 & 0 \end{pmatrix}$, where*

  $m_d = (d^2 - 4)d/24, \quad l_d = (d^2 + 11)d/3 - 2d^2 - 1$

Lemma 12.3. *If S is a nonsingular compact complex surface then*

(a) $c_2(S) = e(S)$

(b) $w_2(S) = c_1(S)\ mod(2)$

(c) $c_1^2(S) = 3\sigma(S) + 2e(S)$

($c_i(S)$, $p_1(S)$, $e(S)$ denote Chern, Pontryagin and Euler classes respectively).

Proof. (a), (b) are standard facts, and $p_1(S) = 3\sigma(S)$ is the index theorem. To prove (c) we use the splitting principle (e.g. [Sh]), which says that it suffices to prove the case when $T(S) = L_1 \oplus L_2$ is a sum of line bundles. By definition, $p_i(S) = (-1)^i c_{2i}(S)$ so:

$$\begin{aligned} p_1(L_1 \oplus L_2) &= -c_2(L_1 \oplus \bar{L}_1 \oplus L_2 \oplus \bar{L}_2) \\ &= -c_2((1 - x_1)(1 + x_1)(1 - x_2)(1 + x_2)) \\ &= -c_2\left(1 - (x_1 + x_2)^2 + 2x_1x_2 + x_1^2x_2^2\right) \\ &= (x_1 + x_2)^2 - 2x_1x_2 \\ &= c_1^2 - 2c_2 \qquad \square \end{aligned}$$

Proof. (of proposition). Recall $H^*(\mathbb{CP}^3) = \mathbb{Z}[h]/\langle h^4\rangle$, where h is the hyperplane class. Let $j : V_d \hookrightarrow \mathbb{CP}^3$ be the inclusion, then deg $(V_d) = d \implies j_*[V_d]$ is the Poincare dual of dh, so the Euler class of the normal bundle $\nu(V_d)$ is $j^*(dh)$. Also, $j^*h^2 = d[V_d]^*$, since

$$\langle j^*h^2, [V_d]\rangle = \langle h^2, j_*[V_d]\rangle = \langle h^2, dh \cap [\mathbb{CP}^3]\rangle = d\langle h^2, h \cap [\mathbb{CP}^3]\rangle = d\langle h^3, [\mathbb{CP}^3]\rangle = d$$

By taking the Chern classes of both sides of $j^*T(\mathbb{CP}^3) = T(V_d) \oplus \nu(V_d)$, we get

$$j^*(1+h)^4 = (1 + c_1 + c_2)(1 + dj^*h)$$

Hence $c_1(V_d) = (4-d)j^*h$, and $c_2(V_d) = (6-4d+d^2)j^*h^2 = (6-4d+d^2)d[V_d]$. So by Lemma 12.3, $w_2(V_d) = 0 \Leftrightarrow d$ is even, and $b_2(V_d) = e(V_d) - 2 = d(6-4d+d^2) - 2$. By using the expression of $c_1(V_d)$ and by (c) of Lemma 12.3, we calculate $\sigma(V_d) = (4-d^2)d/3$. Clearly $\lambda_d = (b_2+\sigma)/2$ and $\mu_d = (b_2-\sigma)/2$ computes the expressions of λ_d and μ_d. □

12.3 Handlebody descriptions of V_d

By using the techniques of Section 11.2 we can draw handlebody pictures of all V_d's. As discussed before, since the smooth manifold structure of V_d is independent of which polynomial equation we choose to describe it, we choose the following simplest one:

$$f(z_0, z_1, z_2, z_3) = z_0^d + z_1^d + z_2^d + z_3^d = 0$$

Though the projection $[z_0, z_1, z_2, z_3] \mapsto [z_0, z_1, z_2]$ is not a well defined map at all points of $\mathbb{CP}^3$, its restriction to V_d gives a well-defined d-fold cyclic branched covering map, $\pi : V_d \to \mathbb{CP}^2$, branched over the degree d complex curve $L_d \subset \mathbb{CP}^2$ given by the equation $z_0^d + z_1^d + z_2^d = 0$. By decomposing $\mathbb{CP}^2 = N \cup B^4$, where N is the tubular neighborhood of $\mathbb{CP}^1$ (Euler class +1 disk bundle over S^2) and its complement B^4 (Figure 12.11), we can describe $L_d \cap N$ as the Milnor fiber $M(d,d)$ (Section 12.1) union $L_d \cap B^4$, which is d disjoint copies of trivial disks. Figure 12.12 demonstrates how L_d sits in $\mathbb{CP}^2 = N \cup B^4$.

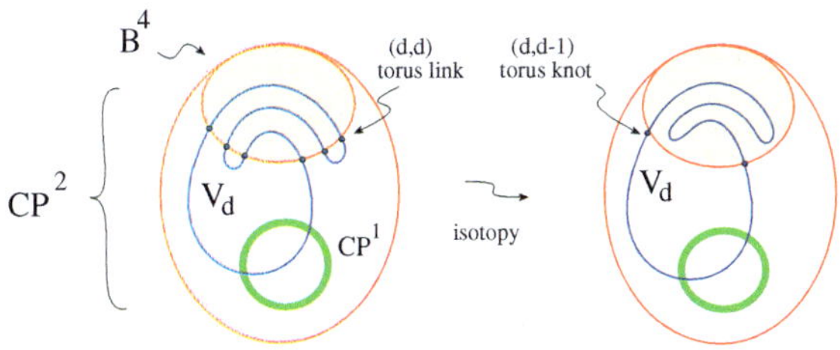

Figure 12.11

Now by a small isotopy of L_d we can move d half bands of $M(d,d)$ from N side to B^4 side, so that $L_d \cap N$ appears as the Milnor fiber $M(d, d-1)$ union $L_d \cap B^4$, which is a single trivial disk $D \subset B^4$ (Figure 12.13). Now by using the technique of Section 11.2, we can take the branched cover of N branched along $L_d \cap N$; the branched cover of B^4 along D is just B^4 (the 4-handle) so we don't need to draw that part of it. To apply this process of course we first need to draw $M(d, d-1)$ as a disk with handles, then proceed.

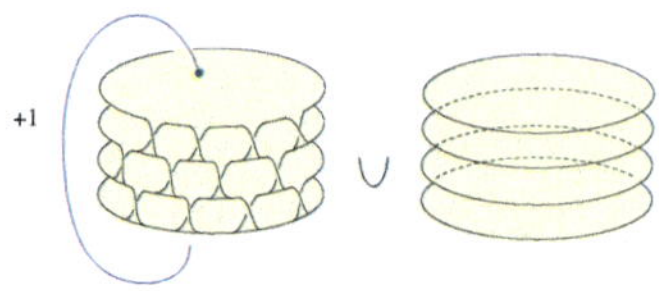

Figure 12.12: $L_d = M(4,4) \cup \sqcup_{j=1}^{4} D_j$

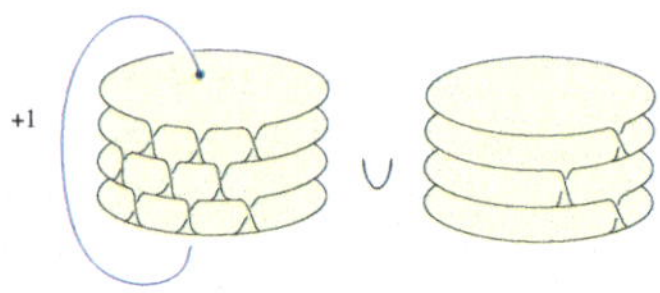

Figure 12.13: $L_d = M(4,3) \cup D$

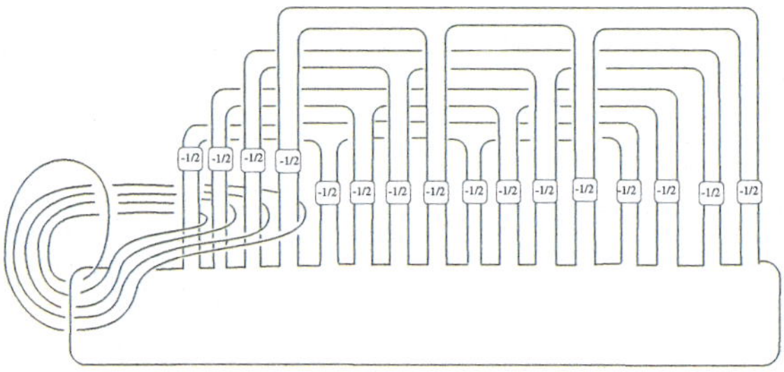

Figure 12.14: M(5,4)

Exercise 12.2. *Show that (a)* $V_2 = S^2 \times S^2$ *and (b)* $V_3 = \mathbb{CP}^2 \# 6\overline{\mathbb{CP}}^2$ *(Hint: Figure 12.15.)*

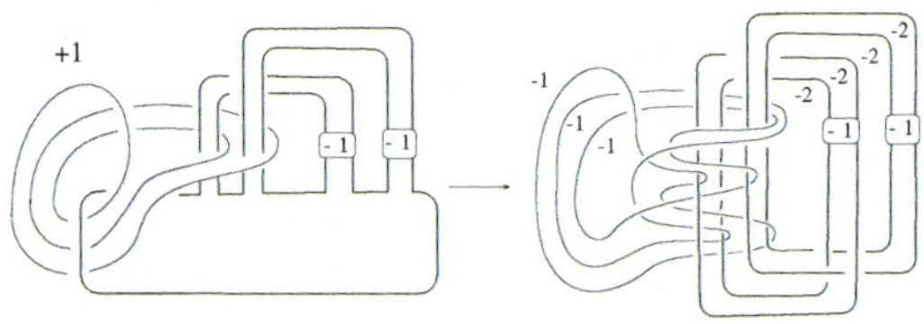

Figure 12.15: Constructing V_3

For example, by this process we get the following handlebody of V_5 (Figure 12.16):

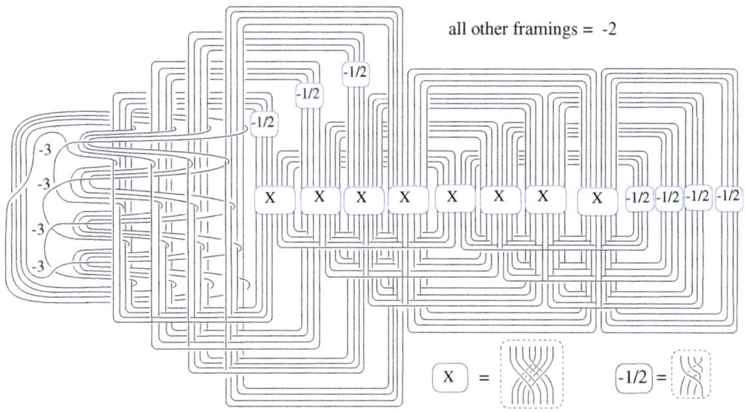

Figure 12.16: V_5

More generally we can construct complex surfaces by taking k-fold branched coverings, $\pi : V_k(d) \to \mathbb{CP}^2$, branched along L_d, where k divides d. For example, it is known that Kummer surface V_4 is $V_2(6)$ (Figure 12.17):

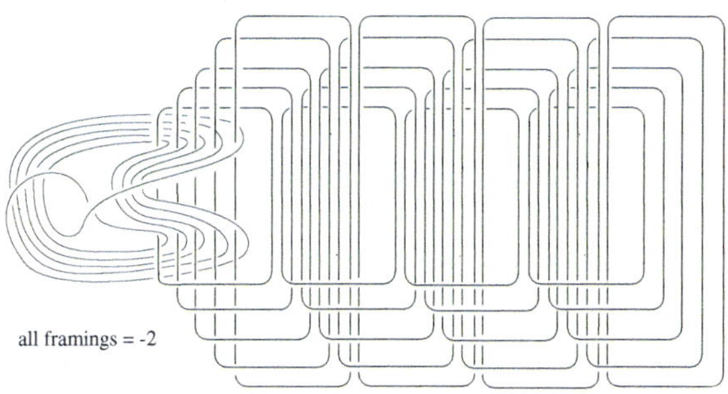

Figure 12.17: V_4

12.4 $\Sigma(a,b,c)$

One way to study Brieskorn manifold $\Sigma(a,b,c)$ is to resolve the singularity of the complex hypersurface $V(a,b,c) = \{(x,y,z) \in \mathbb{C}^3 \mid x^a + y^b + z^c = 0\}$ at the origin, to get a smooth manifold $M(a,b,c)$ which $\Sigma(a,b,c)$ bounds. The topology of resolutions is a rich subject, which is not yet completely understood (e.g. [AKi]), but in this special case there are concrete ways of doing this. A nice approach is explained in [HNK]: The idea is to uniformize the algebraic function $f(x,y) = \sqrt[c]{x^a + y^b}$ by successive blowing ups, $\mathbb{C}^2$ at the origin $\pi : \tilde{\mathbb{C}}^2 \to \mathbb{C}^2$, until $V(a,b) = \{(x,y) \mid x^a + y^b = 0\}$ becomes a smooth curve Γ, with $\pi^{-1}(0)$, consisting of a union of 2-spheres $P(a,b)$, all transversal to each other.

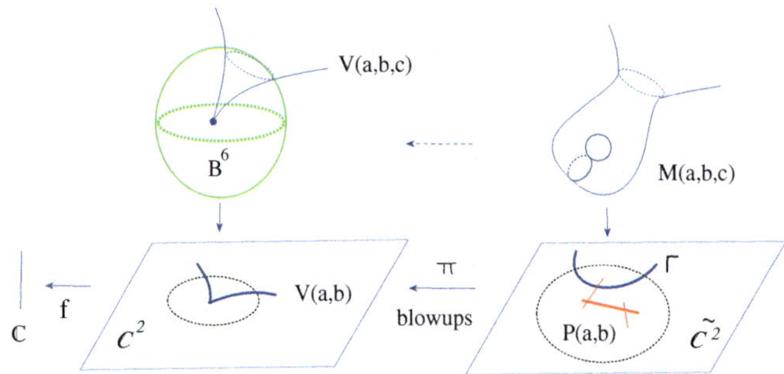

Figure 12.18

This process is continued until the 2-spheres, $\alpha_1, \alpha_2, \ldots$ of $P(a,b)$, on which $f \circ \pi$ has multiplicity 0 (mod c), are disjoint from each other. Also the other spheres, $\beta_1, \beta_2, \ldots$ of $P(a,b)$ and Γ are disjoint from each other, and the framings of, β_j are divisible by c.

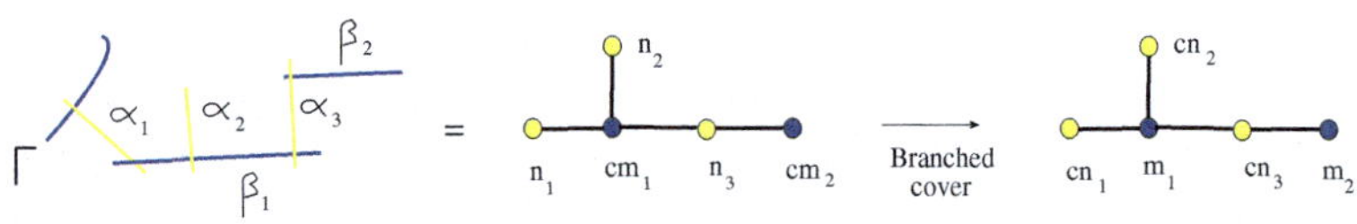

Figure 12.19

This means that the function $z \mapsto \sqrt[c]{z}$ is uniformized on $\tilde{\mathbb{C}}^3$, and we can take the c-fold branched covering of $\tilde{\mathbb{C}}^3$ along $\cup\beta_j \cup \Gamma$. So $M(a,b,c)$ is the branched covering of the plumbed 4-manifold $\cup\alpha_i \cup \beta_j$ along $\cup\beta_j$ (done as in Exercise 11.9).

$M(a,b,c)$ is the plumbing obtained from the plumbing of the spheres $P(a,b)$ by simply by dividing the framings of β_j spheres by c, and multiplying the framings of α_j spheres by c. The idea is basically: $V(a,b,c)$ is the c-fold branched covering space of $\mathbb{C}^2$ branched along $V(a,b)$, and throughout the blowing ups this branched covering information is carried along so that $M(a,b,c)$ is the branched covering space of $\tilde{\mathbb{C}}^2$, branched along $\cup\beta_j \cup \Gamma$.

Let us do the example $\Sigma(2,3,5)$ by hand using the above recipe. Recall that blowing up the (y,z) chart gives two new charts (u,v) and (u',v'); they are related to the old chart by $(y,z) = (uv,u)$ and $(y,z) = (u',u'v')$ (where $u = 0$ and $u' = 0$ is the equation of the new -1 sphere which appeared as the result of blowup). Here we will only trace the pictures on selected charts and guess the other parts. (cf. [Mu], [Lau], [OW], [AKi]).

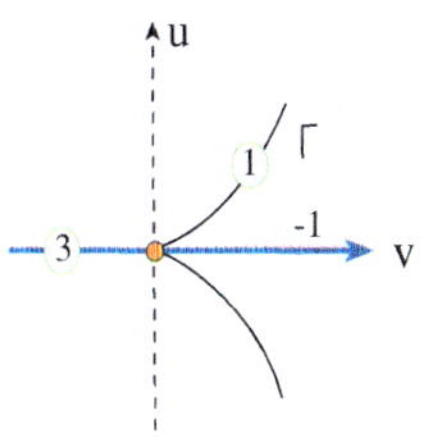

Figure 12.20:
$(y,z) = (uv,u)$
$u^3(v^3 + u^2) = 0$

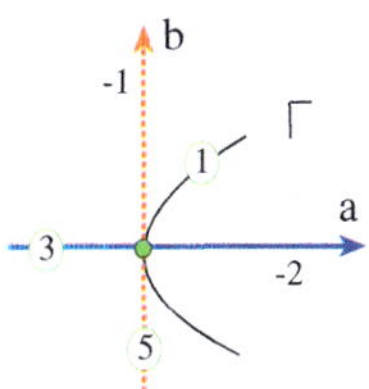

Figure 12.21:
$(u,v) = (ab,a)$
$a^5b^3(a + b^2) = 0$

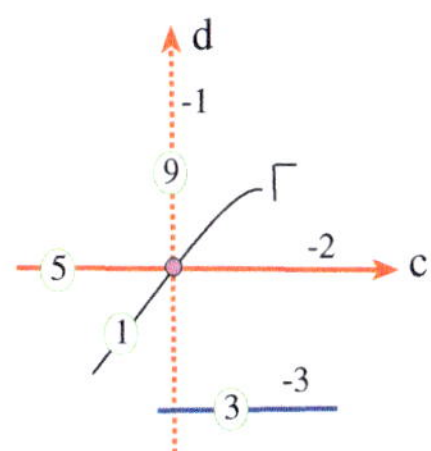

Figure 12.22:
$(a,b) = (cd,c)$
$c^9d^5(d + c) = 0$

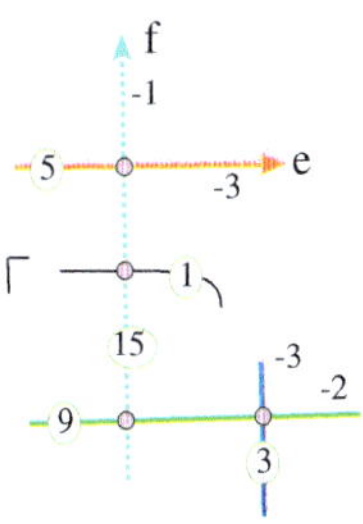

Figure 12.23:
$(c,d) = (e,ef)$
$e^{15}f^5(1 + f) = 0$

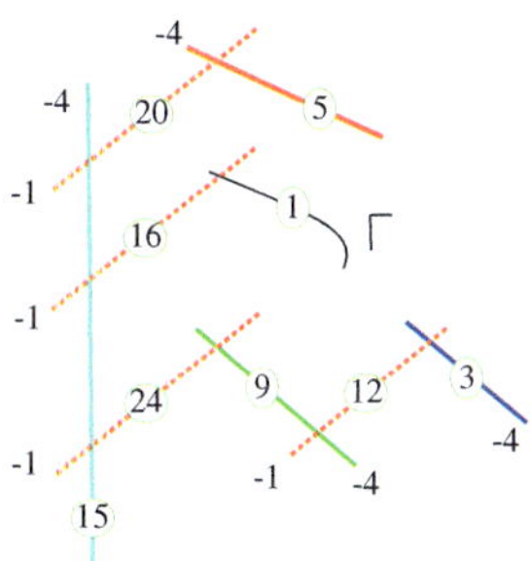

Figure 12.24:
P(3,5)

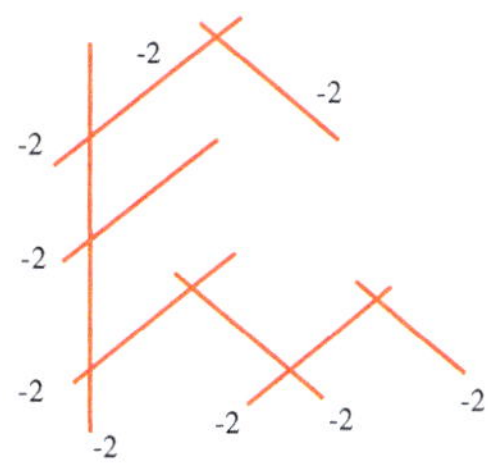

Figure 12.25:
M(2,3,5)

We first blow up the curve $y^3 + z^5 = 0$ at the origin to get Figure 12.20, then by successive blowups at the indicated points we get Figures 12.20, through 12.23. The self-intersections of spheres are indicated by integers, and their multiplicities (which can be read from the equations) are denoted by circled integers. The algebraic curve itself is assigned the multiplicity 1 (as a rule a blowup at a point p results in a new -1 sphere, whose multiplicity is the sum of the multiplicities of the spheres meeting at that point p). Then finally to separate the odd multiplicity spheres from the even ones, we perform the indicated blowups to get Figure 12.24. Then applying the above-mentioned branched covering process, from Figure 12.24 we get Figure 12.25, which is $M(2,3,5)$.

Exercise 12.3. *By resolving singularities verify the following identifications.*

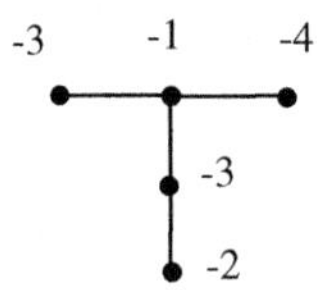

Figure 12.26:
$M(3,4,5)$

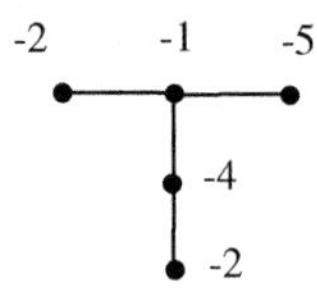

Figure 12.27:
$M(2,5,7)$

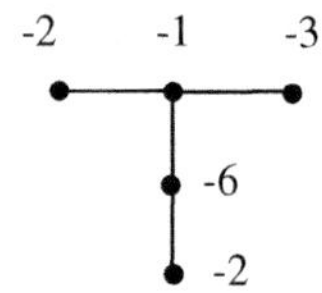

Figure 12.28:
$M(2,3,11)$

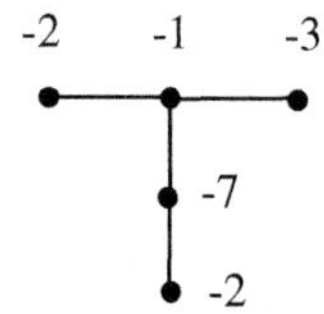

Figure 12.29:
$M(2,3,13)$

In fact, by [HJ], when a_j's pairwise coprime $M(a_1, a_2, a_3)$ is the following plumbing:

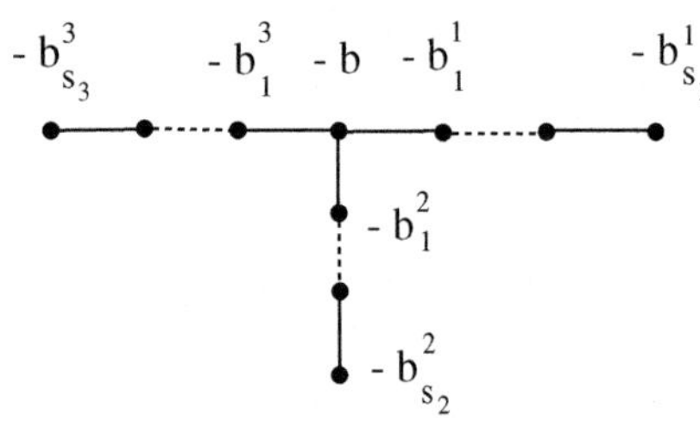

Figure 12.30

where $a_j/b_j = [b^j_1, ..., b^j_{s_j}]$, with $b^j_i \geq 2$, and $ba_1a_2a_3 = 1 + b_1a_2a_3 + a_1b_2a_3 + a_1a_2b_3$, where b_k are given by $0 < b_k < a_k$ and $a_ia_jb_k = -1 \pmod{a_k}$ for $\{i,j,k\} = \{1,2,3\}$ (see also [NR]).

Exercise 12.4. *By using Exercise 12.3 verify the identifications of Figure 2.14 (Hint: Apply blowing up/down and handle sliding operations to the corresponding $M(a,b,c)$).*

Exercise 12.5. *By using Exercise 12.3 show that we have the following identifications with the boundaries of the indicated contractible manifolds defined in Figure 2.11 (Hint: Figure 12.31 indicates a diffeomorphism $\Sigma(2,5,7) \approx \partial W^-(0,3)$, then use Exercise 2.2. The others are similar (cf. [AK2])).*

$$\Sigma(2,5,7) \approx \partial W^+(0,0)$$

$$\Sigma(3,4,5) \approx \partial W^+(-1,0)$$

$$\Sigma(2,3,13) \approx \partial W^+(1,0)$$

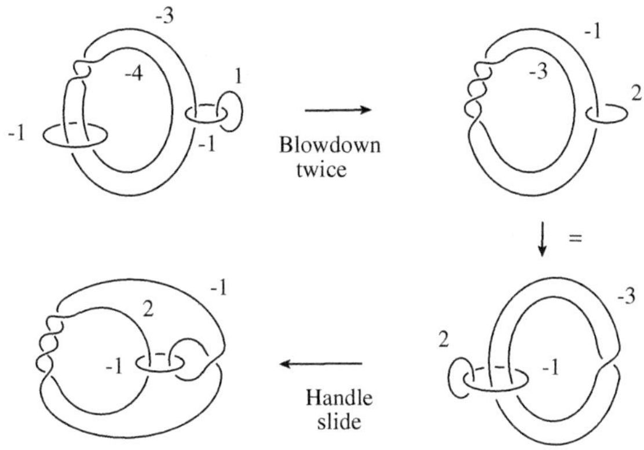

Figure 12.31: $\Sigma(2,5,7) \approx \partial W^-(0,3)$

Remark 12.4. *Proposition 12.2 is a special case of the following (e.g. [Hi], [Ham]): If X is a closed, orientable, simply connected 4-manifold with $b_2(X) > 0$, and $Z^4 \to X^4$ is a d-fold cyclic branched covering map branched along a closed orientable surface $F \subset X$ of genus $g(F)$, then:*

$$b_2(Z) = db_2(X) + 2(d-1)g(F)$$

$$\sigma(Z) = d\sigma(X) - (\frac{d^2-1}{3d})F.F$$

Exercise 12.6. *By identifying the handlebody of $E(n)$ in Figure 7.4 as the handlebody $M(2,3,6n-1)$ (Figure 12.4) plus a pair of 2-handles, and then by imitating the steps of the proof of Proposition 12.1, show that Figure 12.32 gives a handlebody of $E(n)$ (this is done in [GS]).*

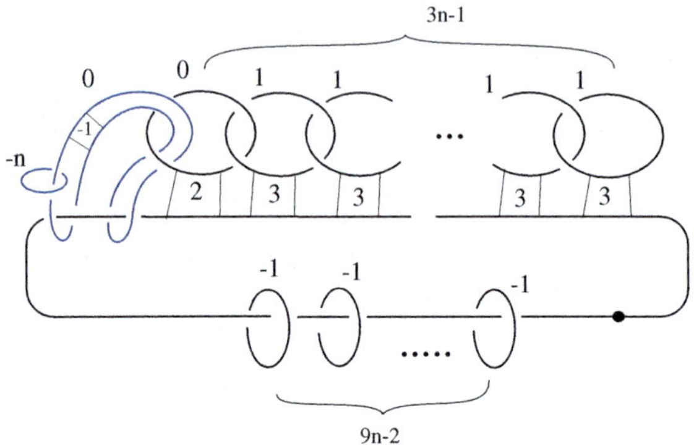

Figure 12.32: $E(n)$

Chapter 13

Seiberg–Witten invariants

Seiberg–Witten invariants are smooth 4-manifold invariants discovered by N. Seiberg and E. Witten; they help to distinguish smooth manifolds. Here we will give an introduction to these invariants (also see [A8], [Mor], [Mar], [Ni], [Sc], [Moo], [FS6]). Let X be a closed smooth 4-manifold with $b_2^+(X) \neq 0$. A *Spin*^{c} structure on X is an integral lift of its second Stiefel-Whitney class $w_2(X) \in H^2(X, \mathbb{Z}_2)$ under the reduction map $H^2(X, \mathbb{Z}) \to H^2(X, \mathbb{Z}_2)$, given by $a \mapsto a \ (mod\ 2)$. When $b_2^+(X) > 1$, the Seiberg–Witten invariant is a function $SW_X : \mathcal{C}(X) \to \mathbb{Z}$, defined on the set of *Spin*^{c} structutes on X:

$$\mathcal{C}(X) = \{L \in H^2(X; \mathbb{Z}) \mid w_2(X) = c_1(L) \ (mod\ 2)\}$$

It turns out that the set of so-called *basic classes* $\mathcal{B}(X) := \{L \in \mathcal{C}(X) \mid SW_X(L) \neq 0\}$ is a finite set. Also, every compact oriented smooth 4-manifold has a *Spin*^{c} structure . This is because the homomorphism $H_2(X; \mathbb{Z}) \to \mathbb{Z}_2$, given by $x \to x^2$ (mod 2), is zero on the torsion elements of $H_2(X; \mathbb{Z})$, so it lifts to some $\gamma : H_2(X; \mathbb{Z}) \to Z$. Let $\alpha \in H^2(X; \mathbb{Z})$ be a preimage of γ. Also recall that the class $w_2 := w_2(X)$ satisfies $x.w_2 = x^2$ for all $x \in H_2(X, \mathbb{Z}_2)$ (Wu relation [MS]). So the class $b = w_2 - \rho(\alpha)$ lies in the kernel of the evaluation map h in the universal coefficient sequence (here ρ denotes the $\mathbb{Z}_2$ reduction):

$$\begin{array}{ccccccc}
0 \to & Ext(H_1(X); \mathbb{Z}_2) & \longrightarrow & H^2(X; \mathbb{Z}_2) & \xrightarrow{h} & Hom(H_2(X); \mathbb{Z}_2) & \to 0 \\
& \uparrow Ext(\rho) & & \uparrow \rho & & \uparrow Hom(\rho) & \\
0 \to & Ext(H_1(X); \mathbb{Z}) & \longrightarrow & H^2(X; \mathbb{Z}) & \longrightarrow & Hom(H_2(X); \mathbb{Z}) & \to 0
\end{array}$$

So b comes from $Ext(H_1(X); \mathbb{Z}_2)$. The universal coefficient sequence is natural with respect to coefficient homomorphism and the first vertical arrow is onto (e.g. [Sp]), hence we get $b = \rho(\beta)$ for some β, which implies that w_2 comes from an integral class.

13.1 Representations

Reader can consult [A22] for a quick background in bundle theory and the basic differential geometric constructions used here. Let us recall some basic identifications: $SU(2) = Spin(3) = S^3 \subset \mathbb{H}$, where $\mathbb{H}$ denotes the quaternions and S^3 are the unit quaternions, we have $SO(3) = SU(2)/\mathbb{Z}_2$ and we also have the following identifications:

$$\begin{aligned} Spin(4) &= SU(2) \times SU(2) \\ Spin^c(4) &= (SU(2) \times SU(2) \times S^1)/\mathbb{Z}_2 = (Spin(4) \times S^1)/\mathbb{Z}_2 \\ SO(4) &= (SU(2) \times SU(2))/\mathbb{Z}_2 \\ Spin^c(3) &= (SU(2) \times S^1)/\mathbb{Z}_2 = U(2) \end{aligned}$$

There is also the idetification $\{(A,B) \in U_2 \times U_2 \mid det(A) = det(B)\} = Spin^c(4)$ via:

$$(A,B) \rightsquigarrow (A.(detA)^{-1/2}\, I\ ,\ B.(detA)^{-1/2}\, I\ ,\ (detA)^{1/2})$$

By projecting the above descriptions to various factors we get fibrations:

$$S^1 \longrightarrow Spin^c(4) \to SO(4)$$

$$\mathbb{Z}_2 \longrightarrow Spin^c(4) \to SO(4) \times S^1$$

The fibrations above naturally extend to fibrations

$$S^1 \to Spin^c(4) \to SO(4) \to K(\mathbb{Z},2) \to BSpin^c(4) \to BSO(4) \to K(\mathbb{Z},3) \tag{13.1}$$

The last map in the sequence is given by the Bockstein of the universal second Steifel–Whitney class $\delta(w_2)$, which is the obstruction to lifting w_2 to an integral class. This explains why this corresponds to a $Spin^c(4)$ structure. We also have the fibration:

$$\mathbb{Z}_2 \to Spin^c(4) \to SO(4) \times S^1 \to K(\mathbb{Z}_2,1) \to BSpin^c(4) \to BSO(4) \times BS^1 \to K(\mathbb{Z}_2,2)$$

The last map is given by $w_2 \times 1 + 1 \times \rho(c_1)$, which clearly vanishes when $\delta(w_2) = 0$. This says that $Spin^c$ can be thought of as a complex line bundle $L \to X$ whose first Chern class is an integral lift of $w_2(X)$. Finally we have the fibration:

$$\mathbb{Z}_2 \to Spin(4) \times S^1 \to Spin^c(4) \to K(\mathbb{Z}_2,1) \to BSpin(4) \times BS^1 \to BSpin^c(4) \to K(\mathbb{Z}_2,2)$$

The last map is given by w_2. This sequence says that locally a $Spin^c(4)$ bundle consists of a pair of a $Spin(4)$ bundle and a complex line bundle. Also recall the identifications:

$$H^2(X;\mathbb{Z}) = [X, K(\mathbb{Z},2)] = [X, BS^1] = \{\text{Complex line bundles } L \to X\} \tag{13.2}$$

Definition 13.1. *A $Spin^c(4)$ structure on X^4 is a complex line bundle $L \to X$ with $w_2(TX) = c_1(L)$ (mod 2) (i.e. L is a characteristic line bundle) . This means a principal $Spin^c(4)$-bundle $P \to X$ such that the associated framed bundles of TX and L satisfy:*

$$P_{SO(4)}(TX) = P \times_{\rho_0} SO(4)$$

$$P_{S^1}(L) = P \times_{\rho_1} S^1$$

where $(\rho_0, \rho_1) : Spin^c(4) \to SO(4) \times S^1$ are the obvious projections.

So we have the following natural projections:

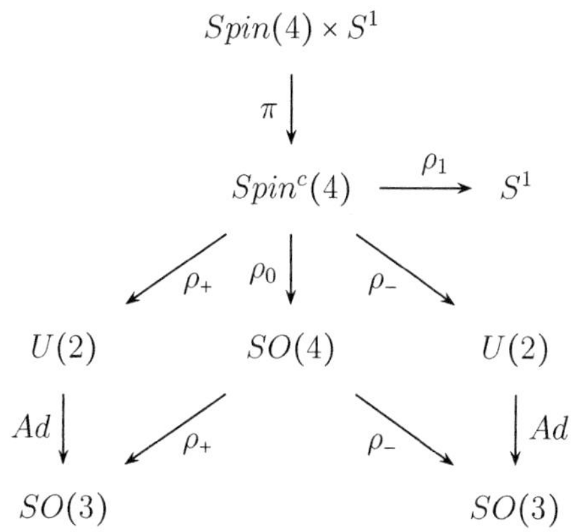

Let $\tilde{\rho}_\pm = Ad \circ \rho_\pm$, also call $\bar{\rho}_\pm = \rho_\pm \circ \pi$. For $x \in \mathbb{H} = \mathbb{R}^4$ we have

$$\begin{array}{lcll}
\rho_1[\, q_+, q_-, \lambda \,] & = & \lambda^2 & \\
\rho_0[\, q_+, q_-, \lambda \,] & = & [\, q_+, q_- \,] & \text{i.e. } x \longmapsto q_+ x q_-^{-1} \\
\rho_\pm[\, q_+, q_-, \lambda \,] & = & [\, q_\pm, \lambda \,] & \text{i.e. } x \longmapsto q_\pm x \lambda^{-1} \\
\tilde{\rho}_\pm[\, q_+, q_-, \lambda \,] & = & Ad \circ q_\pm & \text{i.e. } x \longmapsto q_\pm x q_\pm^{-1} \\
\bar{\rho}_\pm(\, q_+, q_-, \lambda \,) & = & \lambda q_\pm &
\end{array}$$

Besides TX and L, a $Spin^c(4)$ bundle $P \to X$ induces a pair of $U(2)$ bundles:

$$W^\pm = P \times_{\rho_\pm} \mathbb{C}^2 \longrightarrow X$$

In general these are very different $\mathbb{C}^2$-bundles (e.g. [A22]), for example,

$$c_2(W^+)[M] = (c_1^2(L) - 3\sigma(M) - 2\chi(M))/4 \tag{13.3}$$

$$c_2(W^-)[M] = (c_1^2(L) - 3\sigma(M) + 2\chi(M))/4$$

Let $\Lambda^p(X) = \Lambda^p T^*(X)$ be the bundle of exterior p forms over a Riemanian manifold (X, g) (g is metric); we can construct the bundle of self (antiself)-dual 2-forms $\Lambda^2_\pm(X)$, which we abbreviate as $\Lambda^\pm(X)$. We can identify $\Lambda^2(X)$ by the Lie algebra $so(4)$-bundle

$$\Lambda^2(X) = P(T^*X) \times_{ad} so(4) \quad \text{by} \;\; \Sigma\, a_{ij}\, dx^i \wedge dx^j \;\longleftrightarrow\; (a_{ij})$$

where $ad : SO(4) \to so(4)$ is the adjoint representation. The adjoint action preserves the two summands of $so(4) = spin(4) = so(3) \times so(3) = \mathbb{R}^3 \oplus \mathbb{R}^3$. By above identification it is easy to see that the ± 1 eigenspaces $\Lambda^\pm(X)$ of the star operator $* : \Lambda(X) \to \Lambda(X)$ corresponds to these two $\mathbb{R}^3$-bundles; this gives:

$$\Lambda^2_\pm(X) := \Lambda^\pm(X) = P \times_{\tilde{p}_\pm} \mathbb{R}^3 = P_{SO(4)}(TX) \times_{\tilde{p}_\pm} \mathbb{R}^3$$

Chern classes are given by $c_2(\Lambda^\pm(X)) = 4c_2(W^\pm)$. Also if the $Spin^c(4)$ bundle $P \to X$ lifts to the $Spin(4)$ bundle $\bar{P} \to X$ (i.e. when $w_2(X) = 0$), corresponding to the obvious projections $r_\pm : Spin(4) \to SU(2)$, $r_\pm(q_-, q_+) = q_\pm$, we get a pair of $SU(2)$ bundles:

$$V^\pm = P \times_{r_\pm} \mathbb{C}^2$$

Clearly, since $x \longmapsto q_\pm x \lambda^{-1} = q_\pm x\, (\lambda^2)^{-1/2}$, in this case we have:

$$W^\pm = V^\pm \otimes L^{-1/2}$$

13.2 Action of $\Lambda^*(X)$ on $W^\pm$

From the definition of $Spin^c(4)$ structure above we can identify ($\mathbb{H}$ denotes quaternions):

$$T^*(X) = P \times \mathbb{H}/(p, v) \sim (\tilde{p}, q_+ v\, q_-^{-1}) \;\; , \text{ where } \;\; \tilde{p} = p[\, q_+, q_-, \lambda \,]$$

We define left actions (Clifford multiplications), which are well defined by

$$T^*(X) \otimes W^+ \longrightarrow W^- \;\; , \text{ by } \;\; [\, p, v \,] \otimes [\, p, x \,] \longmapsto [\, p, -vx \,]$$

$$T^*(X) \otimes W^- \longrightarrow W^+ \;\; , \text{ by } \;\; [\, p, v \,] \otimes [\, p, x \,] \longmapsto [\, p, vx \,]$$

From the identifications, we can check the well definededness of these actions, e.g.:

$$[p,v\,]\otimes[\,p,x\,]\sim[\,\tilde{p},q_+v\,q_-^{-1}\,]\otimes[\,\tilde{p},q_+x\lambda^{-1}\,]\longmapsto[\,\tilde{p},\,q_-(-\bar{v}x)\lambda^{-1}\,]\sim[\,p,-\bar{v}x\,]$$

Alternatively we can describe these actions by using only complex numbers with the aid of the following obvious fact:

Lemma 13.2. *Let $P\to X$ be a principal G bundle and $E_{\rho_i}\to X$ be vector bundles associated to representations $\rho_i : G\to Aut(V_i)$, $i=1,2,3$. Then a map, $\lambda : V_1\times V_2\to V_3$, induces an action, $E_{\rho_1}\times E_{\rho_2}\to E_{\rho_3}$ if $\lambda(\rho_1(g)v_1,\rho_2(g)v_2)=\rho_3(g)\lambda(v_1,v_2)$.*

By dimension reason complexification of these representations give

$$\rho : T^*(X)_{\mathbb{C}}\xrightarrow{\cong}Hom(W^\pm,W^\mp)\cong W^\pm\otimes W^\mp$$

We can put them together as a single representation (which we still call ρ),

$$\rho : T^*(X)\longrightarrow Hom(W^+\oplus W^-)\ ,\ \text{by}\ \ v\longmapsto\rho(v)=\begin{pmatrix}0 & v\\ -\bar{v} & 0\end{pmatrix}$$

We have $\rho(v)\circ\rho(v)=-|v|^2I$. By the universal property of the Clifford algebra this representation extends to the Clifford algebra $C(X)=\Lambda^*(X)$ (exterior algebra),

$$\begin{array}{ccc}\Lambda^*(X) & & \\ \downarrow & \searrow & \\ T^*(X) & \longrightarrow & Hom\,(W^+\oplus W^-)\end{array}$$

One can construct this extension without the aid of the universal property of the Clifford algebra, for example since

$$\Lambda^2(X)=\left\{\,v_1\wedge v_2=\frac{1}{2}(v_1\otimes v_2-v_2\otimes v_1)\;|\;v_1,v_2\in T^*(X)\,\right\}$$

The action of $T^*(X)$ on $W^\pm$ determines the action of $\Lambda^2(X)=\Lambda^+(X)\otimes\Lambda^-(X)$, and since $2Im\,(v_2\bar{v}_1)=-v_1\bar{v}_2+v_2\bar{v}_1$ we have the action ρ with property

$$\Lambda^+(X)\otimes W^+\longrightarrow W^+\ \text{ is given by }\ [\,p,v_1\wedge v_2\,]\otimes[\,p,x\,]\longrightarrow[\,p,Im\,(v_2\bar{v}_1)x\,]$$

$$\rho : \Lambda^+\longrightarrow Hom(W^+,W^+)$$

$$\rho(v_1 \wedge v_2) = \frac{1}{2}[\,\rho(v_1), \rho(v_2)\,] \qquad (13.4)$$

Let us write the local descriptions of these representations. We first choose a local orthonormal basis of covectors $\{e^1, e^2, e^3, e^4\}$ for $T^*(X)$, then we can take

$$\{\, f_1 = \frac{1}{2}(e^1 \wedge e^2 \pm e^3 \wedge e^4),\ f_2 = \frac{1}{2}(e^1 \wedge e^3 \pm e^4 \wedge e^2),\ f_3 = \frac{1}{2}(e^1 \wedge e^4 \pm e^2 \wedge e^3)\,\}$$

to be a basis for $\Lambda^\pm(X)$. After the local identification, $T^*(X) = \mathbb{H}$, we can take $e^1 = 1$, $e^2 = i$, $e^3 = j$, $e^4 = k$. Let us identify $W^\pm = \mathbb{C}^2 = \{z + jw \mid z, w \in \mathbb{C}\,\}$, then the multiplication by $1, i, j, k$ (action on $\mathbb{C}^2$ as multiplication on the left) induces the representations $\rho(e^i)$, $i = 1, 2, 3, 4$. From this we see that $\Lambda^+(X)$ acts trivially on W^-; and the basis f_1, f_2, f_3 of $\Lambda^+(X)$ acts on W^+ as multiplication by i, j, k, respectively (these are called Pauli matrices),

$$\rho(e^1) = \begin{pmatrix} & & 1 & 0 \\ & & 0 & 1 \\ -1 & 0 & & \\ 0 & -1 & & \end{pmatrix} \qquad \rho(e^2) = \begin{pmatrix} & & i & 0 \\ & & 0 & -i \\ i & 0 & & \\ 0 & -i & & \end{pmatrix}$$

$$\rho(e^3) = \begin{pmatrix} & & 0 & -1 \\ & & 1 & 0 \\ 0 & -1 & & \\ 1 & 0 & & \end{pmatrix} \qquad \rho(e^4) = \begin{pmatrix} & & 0 & -i \\ & & -i & 0 \\ 0 & -i & & \\ -i & 0 & & \end{pmatrix}$$

$$\rho(f_1) = \begin{pmatrix} i & 0 \\ 0 & -i \end{pmatrix} \quad \rho(f_2) = \begin{pmatrix} 0 & -1 \\ 1 & 0 \end{pmatrix} \quad \rho(f_3) = \begin{pmatrix} 0 & -i \\ -i & 0 \end{pmatrix}$$

In particular we get an isomorphism, $\rho : \Lambda^+(X) \longrightarrow su\,(W^+)$ (traceless skew adjoint endomorphism of W^+). After complexifying, this isomorphism extends to an isomorphism $\rho : \Lambda^+(X)_{\mathbb{C}} \cong sl\,(W^+)$ (traceless endomorphism of W^+). We will also define a map σ,

$$\begin{array}{ccccc} & & \Lambda^+(X) & \xrightarrow{\cong} & su\,(W^+) \\ & & \cap & & \cap \\ W^+ & \xrightarrow{\sigma} & \Lambda^+(X)_{\mathbb{C}} & \xrightarrow{\rho} & sl\,(W^+) \end{array}$$

Remark 13.3. *$Hom(W^+, W^+) \cong W^+ \otimes (W^+)^*$, and the dual space $(W^+)^*$ can be identified with $\bar{W}^+$ ($= W^+$ with scalar multiplication $c.v = \bar{c}v$) by the pairing $W^+ \otimes \bar{W}^+ \to \mathbb{C}$, given by $z \otimes w \to z\bar{w}$. Usually $sl\,(W^+)$ is denoted by $(W^+ \otimes \bar{W}^+)_0$ and the trace map gives the identification*

$$W^+ \otimes \bar{W}^+ = (W^+ \otimes \bar{W}^+)_0 \oplus \mathbb{C}$$

Now define $\sigma : W^+ \to \Lambda^+(X)$ by $[\,p, x\,] \longmapsto [\,p,\ \frac{1}{2}(xi\,\bar{x})\,]$ (multiplications in $\mathbb{H}$). By local identification as above $W^+ = \mathbb{C}^2$ and $\Lambda^+(X) = \mathbb{R} \oplus \mathbb{C}$, we see σ corresponds to

$$\sigma(z,w) = i\left(\frac{|z|^2 - |w|^2}{2}\right) - k\ Re(z\bar{w}) + j\ Im(z\bar{w}) \tag{13.5}$$

Here i, j, k correspond to the basis elements f_1, f_2, f_3 of $\Lambda^+(X)$. By applying the isomorphism ρ we make the composition $\boldsymbol{\sigma} := \rho \circ \sigma$ take values in $sl\,(W^+)$,

$$\boldsymbol{\sigma}(z,w) \quad = \quad \rho\,[\,\frac{|z|^2 - |w|^2}{2}\,f_1 + Im(z\bar{w})\,f_2 - Re(z\bar{w})\,f_3\,]$$

By plugging in the values of the Pauli matrices $\rho(f_1)$, $\rho(f_2)$, $\rho(f_3)$ we calculate

$$(z,w) \longmapsto \boldsymbol{\sigma}(z,w) = i\begin{pmatrix} (\,|z|^2 - |w|^2)/2 & z\bar{w} \\ \bar{z}w & (\,|w|^2 - |z|^2)/2 \end{pmatrix} \tag{13.6}$$

Put another way, $\boldsymbol{\sigma}$ is the projection of the diagonal elements of $W^+ \otimes \bar{W}^+$ onto the subspace $(W^+ \otimes \bar{W}^+)_0$, that is, if we write $x = z + jw$ we get $\boldsymbol{\sigma}(z,w) = i(z\bar{w}^t)_0$, and

$$|\,\rho(u)\,| \quad = \quad \sqrt{2}\,|\,u\,| \tag{13.7}$$

$$|\,\boldsymbol{\sigma}(\psi)\,| \quad = \quad \frac{1}{2}|\,\psi\,|^2 \tag{13.8}$$

$$\langle\,\boldsymbol{\sigma}(\psi)\,\psi, \psi\,\rangle \quad = \quad \frac{i}{2}|\,\psi\,|^4 \tag{13.9}$$

$$\langle \rho(u)\,\psi\,,\ \psi\rangle \quad = \quad 2i\,\langle \rho(u)\,, \boldsymbol{\sigma}(\psi)\rangle \tag{13.10}$$

Here the norm in $su(2)$ is induced by the inner product $\langle A, B\rangle = -trace(AB)$.

By renaming $\boldsymbol{\sigma}(\psi,\psi) := \boldsymbol{\sigma}(\psi)$ we can extend the definition of $\boldsymbol{\sigma}$ to $W^+ \otimes \bar{W}^+$ by

$$\langle \rho(u)\,\psi\,,\,\varphi\rangle = i\,\langle \rho(u)\,,\boldsymbol{\sigma}(\psi,\varphi)\rangle$$

$$\begin{array}{ccc} W^+ & \xrightarrow{\sigma} & su(W^+) \\ \cap & & \cap \\ W^+\otimes \bar{W}^+ & \xrightarrow{\sigma} & sl\,(W^+) \;\; = (W^+\otimes \bar{W}^+)_0 \end{array}$$

Remark 13.4. *A $Spin^c(4)$ structure can also be defined as a pair of $\mathbb{C}^2$ bundles:*

$$W^\pm \longrightarrow X \;\; \text{with} \;\; det(W^+) = det(W^-) \longrightarrow X \;\; \text{(a complex line bundle)},$$

$$\text{and an action } c_\pm : T^*(X) \longrightarrow Hom(W^\pm, W^\mp) \;\; \text{with} \;\; c_\pm(v)c_\mp(v) = -|v|^2 I$$

In particular this means that the associated bundles $P(W^\pm) \to X$ have $U(2)$ structures. The first definition of $Spin^c(4)$ structure clearly implies this, and conversely we can obtain the first definition by letting the principal $Spin^c(4)$ bundle be:

$$P = \{(p_+, p_-) \in P(W^+) \times P(W^-) \mid det(p_+) = det(p_-)\}$$

Clearly, $Spin^c(4) = \{(A,B) \in U_2 \times U_2 \mid det(A) = det(B)\}$ acts on P freely. This definition generalizes and gives way to the following definition:

Definition 13.5. *A Dirac bundle $W \longrightarrow X$ is a Riemannian vector bundle with an action $\rho : T^*(X) \longrightarrow Hom(W,W)$, satisfying $\rho(v)\circ\rho(v) = -|v|^2 I$. W is also equipped with a connection D satisfying*

$$\langle \rho(v)x, \rho(v)y\rangle = \langle x, y\rangle$$

$$D_Y(\rho(v)s) = \rho(\nabla_X v)s + \rho(v)D_Y(s)$$

where ∇ is the Levi-Civita connection on $T^*(X)$, and Y is a vector field on X.

An example of a Dirac bundle is $W = W^+ \oplus W^- \longrightarrow X$ and $D = d + d^*$ with $W^+ = \oplus\Lambda^{2k}(X)$ and $W^- = \oplus\Lambda^{2k+1}(X)$, where $\rho(v) = v \wedge . + v \perp$. (exterior + interior product with v). In this case $\rho : W^\pm \to W^\mp$ In the next section we will discuss the natural connections D for $Spin^c$ structures $W^\pm$.

13.3 Dirac operator

Let $\mathcal{A}(L)$ denote the space of connections on a $U(1)$ bundle $L \to X$ over a smooth 4-manifold with $w_2(X) = c_1(L)\ (mod\ 2)$. Note that after choosing a base connection a_0 we can identify $\mathcal{A}(L) = \{a_0\} + \Omega^1(X)$. Any $A \in \mathcal{A}(L)$ and the Levi-Civita connection A_0 on the tangent bundle coming from Riemanian metric of X define a product connection on $P_{SO(4)} \times P_{S^1}$. Since $Spin^c(4)$ is the 2-fold covering of $SO(4) \times S^1$, they have the same Lie algebras, $spin^c(4) = so(4) \oplus i\ \mathbb{R}$. Hence we get a connection $\tilde{A}$ on the $Spin^c(4)$ principal bundle $P \longrightarrow X$. In particular the connection $\tilde{A}$ defines connections to all the associated bundles of P, giving back A, A_0 on L, $T(X)$ respectively, and two new connections $A^{\pm}$ on bundles $W^{\pm}$. We denote the corresponding covariant derivative by

$$\nabla_A : \Gamma(W^+) \to \Gamma(T^*X \otimes W^+)$$

Locally, by choosing orthonormal tangent vector field $e = \{e_i\}_{i=1}^4$ and the dual basis of 1-forms $\{e^i\}_{i=1}^4$ in a neighborhood U of a point $x \in X$, we can write

$$\nabla_A = \sum e^i \otimes \nabla_{e_i}$$

where $\nabla_{e_i} : \Gamma(W^+) \to \Gamma(W^+)$ is the covariant derivative ∇_A along e_i. Composing this map with the map $\Gamma(T^*X \otimes W^+) \to \Gamma(W^-)$, induced by the Clifford multiplication (Section 13.2), gives the Dirac operator $\not{D}_A : \Gamma(W^+) \to \Gamma(W^-)$, i.e.

$$\not{D}_A = \sum \rho(e^i)\nabla_{e_i}$$

Locally, $W^{\pm} = V^{\pm} \otimes L^{-1/2}$, and the connection A gives the untwisted Dirac operator

$$\not{\partial} : \Gamma(V^+) \to \Gamma(V^-)$$

Notice that as in $W^{\pm}$, forms $\Lambda^*(X)$ act on $V^{\pm}$. Now let $\omega = (\omega_{ij})$ be the Levi-Civita connection 1-form, which is an $so(4) = su(2) \oplus su(2)$ valued *equivariant* 1-form on $P_{SO(4)}(X)$. Let $e^*(\omega) := \tilde{\omega} = (\tilde{\omega}_{ij}) \oplus (\tilde{\omega'}_{ij})$ be the pull-back 1-form on U. Since $P_{SO(4)}(U) = P_{Spin(4)}(U)$, the orthonormal basis $e \in P_{SO(4)}(U)$ determines an orthonormal basis $\sigma = \{\sigma^k\} \in P_{SU_2}(U)$, then (e.g. [LaM] p.110)

$$\not{\partial}(\sigma^k) = \sum_{i<j} \rho(\tilde{\omega}_{ij})\, \sigma^k$$

A metric on $T(X)$ give metrics on $W^{\pm}$ and $T^*(X) \otimes W^{\pm}$, hence we can define the adjoint $\nabla_A^* : \Gamma(T^*X \otimes W^-) \to \Gamma(W^+)$,

$$\nabla_A^* = -\sum e^i \lrcorner \nabla_{e_i}$$

Similarly we define $\not{D}_A : \Gamma(W^-) \to \Gamma(W^+)$, which turns out to be the adjoint of the previous $\not{D}_A$ and makes the following commute (vertical maps are Clifford multiplications):

$$\begin{array}{ccccc}
\Gamma(W^+) & \xrightarrow{\nabla_A} & \Gamma(T^*X \otimes W^+) & \xrightarrow{\nabla_A} & \Gamma(T^*X \otimes T^*X \otimes W^+) \\
\| & & \downarrow & & \downarrow \\
\Gamma(W^+) & \xrightarrow{\not{D}_A} & \Gamma(W^-) & \xrightarrow{\not{D}_A} & \Gamma(W^+)
\end{array}$$

Let $\mathbf{F}_A = iF_A \in \Lambda^2(X; i\mathbb{R})$ be the curvature of the connection A on L and $F_A^+ \in \Lambda^+(X)$ be the self dual part of F_A and s be the scalar curvature of X. Weitzenbock formula says

$$\not{D}_A^2(\psi) \; = \; \nabla_A^* \nabla_A \psi + \frac{s}{4}\psi - \frac{i}{2}\rho(F_A^+)\psi \tag{13.11}$$

To check this we we can assume $\nabla_{e_i}(e^j) = 0$ at the point x

$$\begin{aligned}
\not{D}_A^2 \psi &= \sum \rho(e^i).\, \nabla_{e_i} [\, \sum \rho(e^j).\, \nabla_{e_j} \psi \,] \\
&= \nabla_A^* \nabla_A \psi + \sum_{i<j} \rho(e^i)\, \rho(e^j)\, (\nabla_{e_i} \nabla_{e_j} - \nabla_{e_j} \nabla_{e_i})\, \psi \\
&= \nabla_A^* \nabla_A \psi + \frac{1}{2} \sum_{i,j} \rho(e^i)\, \rho(e^j)\, \Omega_{ij}^A \, \psi
\end{aligned}$$

$\Omega_{ij}^A = R_{ij} - \frac{i}{2} F_{ij}$ is the curvature on $V^+ \otimes L^{-1/2}$ (endemorphism of W^+ valued 2-form). R_{ij} is the Riemannian curvature, and the imaginary valued 2-form $\mathbf{F}_A = \sum i F_{ij} e^i \wedge e^j$ is the curvature of A of the complex line bundle L. So if $\psi = \sigma \otimes \alpha \in \Gamma(V^+ \otimes L^{-1/2})$, then

$$\begin{aligned}
\frac{1}{2} \sum_{i,j} \rho(e^i)\, \rho(e^j)\, \Omega_{ij}^A \, (\sigma \otimes \alpha) &= \frac{1}{2} (\sum \rho(e^i)\, \rho(e^j)\, R_{ij}\, \sigma) \otimes \alpha \\
&\quad - \frac{i}{4} \sum \rho(e^i)\, \rho(e^j) \sigma \otimes (F_{ij} \alpha) \\
&= \frac{1}{8} \sum \rho(e^i)\, \rho(e^j)\, \rho(e^k)\, \rho(e^l)\, R_{ijkl}\, (\psi) \\
&\quad - \frac{i}{2} \rho\, (\sum_{i<j} F_{ij}\, e^i \wedge e^j)\, (\psi)
\end{aligned}$$

It is a standard calculation that the first term is $s/4$ (e.g. [LaM] pp. 161, 398), and since $\Lambda^-(X)$ act as zero on W^+ the second term can be replaced by

$$-\frac{i}{2} \rho\, (\sum F_{ij}^+\, e^i \wedge e^j)\, \psi = -\frac{i}{2} \rho\, (F_A^+)\, \psi = -\frac{1}{2} \rho\, (\mathbf{F}_A^+)\, \psi$$

13.4 A special calculation

In coming sections we need some a special case of (13.11). For this, suppose

$$V^+ = L^{1/2} \oplus L^{-1/2}$$

where $L^{1/2} \longrightarrow X$ is some complex line bundle with $L^{1/2} \otimes L^{1/2} = L$. Hence $W^+ = (L^{1/2} \oplus L^{-1/2}) \otimes L^{-1/2} = L^{-1} \oplus \mathbb{C}$. In this case there is a unique connection $\frac{1}{2}A_0$ in $L^{-1/2} \to X$ such that the induced Dirac operator D_{A_0} on W^+ restricted to the trivial summand $\underline{\mathbb{C}} \to X$ is the exterior derivative d. This is because for $\sigma_\pm \in \Gamma(L^{\pm 1/2})$, the following determines $\nabla_{\frac{A_0}{2}}(\sigma_-)$:

$$d(\sigma_+ \otimes \sigma_-) = \nabla_{A_0}[(\sigma_+ + 0) \otimes \sigma_-] \;=\; \partial\!\!\!/(\sigma_+ + 0) \otimes \sigma_- + (\sigma_+ + 0) \otimes \nabla_{\frac{A_0}{2}}(\sigma_-)$$

The following is essentially the Leibniz formula for a Dirac operator, it is a version of the Weitzenbock formula (13.11).

Proposition 13.6. *Let $A, A_0 \in \mathcal{A}(L^{-1})$ and $A = A_0 + ia$. Let $\nabla_a = d + ia$ be the corresponding covariant derivative of the trivial bundle, $\underline{\mathbb{C}} \longrightarrow X$ and $\alpha : X \to \mathbb{C}$. Let u_0 be a section of the bundle $W^+ = L^{-1} \oplus \mathbb{C}$, with a constant $\mathbb{C}$ component and $D\!\!\!\!/_{A_0}(u_0) = 0$, then*

$$D\!\!\!\!/^2_A(\alpha u_0) = (\nabla_a^* \nabla_a \alpha) u_0 + \frac{1}{2}\rho(F_a)\,\alpha.\,u_0 - 2 < \nabla_a \alpha\,, \nabla_{A_0}(u_0) > \tag{13.12}$$

Proof: By abusing notation, $\nabla_A = \nabla^A$ or $\nabla_a = \nabla^a$, for the sake of not cluttering formulas, and abbreviating $\nabla_{e_j} = \nabla_j$ and leaving out summation signs for repeated indices (Einstein convention), we calculate

$$\begin{aligned} \nabla^A(\alpha u_0) &= (d\alpha)u_0 + \alpha \nabla^A(u_0) \\ &= (d\alpha)u_0 + \alpha(\nabla^{A_0}(u_0) + ia u_0) \\ &= (d\alpha + ia\alpha)u_0 + \alpha \nabla^{A_0}(u_0) \\ &= \nabla^a(\alpha)u_0 + \alpha \nabla^{A_0}(u_0) \Rightarrow \\ D\!\!\!\!/_A(\alpha u_0) &= \rho(e^j)\,\nabla^a_j(\alpha)\,u_0 + \alpha\, D\!\!\!\!/_{A_0}(u_0) = \rho(e^j)\,\nabla^a_j(\alpha)\,u_0 \end{aligned}$$

By abbreviating $\mu = \nabla^a_j(\alpha)$ and recalling $d\mu = e^k \otimes \nabla_k(\mu)$, we calculate

$$\begin{aligned} \nabla^A(\rho(e^j)\,\mu\,u_0) &= e^k \otimes \rho(e^j)\,\nabla_k(\mu)u_0 + e^k \otimes\, \rho(e^j)\,\mu\,(\nabla^{A_0}_k(u_0) + i\,a_k\,u_0) \\ &= e^k \otimes \rho(e^j)\,\nabla^a_k(\mu)\,u_0\; + e^k \otimes \rho(e^j)\,\mu\,\nabla^{A_0}_k(u_0) \Rightarrow \end{aligned}$$

$$\begin{aligned}
\not{D}_A(\rho(e^j)\,\mu\,u_0) &= \rho(e^k)\rho(e^j)\,\nabla_k^a(\mu)\,u_0 + \rho(e^k)\rho(e^j)\,\mu\,\nabla_k^{A_0}(u_0) \\
&= -\nabla_j^a(\mu)\,u_0 + \frac{1}{2}\sum_{k<j}\rho(e^k)\rho(e^j)(\nabla_k^a(\mu) - \nabla_j^a(\mu))u_0 \\
&\quad -\mu\,\nabla_j^{A_0}(u_0) - \mu\,\rho(e^j)\sum_{k\neq j}\rho(e^k)\nabla_k^{A_0}(u_0)
\end{aligned} \tag{13.13}$$

But since $0 = \not{D}_{A_0}(u_0) = \sum \rho(e^k)\nabla_k^{A_0}(u_0)$, the last term of (13.12) is $-\mu\,\nabla_j^{A_0}(u_0)$.

Now by plugging in the value $\mu = \nabla_j^a(\alpha)$ in (13.12) and summing over j, we get

$$\not{D}_A^2(\alpha u_0) = -\nabla_j^a\nabla_j^a(\alpha)u_0 + \frac{1}{2}\rho(\sum F_{k,j}^a\, e^k \wedge e^j)\;\alpha.\,u_0 - 2\sum \nabla_j^a(\alpha)\nabla_j^{A_0}(u_0) \quad \square$$

Remark 13.7. *Since u_0 has a constant $\mathbb{C}$ component, and ∇^{A_0} restricts to the usual d on the $\mathbb{C}$ component, the term $< \nabla^a\alpha\,,\nabla^{A_0}(u_0) >$ lies entirely in the L component of W^+*

13.5 Seiberg–Witten invariants

Let X be a closed oriented Riemannian manifold, and $L \to X$ a characteristic complex line bundle. Seiberg–Witten equations are defined for $(A,\psi) \in \mathcal{A}(L)\times\Gamma(W^+)$,

$$\not{D}_A(\psi) = 0 \tag{13.14}$$
$$\mathbf{F}_A^+ = \sigma(\psi) \tag{13.15}$$

The curvature of L is an imaginary valued 2-form $\mathbf{F}_A = iF_A$. Sometimes (13.15) is written as $F_A^+ = -i\sigma(\psi)$, or $\rho(\mathbf{F}_A^+) = \boldsymbol{\sigma}(\psi)$ as an equation in $sl(W^+)$. When convenient, we can abuse notation by denoting the boldface $\mathbf{F}_A^+$ as F_A^+ and assuming it to be imaginary valued.

Definition 13.8. *Gauge group $\mathcal{G}(L) = Map(X,S^1)$ acts on $\tilde{\mathcal{B}}(L) = \mathcal{A}(L)\times\Gamma(W^+)$ as follows: for $s = e^{if} \in \mathcal{G}(L)$,*

$$s^*(A,\psi) = (s^*A, s^{-1}\psi) = (A + s^{-1}ds\,,\, s^{-1}\psi) = (A + i\,df\,,\, s^{-1}\psi) \tag{13.16}$$

From this, and by using $s^{-1}ds = idf$ hence $ds^{-1} = -idf s^{-1}$ we calculate

$$\begin{aligned}
\not{D}_{s^*A}(s^{-1}\psi) &= \not{D}_A(s^{-1}\psi) + \rho(idf)\,(s^{-1}\psi) \\
&= \rho(ds^{-1})\psi + s^{-1}\not{D}_A(\psi) + \rho(idf)\,\rho(s^{-1}\psi) \\
&= s^{-1}\not{D}_A(\psi) \\
\mathbf{F}_{s^*A}^+ &= s^{-1}\mathbf{F}_A^+\,s = \mathbf{F}_A^+ = \sigma(\psi) = \sigma(s^{-1}\psi)
\end{aligned}$$

Hence the solution set $\tilde{\mathcal{M}}(L) \subset \tilde{\mathcal{B}}(L)$ of Seiberg–Witten equations is preserved by the action $(A, \psi) \longmapsto s^*(A, \psi)$ of $\mathcal{G}(L)$ on $\tilde{\mathcal{M}}(L)$. Define

$$\mathcal{M}(L) = \tilde{\mathcal{M}}(L)/\mathcal{G}(L) \quad \subset \quad \mathcal{B}(L) = \tilde{\mathcal{B}}(L)/\mathcal{G}(L)$$

We call a solution (A, ψ) of (13.15) or (13.22) an *irreducible* solution if $\psi \neq 0$ (otherwise *reducible*). $\mathcal{G}(L)$ acts on the subset $\tilde{\mathcal{M}}^*(L)$ of the irreducible solutions freely, we define

$$\mathcal{M}^*(L) = \tilde{\mathcal{M}}^*(L)/\mathcal{G}(L)$$

Any solution (A, ψ) of Seiberg–Witten equations satisfies the C^0 bound

$$|\psi|^2 \leq \max(0, -s) \tag{13.17}$$

where s is the scalar curvature function of X. To see this we first plug (13.15) into Weitzenbock formula (13.11) and get

$$\not{D}_A^2(\psi) = \nabla_A^* \nabla_A \psi + \frac{s}{4}\psi - \frac{i}{2}\boldsymbol{\sigma}(\psi)\psi \tag{13.18}$$

Then at the points of X^4 where $|\psi|^2$ is maximum we calculate $(d^* = - * d*)$:

$$\begin{aligned}
0 \leq \frac{1}{2}\Delta|\psi|^2 &= \frac{1}{2}d^*d < \psi, \psi >= \frac{1}{2}d^*(< \nabla_A\psi, \psi > + < \psi, \nabla_A\psi >) \\
&= \frac{1}{2}d^*(\overline{< \psi, \nabla_A\psi >} + < \psi, \nabla_A\psi >) = d^* < \psi \,, \nabla_A\psi >_{\mathbb{R}} \\
&= < \psi, \nabla_A^*\nabla_A\psi > -|\nabla_A\psi|^2 \leq < \psi, \nabla_A^*\nabla_A\psi > \\
&\leq -\frac{s}{4}| \psi |^2 - \frac{1}{4}| \psi |^4
\end{aligned}$$

The last step follows from (13.18) and (13.9) and gives the required bound.

Proposition 13.9. *$\mathcal{M}(L)$ is compact.*

Proof: Given a sequence $[A_n, \psi_n] \in \mathcal{M}(L)$ we claim that there is a convergent subsequence (which we will denote by the same index), i.e. there is a sequence of gauge transformations $g_n \in \mathcal{G}(L)$ such that $g_n^*(A_n, \psi_n)$ converges in C^∞. Let A_0 be a base connection. By Hodge theory of the elliptic complex,

$$\Omega^0(X) \xrightarrow{d^0} \Omega^1(X) \xrightarrow{d^+} \Omega^2_+(X)$$

$$A - A_0 = h_n + a_n + b_n \in \mathcal{H} \oplus im(d^+)^* \oplus im(d)$$

where $\mathcal{H}$ are the harmonic 1-forms. After applying gauge transformation g_n we can assume that $b_n = 0$, i.e. if $b_n = i\, df_n$ we can let $g_n = e^{if}$. Also,

$$h_n \in \mathcal{H} = H^1(X;\mathbb{R}) \text{ and a component of } \mathcal{G}(L) \text{ is } H^1(X;\mathbb{Z})$$

Hence after a gauge transformation we can assume $h_n \in H^1(X;\mathbb{R})/H^1(X;\mathbb{Z})$ so h_n has convergent subsequence. Consider the first order elliptic operator:

$$D = d^* \oplus d^+ : \Omega^1(X)_{L^p_k} \longrightarrow \Omega^0(X)_{L^p_{k-1}} \oplus \Omega^2_+(X)_{L^p_{k-1}}$$

The kernel of D consists of harmonic 1-forms, hence by Poincare inequality if a is a 1-form orthogonal to the harmonic forms, then for some constant C,

$$||a||_{L^p_k} \le C||D(a)||_{L^p_{k-1}}$$

Now $a_n = (d^+)^*\alpha_n$ implies that they have to be co-closed i.e. $d^*(a_n) = 0$. Since a_n is orthogonal to harmonic forms, and by calling $A_n = A_0 + a_n$, we see

$$||a_n||_{L^p_1} \le C\ ||D(a_n)||_{L^p} \le C||d^+a_n||_{L^p} = C\ ||F^+_{A_n} - F^+_{A_0}||_{L^p}$$

Here we use C for a generic constant. By (13.15), (13.8), and along with the bound on $||\psi_n||$ given by (13.17) there is a C depending only on the scalar curvature s with

$$||a_n||_{L^p_1} \le C \tag{13.19}$$

By iterating this process we get $||a_n||_{L^p_k} \le C$ for all k , hence $||a_n||_\infty \le C$. From the elliptic estimate and $\not{D}_{A_n}(\psi_n) = 0$:

$$\begin{aligned} ||\psi_n||_{L^p_1} &\le C(\ ||\not{D}_{A_0}\psi_n\ ||_{L^p} + ||\psi_n||_{L^p}) = C(\ ||a_n\psi_n\ ||_{L^p} + ||\psi_n||_{L^p}) \\ ||\psi_n||_{L^p_1} &\le C(\ ||a_n||_\infty||\psi_n||_{L^p} + ||\psi_n||_{L^p}) \le C \end{aligned} \tag{13.20}$$

By repeating this (bootstrapping) process we get $||\psi_n||_{L^p_k} \le C$, for all k, where C depends only on the scalar curvature s and A_0. By the Rellich theorem we get a convergent subsequence of (a_n, ψ_n) in L^p_{k-1} norm, for all k. So we get this to be a C^∞ convergence. For the elliptic theory used in this proof, the reader can consult [Mc], [Sal2] □

Remark 13.10. *Compactness holds when X has a nonempty boundary. In this case in the above proof we need to take the gauge fixing condition for the 1-forms a_n not just to be co-closed, but also annihilate the normal vectors at the boundary ([KM]).*

It is not clear that the solution set of Seiberg–Witten equations is a smooth manifold. However we can perturb the Seiberg–Witten equation (13.15) by a generic self dual 2-form $\delta \in \Omega^+(X)$, in a gauge invariant way, to obtain a new set of equations with a perturbation term whose solutions set is a smooth manifold.

Definition 13.11. *The parametrized Seiberg–Witten equations with $\delta \in \Omega_2^+(X)$ are:*

$$\not{D}_A(\psi) = 0 \tag{13.21}$$
$$\mathbf{F}_A^+ = \sigma(\psi) + i\,\delta \tag{13.22}$$

Denote this solution space by $\tilde{\mathcal{M}}_\delta := \tilde{\mathcal{M}}_\delta(L)$, and the parametrized solution space by

$$\tilde{\mathcal{M}} = \bigcup_{\delta\in\Omega^+} \tilde{\mathcal{M}}_\delta \times \{\,\delta\,\} \subset \mathcal{A}(L) \times \Gamma(W^+) \times \Omega^+(X)$$

$$\mathcal{M}_\delta = \tilde{\mathcal{M}}_\delta \,/\mathcal{G}(L) \;\subset\; \mathcal{M} = \tilde{\mathcal{M}} \,/\mathcal{G}(L) \tag{13.23}$$

$\mathcal{M}_\delta$ is compact (proof similar to Proposition 13.9). Let $\tilde{\mathcal{M}}_\delta^* \subset \tilde{\mathcal{M}}^*$ be the corresponding irreducible solutions (i.e. $\psi \neq 0$ solutions), and also let $\mathcal{M}_\delta^* \subset \mathcal{M}^*$ be their quotients by the gauge group. We will prove that when $b_2(X) > 1$ for a generic choice of δ, $\mathcal{M}_\delta = \mathcal{M}_\delta^*$. But first we will show that for generic δ the set $\mathcal{M}_\delta^*$ is a closed smooth manifold.

Proposition 13.12. *$\tilde{\mathcal{M}}$ is a smooth manifold. The projection $\pi : \tilde{\mathcal{M}} \longrightarrow \Omega^+(X)$ is a proper surjection of the Fredholm index: (here we identify $c_1(L)^2 =< c_1(L)^2, [X] >$)*

$$d(L) = \frac{1}{4}[\, c_1(L)^2 - (2\chi + 3\sigma)\,] \tag{13.24}$$

where χ and σ are Euler characteristic and the signature of X.

Proof: $\tilde{\mathcal{M}} = \mathcal{P}^{-1}(0)$, where $\mathcal{P}$ is the map $(A,\psi,\delta) \longmapsto (F_A^+ - \sigma(\psi) - i\delta\;, \not{D}_A(\psi))$. Linearization of P at (A_0,ψ_0,δ_0), for $\psi_0 \neq 0$ is given by ($\sigma(\psi)$ is a quadratic function):

$$P : \Omega^1(X) \oplus \Gamma(W^+) \oplus \Omega^+(X) \longrightarrow \Omega^+(X) \oplus \Gamma(W^-) \tag{13.25}$$

$$P(a,\psi,\epsilon) = (d^+a - 2\sigma(\psi,\psi_0) - i\epsilon\;,\; \not{D}_{A_0}\psi + \rho(a)\psi_0) \tag{13.26}$$

To see that this is onto we pick $(\kappa,\theta) \in \Omega^+(X) \oplus \Gamma(W^-)$, by varying ϵ we can see that $(\kappa, 0)$ is in the image of P. To see $(0,\theta)$ is in the image of P, we will show that if it is orthogonal to $image(P)$ then it has to be $(0,0)$; so assume for all a and ψ we have

$$< \not{D}_{A_0}\psi, \theta > + < \rho(a)\psi_0, \theta >= 0$$

By choosing $a = 0$ and by the self adjointness of the Dirac operator we get $\not{D}_{A_0}\theta = 0$. Then by choosing $\psi = 0$ and a to be a "bump" form supported in U, we see that $<\rho(a)\psi_0, \theta> = 0$ for all such 1-forms a. This implies $\theta = 0$ pointwise on U. Here we have used $T^*(X)_{\mathbb{C}} \cong W^+ \otimes W^-$. Since we can get $\theta = 0$ at any sequence of points in U, by unique continuation $\theta = 0$ ([Aro]).

By the implicit function theorem $\tilde{\mathcal{M}}$ is a smooth manifold, and by the Smale–Sard theorem $\tilde{\mathcal{M}}_\delta$ are smooth manifolds, for generic choice of δ's (to be precise, for this we need to set up Banach space structures to write the Seiberg–Witten as the equations for the vanishing of a smooth section of a Banach space bundle over a Banach manifold; these technicalities will be ignored here). Hence their free quotients $\mathcal{M}$ and $\mathcal{M}_\delta$, are smooth manifolds. Also since the derivative of the gauge group action $s \mapsto s^*(A_0)$,

$$\mathcal{G}(L) = Map(X, S^1) \to \mathcal{A}(L) \tag{13.27}$$

at $s = I$, is just $d : \Omega^0(X) \to \Omega^1(X)$, the tangent space of $\mathcal{M}_\delta = \tilde{\mathcal{M}}_\delta / \mathcal{G}(L)$ is obtained by intersecting the tangent space of $\tilde{\mathcal{M}}_\delta$ with $im(d)^\perp = kernel(d^*)$, which is called "gauge fixing". So after taking gauge fixing into account, the dimension of $\mathcal{M}_\delta(L)$ is given by the index of $P + d^*$ (c.f. [DK]). $P + d^*$ is the compact perturbation of

$$S : \Omega^1(X) \oplus \Gamma(W^+) \longrightarrow [\, \Omega^0(X) \oplus \Omega^2_+(X)\,] \oplus \Gamma(W^-) \tag{13.28}$$

$$S = \begin{pmatrix} d^* \oplus d^+ & 0 \\ 0 & \not{D}_{A_0} \end{pmatrix}$$

By the Atiyah–Singer index theorem:

$$\begin{aligned} d(L) = \dim \mathcal{M}_\delta(L) = \text{ind}(S) &= \text{index}(d^* \oplus d^+) + \text{index}_{\mathbb{R}} \not{D}_{A_0} \\ &= -\frac{1}{2}(\chi + \sigma) + \frac{1}{4}(c_1(L)^2 - \sigma) \\ &= \frac{1}{4}[\, c_1(L)^2 - (2\chi + 3\sigma)\,] \qquad (13.29) \\ &= \frac{c_1(L)^2 - \sigma}{4} - (1 - b_1 + b^+) \end{aligned}$$

where b_1 is the first Betti number of X, and b^+ is the dimension of the positive define part H^2_+ of $H^2(X;\mathbb{Z})$. Clearly $\text{ind}(S)$ is even; and if $b^+ - b_1$ is odd, this expression is even, then for the characteristic line bundle L we must have $c_1(L)^2 = \sigma \bmod 8$. □

Now assume that $H^1(X) = 0$, then $\mathcal{G}(L) = K(\mathbb{Z}, 1)$. Therefore being a free quotient of a contractible space by $\mathcal{G}(L)$, we have

$$\mathcal{B}^*(L) = K(\mathbb{Z}, 2) = \mathbb{CP}^\infty$$

By (13.25) the orientation of H^2_+ gives an orientation to $\mathcal{M}_\delta(L)$. Now by Proposition 13.12, when b^+ is odd $\mathcal{M}_\delta(L) \subset \mathcal{B}^*(L)$ is an even $2d = 2d(L)$ dimensional smooth closed oriented submanifold, then we can define Seiberg–Witten invariants as

$$SW_X(L) = \langle \mathcal{M}_\delta(L)\ ,\ [\ \mathbb{CP}^d\ ]\rangle \tag{13.30}$$

Even though $\mathcal{M}_\delta(L)$ depends on the perturbation term δ and on the metric, when $b_2^+(X) \geq 2$, the invariant $SW_L(X)$ is independent of these choices. To see this let

$$\pi_+ : \Omega^2_+(X) \to \mathcal{H}^2_+ \text{ be the projection to self dual harmonic forms}$$

Lemma 13.13. *The set of $\delta \in \Omega^2_+(X)$ such that $\mathcal{M}_\delta(L)$ contains a reducible solution ($\psi = 0$) is given by $\mathcal{Z} = \{\delta \in \Omega^2_+(X) \mid \pi_+(\delta) = -2\pi c_1(L)^+\}$. Furthermore, $\mathcal{Z}$ is a codimension $b_2^+(X)$ affine subspace of $\Omega^2_+(X)$, which we can identify as $\mathcal{Z}^\perp = \mathcal{H}^2_+$.*

Proof. If $\mathcal{M}_\delta(L)$ contains a reducible solution, then by (13.22) $F_A^+ - i\delta = 0$ for some A. But since $c_1(L) = [\frac{i}{2\pi}F_A]$, we get $\pi_+(\delta) = -2\pi c_1(L)^+$, which means $\delta \in \mathcal{Z}$. In particular, for any pair of connections A_0 and A the form $F^+_{A_0} - F^+_A \in \text{image}\,(d^+)$ since $[F_{A_0}] = [F_A]$.

Fix a connection A_0 with $F^+_{A_0} - i\delta_0 = 0$, then for any other solution $F^+_A - i\delta = 0$ we get $\delta - \delta_0 = i(F^+_{A_0} - F^+_A) = d^+(\sigma)$ for some $\sigma \in \Omega^1(X)$. This gives an affine identification $\mathcal{Z} = im(d^+)$. Now we claim $\Omega^2_+ = \mathcal{H}^2_+ \oplus im(d^+)$, where $\mathcal{H}^2_+$ is the self-intersection positive harmonic 2-forms. Clearly this claim implies the result. To prove the claim we need to show that $coker(d^+) = \mathcal{H}^2_+$ where $d^+ : \Omega^1 \to \Omega^2_+$. If $\langle d^+\beta, \alpha\rangle = 0\ ,\ \forall\beta \Rightarrow \langle \beta, d^*\alpha\rangle = 0 \Rightarrow d^*\alpha = 0$. Also since $*\alpha = \alpha \Rightarrow d\alpha = 0$, so $\alpha \in \mathcal{H}^2_+$ □.

Therefore when $b_2^+(X) \geq 1$ we can choose a generic $\delta \in \mathcal{Z}^\perp \subset \Omega^2_+(X)$ to make $\mathcal{M}_\delta(L) = \mathcal{M}^*_\delta(L)$ smooth, and when $b_2^+(X) \geq 2$ any two such choices for δ can be connected by a path in $\delta \in \mathcal{Z}^\perp$, which has an affect of connecting the corresponding manifolds $\mathcal{M}_\delta(L)$ by a cobordism. $\mathcal{M}_\delta(L)$ also depends on the metric on X, similarly by connecting any two choices by a path of metrics will result in changing $\mathcal{M}_\delta(L)$ by a cobordism in $\mathbb{CP}^\infty$. So if $b_2^+(X) \geq 2$, Definition 13.30 does not depend on generic choice of (g, δ).

Also by (13.17) if X has non-negative scalar curvature then all the solutions are reducible, i.e. $\psi = 0$, so for $b^+ \geq 2$ and for a generic metric $\tilde{\mathcal{M}} = \varnothing$, which implies $SW_L(X) = 0$. The right hand side of (13.30) can be thought as the oriented count of the intersection points $\mathcal{M}_\delta(L) \cap V_1 \cap .. \cap V_d$, where $V_j \subset \mathbb{CP}^\infty$ are generic copies of the divisors representing $H^1(\mathbb{CP}^\infty; \mathbb{Z})$.

As we saw in Definition 13.1, we can identify $Spin^c$ structures with characteristic line bundles $\mathcal{C}(X)$, so we can assume SW_X as a function defined on $\mathcal{C}(X)$,

$$SW_X : \mathcal{C}(X) = \{L \in H^2(X;\mathbb{Z}) \mid w_2(X) = c_1(L)\ (mod\ 2)\} \to \mathbb{Z} \tag{13.31}$$

Lemma 13.14. *For any closed oriented smooth 4-manifold X with $b_2^+(X) \geq 2$, there are finitely many $L \in \mathcal{C}(X)$ for which $SW_X(L) \neq 0$.*

Proof. It is enough to show that in order to have a nonempty solution set $\mathcal{M}_\delta$ the integral class $c_1(L)$ must lie in a fixed bounded subset of $\mathcal{H}^2$. We know $c_1(L) = [F]$, where $F = \frac{i}{2\pi} F_A \in \Omega^2(X)$, write $F = F^+ + F^-$, hence $||F||^2 = ||F^+||^2 + ||F^-||^2$, and by (13.29)

$$c_1(L)^2 = ||F^+||^2 - ||F^-||^2 \geq 2\chi(X) + 3\sigma(X) \tag{13.32}$$

Also by (13.22), (13.8), and (13.17) $||F^+||^2$ is bounded in terms of the scalar curvature of X, and hence by (13.32) $||F^-||^2$ is bounded as well, so $||F||$ is bounded. □

We can combine Seiberg–Witten invariants as a single element (13.33) in the group ring $\mathbb{Z}H^2(X;\mathbb{Z}) \cong \mathbb{Z}H_2(X;\mathbb{Z})$, where for $L \in \mathcal{C}(X)$ the symbol t^L denotes the corresponding element in the group ring satisfying the formal property $t^L t^K = t^{K+L}$,

$$SW_X = \sum_{L \in \mathcal{C}(X)} SW_X(L).t^L \tag{13.33}$$

If (X_i, L_i), $i = 1, 2$ are two oriented compact smooth manifolds with $Spin^c$ structures such that $b^+(X_i) > 0$ and with common boundary, which is a 3-manifold Y^3 with a positive scalar curvature; then gluing these manifolds together along their boundaries produces a manifold $X = X_1 \smile_\partial X_2$ with vanishing Seiberg-Witten invariants (cf. [FS3]). This is because by (13.17) the restriction of Seiberg–Witten solutions to Y would be reducible $\psi = 0$, violating the dimension formula. For example, when $Y = S^3$ then $X = \bar{X}_1 \# \bar{X}_2$, where each $\bar{X}_i$ denotes the closed manifold obtained by capping the boundary of X_i). Then by (13.29) the dimension of the Seiberg–Witten solution space is

$$d(X) = d(\bar{X}_1) + d(\bar{X}_2) + 1 \tag{13.34}$$

Then if X is a manifold of simple type i.e. $d(X) = 0$, by making a metrically long neck at the connected sum region we see that the Seiberg–Witten solutions on X would converge to solutions to each connected sum regions with one of $d(\bar{X}_i)$ negative, which is a contradiction (see also [Sal1]).

Definition 13.15. *A smooth manifold X^4 is said to be of simple type if only the 0-dimensional moduli spaces $\mathcal{M}_\delta(L)$ give nonzero invariants $SW_L(X)$.*

Remark 13.16. *Symplectic manifolds are of simple type [Ta2]. Also any closed smooth manifold X^4 with $b_2^+(X) > 1$ and $b_1(X) = 0$, containing a genus $g > 0$ imbedded surface $\Sigma \subset X$ representing a non-torsion homology class with $\Sigma.\Sigma = 2g - 2$, is of simple type (Thm 1.4 of [Oz2]). Currently there are no known examples of 4-manifolds which are not simple type. It is conjecture that all manifolds with $b^+ > 1$ are of simple type.*

13.6 S–W when $b_2^+(X) = 1$

In case (X^4, g) is a closed Riemannian manifold with $b_2^+(X) = 1$, Seiberg–Witten invariant depends on the perturbation term $SW_X = SW_{X,g,\delta}$. Let ω_g be the unique harmonic form in Ω_+^2 representing $1 \in \mathcal{H}_+^2(X) \cong \mathbb{R}$. By Lemma 13.13 the condition that $\mathcal{M}_\delta(L)$ cannot contain any reducible solution ($\psi \neq 0$) is $(2\pi c_1(L) + \delta)^+ \neq 0$, which can be expressed by $(2\pi c_1(L) + \delta).\, \omega_g \neq 0 \Rightarrow (g, \delta)$ lies in one of the (plus/minus) *chambers*:

$$C_\pm = \{(g, \delta) \mid \pm (2\pi c_1(L) + \delta).\, \omega_g > 0\}$$

Varying (g, δ) in the same chamber does not change $SW_{X,g,\delta}$. Let $H = H_g \in H_2(X, \mathbb{R})$ be the Poincare dual of ω_g, define $SW_{X,H}^\pm = SW_{X,g,\delta}$ if $(g, \delta) \in C_\pm$. By [KM] and [LL],

$$SW_{X,H}^+(L) - SW_{X,H}^-(L) = (-1)^{1+d(L)} \tag{13.35}$$

So for any basic class L at least one of $SW_X^\pm(L)$ is nonzero. Also we can reduce the dependence on δ by taking it to be an arbitrarily small generic perturbation. Let L denotes the Poincare dual of $c_1(L)$, and define a "small" Seiberg–Witten invariant to be

$$SW_{X,H}(L) = \begin{cases} SW_{X,H}^+(L) & \text{if } L.H > 0 \\ SW_{X,H}^-(L) & \text{if } L.H < 0 \end{cases} \tag{13.36}$$

This quantity depends on the metric g, since $H = H_g$, but in fact it depends only on

$$P = \{h \in H_2(X; \mathbb{R}) \mid h^2 > 0\}$$

P has two components, the component containing H is given by $P_H = \{h \in P \mid h.H > 0\}$. Any $H', H'' \in P_H$ with $L.H' < 0$ and $L.H'' > 0$ gives the so-called "wall crossing relation":

$$SW_{X,H''}(L) - SW_{X,H'}(L) = (-1)^{1+d(L)}$$

Any nonzero class H with $H.H \geq 0$ orients all other non-zero classes with $Q.Q \geq 0$. This is because $H.Q \neq 0$ since $b_2^+(X) = 1$. Hence Q can be oriented by the rule $H.Q > 0$.

Exercise 13.1. *(Light cone lemma) Let X^4 be a closed smooth manifold with $b_2^+(X) = 1$, then show that if $H_1, H_2 \in H_2(X) - \{0\}$ with $H_1^2 \geq 0, H_2^2 \geq 0$ and $H_1.H_2 = 0$, then $H_2 = \lambda H_1$ for some $\lambda > 0$. (Hint: write them in terms of a basis and use the Schwarz inequality.)*

Remark 13.17. *When $b_2^-(X) \leq 9$ then $SW_{X,H}(L)$ doesn't depend on the choice of H in P. This is because to have a solution to Seiberg–Witten equations we must have $d(L) \geq 0$, so $c_1(L)^2 - 3\sigma(X) - 2\chi(X) \geq 0 \implies c_1^2 - 9 + b_2^- \geq 0 \implies c_1(L)^2 \geq 0$. The set P is closed under convex linear combinations, so if $L.H_1 > 0$ and $L.H_2 < 0$ for some $H_1, H_2 \in P$, then $L.H_3 = 0$ for some $H_3 \in P$. So by the light cone lemma $H_3 = \lambda L$, which implies $H_3^2 = 0$, a contradiction. Hence if $L.H_1 > 0$ then $L.H_2 > 0$ for every other element $H_2 \in P$.*

13.7 Blowup formula

The blowup formula [Wi] relates the Seiberg–Witten invariants of $X\#\bar{\mathbb{CP}}^2$ to that of X (e.g. Exercise 13.3). We have $H^2(X\#\bar{\mathbb{CP}}^2;\mathbb{Z}) \cong H^2(X;\mathbb{Z}) \oplus H^2(\bar{\mathbb{CP}}^2;\mathbb{Z})$. Let E denote the generator of $H^2(\mathbb{CP}^2;\mathbb{Z}) = \mathbb{Z}$, which is represented by a sphere $\mathbb{CP}^1$. When X is a complex surface with canonical class K, then the canonical class of its blowup is $K+E$. When X has simple type (Definition 13.15), that is, for each basic class L, the dimension, $d(L)$, of the associated Seiberg–Witten moduli space is 0, then

$$SW_{X\#\bar{\mathbb{CP}}^2} = SW_X(t^E + t^{-E})$$

More specifically, this means if $\{B_1,...,B_k\}$ are the basic classes of X, then the basic classes of $X\#\bar{\mathbb{CP}}^2$ are given by $\{B_1 \pm E,...,B_k \pm E\}$ and $SW_{X\#\bar{\mathbb{CP}}^2}(B_j \pm E) = SW_X(B_j)$.

13.8 S–W for torus surgeries

Let X be a closed oriented smooth 4-manifold with $b_2^+ \geq 1$. The following theorem of [MMS] relates the Seiberg–Witten invariants of manifolds obtained by torus surgeries to X, along a given $T^2 \times B^2 \subset X$. Let M be the closed complement $X - T^2 \times B^2$, so that $\partial M = T^3$. Fix generators of $\mathbf{a},\mathbf{b},\mathbf{c}$ of $H_1(\partial M;\mathbb{Z})$ (for example this could be the standard generators of $H_1(T^2\times S^1)$ induced from the imbedding). For each indivisible element, $\gamma = r\mathbf{a} + q\mathbf{b} + p\mathbf{c}$, let M_γ be the manifold obtained by gluing $T^2 \times B^2$ back to M via a diffeomorphism $\phi_\gamma : T^2 \times S^1 \to \partial M$, with the property $\phi_*[pt \times \partial B^2] = \gamma$ (check that the diffeomorphism type of M_γ depends only on γ, not on the choice of ϕ_γ). Note that p-log transformations of Section 6.3 are a special case of this. By an application of the adjunction formula, it follows that any $Spin^c$ structure L on M_γ which restricts nontrivially to $T^2\times B^2$, then $SW_{M_\gamma}(L) = 0$ (for simplicity in first reading assume $b_2^+ \geq 2$).

For each $Spin^c$ structure $L_0 \to M$ which restricts trivially to ∂M, let $\mathcal{C}_0(M_\gamma)$ denote the set of all $Spin^c$ structures on M_γ whose restriction to M equals L_0, define

$$SW^0_{M_\gamma} = \sum_{L\in\mathcal{C}_0(M_\gamma)} SW_{M_\gamma}(L) \tag{13.37}$$

Theorem 13.18. *([MMS]) The following holds:*

$$SW^0_{M_\gamma} = rSW^0_{M_\mathbf{a}} + qSW^0_{M_\mathbf{b}} + pSW^0_{M_\mathbf{c}}$$

where it is understood that in the $b_2^+ = 1$ case, the invariants of this formula are computed in corresponding chambers (Section 13.6).

13.9 S–W for manifolds with T^3 boundary

In [Ta3] Taubes defined Seiberg–Witten invariants for 4-manifolds whose boundary is a disjoint union of 3-tori, provided that there is $\tau \in H^2(X;\mathbb{R})$ which restricts nontrivially to each boundary component. This invariant is an element of the group ring $\mathbb{Z}H^2(X,\partial X;\mathbb{Z})$, which can be identified with $\mathbb{Z}H_2(X;\mathbb{Z})$ (Poincare duality), for example

$$SW_{T^2\times D^2} = (t^{-1} - t)^{-1} = t + t^3 + t^5 + \dots$$

where $t = t^T$ is the corresponding element of a generator $T \in H_2(T^2 \times D^2)$. Here an implicit orientation for H_2 has been chosen to be able to distinguish t from t^{-1}.

Theorem 13.19. *([Ta3]) Let X be a compact, connected, oriented 4-manifold with $b_2^+ \geq 1$, and with boundary consisting of a disjoint union of 3-dimensional tori. Let $T^3 \subset X$ be an imbedding such that there is $\tau \in H^2(X;\mathbb{R})$ which restricts nontrivially to T^3 and to the each boundary component of X, then the following hold:*

(a) If T^3 splits X as a pair $X = X_+ \cup_{T^3} X_-$ of 4-manifolds with boundary T^3, and $j_*^{\pm} : \mathbb{Z}H_2(X_\pm) \to \mathbb{Z}H_2(X)$ is the map induced by inclusion, then:

$$SW_X = j_*^- SW_{X_-} \,.\, j_*^+ SW_{X_+} \tag{13.38}$$

(b) If T^3 does not split X, and X_1 is the complement of a tubular neighborhood of T^3, and $j_* : \mathbb{Z}H_2(X_1) \to \mathbb{Z}H_2(X)$ is the map induced by inclusion, then,

$$SW_X = j_* SW_{X_1} \tag{13.39}$$

In the case $b_2^+(X) = 1$ the choice of τ supplies an orientation for $H_+^2(X;\mathbb{R})$. This is because the restriction $i^*(\tau) \in H^2(T^3;\mathbb{R})$ is nonzero, hence there is a class in $H_2(T^3;\mathbb{Z})$ whose push forward $H \in H_2(X;\mathbb{Z})$ is nonzero and pairs positively with τ. This homology class has self intersection zero, so its dual class in $H^2(X;\mathbb{Z})$ lies in the "light cone", thus it defines an orientation to any line in $H^2(X;\mathbb{R})$ on which the cup product pairing is positive definite. With this H we define the Seiberg–Witten invariant. (Section 13.6).

Example 13.1. *(Removing tori) ([Ta3], [MMS]) Let X be a closed smooth 4-manifold with $H_1(X) = 0$, and let $N \approx T \times D^2 \subset X$ be a tubular neighborhood of a homologically nontrivial imbedded torus $T^2 \approx T \subset X$, which splits X into two pieces $X_+ = X - N$ and $X_- = N$. Then by (13.38), $SW_X = j_* SW_{X-N} \,.\, (t^{-1} - t)^{-1}$, where t corresponds to T, so*

$$j_* SW_{X-N} = SW_X \,.\, (t^{-1} - t) \tag{13.40}$$

Here $H_2(N) \to H_2(X)$ and $H_2(\partial N) \to H_2(X - N)$ are injective maps, but the map $j_* : H_2(X - N) \to H_2(X)$ is not injective, it has 2- dimensional kernel $\Lambda = \langle \alpha_1, \alpha_2 \rangle$ generated by the so-called "rim tori" $\alpha_1 = a_1 \times m$, $\alpha_2 = a_2 \times m$, where a_1, a_2 are the circle generators of T, and m is the meridian of the torus T in X. Then (13.40) implies the following (for simplicity we are identifying classes in $X_\pm$ and X when it makes sense):

$$SW_{X-N} = SW_X\,(t^{-1} - t) + r(1 - t^{\alpha_1}) + s(1 - t^{\alpha_2}) \tag{13.41}$$

where $r, s \in SW_{X-N}$, and j_* maps $t \to t$, $\alpha_i \to 0$. We note that when $X = E(n)$, $n \geq 2$ and T is the fiber torus, only the first term of (13.41) is nonzero. We see this by using $E(2n) = E(n) \natural E(n)$, and applying (13.44) to both sides of $SW_{E(2n)} = (SW_{E(n)-N})^2$.

Now let N_1, N_2 be two disjoint parallel copies of N in X and $\mathcal{N} = N_1 \cup N_2$. Let X_0 be the manifold obtained by identifying the two T^3 boundary components of $X - \mathcal{N}$,

$$X_0 = (T^3 \times [0,1]) \sqcup (X - \mathcal{N}) \Big/ \begin{array}{ccc} (x,0) & \sim & x \in \partial N_1 \\ (x,1) & \sim & x \in \partial N_2 \end{array} \tag{13.42}$$

By using (13.41) twice, we can compute $SW_{X-\mathcal{N}}$, then by appying (13.39) we obtain:

$$SW_{X_0} = SW_X\,(t^{-1}-t)^2 + (t^{-1}-t)(1-t^{\alpha_1})r + (t^{-1}-t)(1-t^{\alpha_2})s + (1-t^{\alpha_1})r' + (1-t^{\alpha_2})s' \tag{13.43}$$

where $r, s, r', s' \in SW_{X-\mathcal{N}}$, and t is the image of the core torus. Notice that the map $H_2(X_0 - \mathcal{N}) \to H_2(X_0)$ is injective, since the restriction map $H^1(X_0) \to H^1(T^3)$ is onto.

Remark 13.20. *In some special cases, such as when $X = E(n)$ $n \geq 2$ and T is the fiber torus, only the first term of the formula (13.43) is nonzero, for example this occurs when the adjunction inequality (8.1) rules out the possibility of the rim tori being basic classes (so in (13.43) we can set $\alpha_i = 0$). For example, we can view the rim torus $\alpha_1 = a_1 \times m \subset T^3 \times 1/2$, then the circle $a_2 \subset T^3 \times 1/2$ meets α_1 transversally at one point, hence if we can cap the boundaries of the cylinder $a_2 \times [0,1]$ in $X - \mathcal{N}$ (this happens when $\mathcal{N}$ has vanishing cycles), we get some closed surface $\Sigma \subset X_0$ of some genus $g > 0$, intersecting the torus α_1 at a single point. Then if we violate the adjunction inequality $2g - 2 > \Sigma.\Sigma + 1$, α_1 cannot be a basic class. Here "N has vanishing cycles" means that there are 2-handles in $X - \mathcal{N}$ which are attached to the loops a_1, a_2 with -1 framing. When this happens we say the core torus T in N "is contained in a node neighborhood", in which case we get Σ to be a self-intersection -2 sphere. By replacing this Σ with $\Sigma + \alpha_1$ (and smoothing corners) we can turn Σ into a self-intersection 0 torus which meets α_1 at a single point, hence the adjunction inequality applies and we get a violation. Reader can draw the handlebody of X_0 by using Figures 7.7 and 7.8 and techniques of Chapter 3.*

13.10 S–W for logarithmic transforms

By using Theorem 13.19, Fintushel and Stern ([FS6], [Str]) computed the Seiberg–Witten invariants of elliptic surfaces and their logarithmic transforms as follows: Recall that $E(n) \to S^2$ is a Lefschetz fibration with regular fiber $F = T^2$, and $E(n)$ is the fiber sum $E(n-1) \natural E(1)$ (Exercise 7.2), and also $SW_{E(2)} = 1$ (Theorem 13.28), so

$$1 = SW_{E(2)} = (SW_{E(1)-N})^2$$

We choose homology orientation so that $SW_{E(1)-N} = -1$. Also again by using (13.38) we get $SW_{E(2)} = SW_{E(2)-N} \,.\, SW_N$ and so $SW_{E(2)-N} = t^{-1} - t$, and hence

$$SW_{E(3)} = SW_{E(2)-N} \,.\, SW_{E(1)-N} = t - t^{-1}$$

By continuing this way, inductively we get (notice t^{n-2} corresponds to $(n-2)F$)

$$SW_{E(n)} = (t - t^{-1})^{n-2} \qquad (13.44)$$

Now let $X \mapsto X_p = (X - N) \cup_{j_p} N$ be a log transform (Section 6.3), where T^2 is a homologically essential torus in X imbedded with trivial normal bundle $N = T^2 \times D^2 \subset X$ (notice that j_p maps the torus T by degree p map). by abbreviating $\tau = t^{\tau}$ (here τ corresponds to the torus in $\partial(X - N)$). Now the new induced map $(j_p)_*(\tau) = t^p$ modulo the classes coming from the 2-dimensional kernel Λ, so when $(j_p)_*|\Lambda = 0$, we can write

$$\begin{aligned} SW_{X_p} &= (j_p)_* \left[SW_{X-N}\right] . SW_N = (j_p)_* \left[SW_X \,.\, (\tau^{-1} - \tau)\right] . (t^{-1} - t)^{-1} \\ &= (j_p)_* \, SW_X \,.\, \frac{(t^p - t^{-p})}{(t - t^{-1})} = SW_X \,.\, (t^{p-1} + t^{p-3} + \ldots + t^{-(p-1)})\Big|_{\tau = t^p} \end{aligned} \qquad (13.45)$$

For example, in the case of $SW_{E(n)} = (\tau - \tau^{-1})^{n-2}$ the above calculation gives

$$SW_{E(n)_p} = (t^p - t^{-p})^{n-1} / (t - t^{-1}) \qquad (13.46)$$

Also, for relatively prime integers p and q if $E(n)_{p,q}$ is the log transform of order p and q of $E(n)$ along two of its parallel fibers, then by the above formula we calculate

$$SW_{E(n)_{p,q}} = (\tau - \tau^{-1})^{n-2} \frac{(t^p - t^{-p})}{(t - t^{-1})} \frac{(s^q - s^{-q})}{(s - s^{-1})} \Big| \begin{array}{l} \tau = t^p \\ \tau = s^q \end{array} = \frac{(\tau - \tau^{-1})^n}{(t - t^{-1})(s - s^{-1})} \Big| \begin{array}{l} \tau = t^p \\ \tau = s^q \end{array}$$

Since p and q are relatively prime, and the group ring $\mathbb{Z}[G]$ is UFD when G is a torsion-free Abelian group (e.g.[Tan]) we can conclude that $t = r^q$ and $s = r^p$ for some r hence,

$$SW_{E(n)_{p,q}} = \frac{(r^{pq} - r^{-pq})^n}{(r^p - r^{-p})(r^q - r^{-q})} \tag{13.47}$$

From the handlebody pictures (e.g. Section 6.3, Figures 7.11 and 12.32) it is easy to see that $E(n)_{p,q}$ are simply connected, hence they are exotic copies of $E(n)$.

Exercise 13.2. *Show that if $H_1(X) = 0$ and the loops $b, c \subset \partial N$ in Figure 13.1 are null homotopic in X, then $H^1(X_p) = 0$ and $(j_p)_*|\Lambda = 0$ (consider homology of $(X_p, X_p - N)$).*

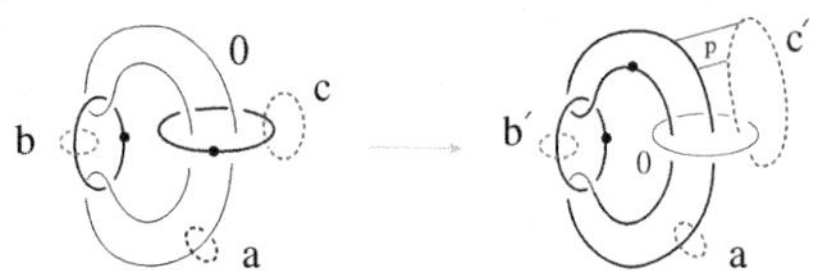

Figure 13.1: p-log transform

13.11 S–W for knot surgery X_K

Here we will give a proof of Theorem 6.3 from [FS1], there is also another proof by Meng and Taubes [MeT] which we will outline in Section 13.12. Theorems 13.19 and 13.18 are main ingredients in these proofs. Gauge theory content of Theorem 6.3 is the following:

Proposition 13.21. *Let X be a closed smooth 4-manifold such that $H_1(X) = 0$, and $b_2^+(X) > 1$, and let $T \subset X$ be a smoothly imbedded torus, such that $[T] \neq 0$, and $K \subset S^3$ a knot, then the Seiberg–Witten invariant of X_K is given by $SW_{X_K} = SW_X \cdot \Delta_K(t^2)$, where $\Delta_K(t)$ is the (symmetrized) Alexander polynomial of the knot $K \subset S^3$, and $t = t_T$.*

Proof. The proof in [FS1] needs the assumption that T is contained in a node neighborhood. Let $K \subset Y$ be either a knot, or a link with 2-components. Let Y_K denote the 0-surgered Y long K (Remark 1.2) if K is a knot, or the round surgered Y along K (Remark 3.1) if K is a link. Note that in both cases there is the distinguished meridian $m \subset Y_K$ coming from the meridian(s) of K. Let $N(m)$ be its tubular neighborhood of m. Now let us recall the knot surgery operation from Section 6.5; here we will state it slightly differently to be able to apply it to more general situations. Let X be a smooth 4-manifold, and $T^2 \times B^2 \subset X$ be an imbedded torus with trivial normal bundle. Then the knot surgery operation is the operation of replacing $T^2 \times B^2$ with $(Y_K - N(m)) \times S^1$, so that the meridian $p \times \partial B^2$ of the torus coincides with the meridian of m in Y_K:

$$X \rightsquigarrow X_{(Y,K)} = (X - T^2 \times B^2) \cup [\, Y_K - N(m) \,] \times S^1$$

When $Y = S^3$ we will abbreviate $X_K = X_{(Y,K)}$. Notice that in the case where K is a knot this definition is consistent with the definition of X_K in Section 6.5.

Now assume $K \subset S^3$ is a knot or a link of two components (visible in the picture), and let $K = K_+$, K_-, and K_0 be links obtained from K, with projections differing from each other by a single crossing change, as shown in Figure 13.2,

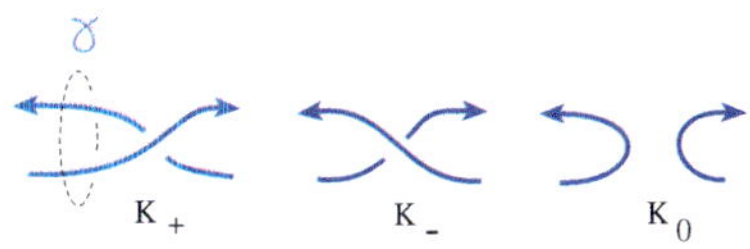

Figure 13.2

Notice if K is a knot then K_- is a knot and K_0 is a link of two components; and if K is a link of two components then K_- is a link of two components and K_0 is a knot. Let $\gamma \subset Y_K$ be the circle, as shown in Figure 13.2, where $K = K_\pm$ or K_0, and let $S^3_K(\gamma^r)$ denote the manifold obtained by surgering S^3_K along γ with framing r, then we claim:

(a) $S^3_{K_+}(\gamma^{-1}) = S^3_{K_-}$ if K_+ is a knot or a link

(b) $S^3_{K_+}(\gamma^0) = S^3_{K_0}$ if K_+ is a knot

(c) $S^3_{K_+}(\gamma^0) = (S^3_{K_0})_C$ if K_+ is a link, $C = \{C_1, C_2\}$ are 2 copies of the meridian of K_0

The proof of (a) is straightforward, the proofs of (b) and (c) follow from the following pictures of Figure 13.3 and 13.4 (also recall Figure 3.10 and the remarks that follow it).

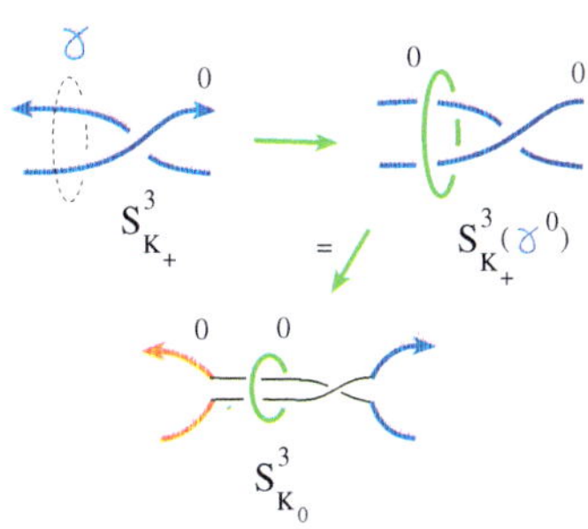

Figure 13.3: If K_+ is a knot

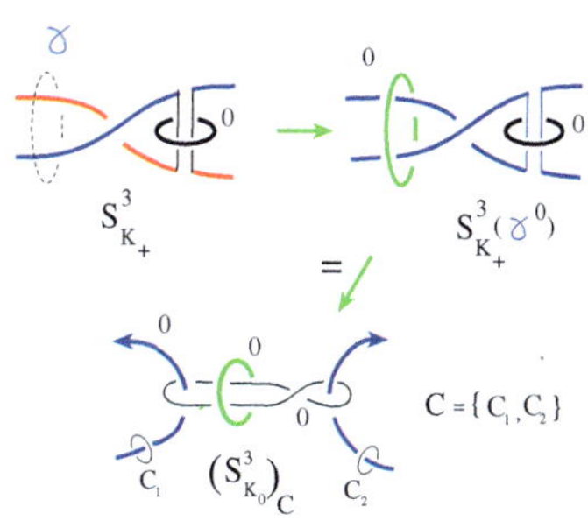

Figure 13.4: If K_+ is a link

Let $M = X_{K_+} - N(\gamma) \times S^1$, then $\partial M = T^3$ is a 3-torus with circle generators $\{\mathbf{a}, \mathbf{b}, \mathbf{c}\}$, as in Figure 13.5 ($c$ is the S^1 direction). Now by using the notations of Section 13.8 let $\gamma = \mathbf{a} + \mathbf{b}$, then from above discussion it follows that $M_{\mathbf{a}} = X_{K_+}$, $M_\gamma = X_{K_-}$, and also

$$M_b = \begin{cases} X_{K_0} & \text{if } K \text{ is a knot} \\ X_{(S^3_K, C)} & \text{if } K \text{ is a link} \end{cases}$$

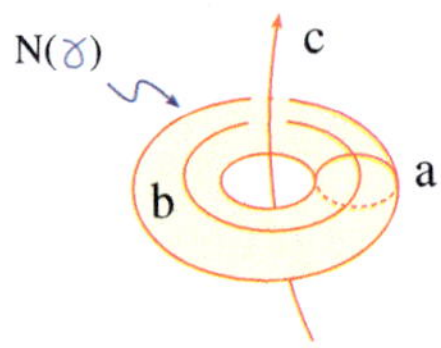

Figure 13.5: $N(\gamma) \times S^1$

Figure 13.6: Surface γ bounds in Y_K

Recall that, if $\gamma \subset T^3$ is a loop, $M_\gamma = M \cup N$ is the compactification of M with $N = T^2 \times D^2$ (∂D^2 coinciding with γ), and let $\mathcal{C}_0(M_\gamma)$ be the set of all *Spin*^{c} structures on M_γ whose restriction to M is equal to a fixed one. Counting $\mathcal{C}_0(M_\gamma)$ is equivalent to counting liftings of a fixed element $H^2(M_\gamma - N)$ to $H^2(M_\gamma)$, dually the liftings of $j_* : H_2(M_\gamma) \to H_2(M_\gamma, N)$. Hence if $H_2(N) \to H_2(M_\gamma)$ is the zero map (e.g. when $\gamma = a$ or $a+b$) then the set $\mathcal{C}_0(M_\gamma)$ has single element $\{\alpha\}$. Now if K_+ is a knot (and $\gamma = \mathbf{a}+\mathbf{b}$), then Theorem 13.18 gives

$$SW_{X_{K_-}}(\alpha) = SW_{X_{K_+}}(\alpha) + \sum_i SW_{X_{K_0}}(\alpha + 2i[T_0])$$

Here T_0 is the torus in the kernel of j_*, as shown in Figure 13.6 (the shaded region union the disk which goes over the 2-handle γ^0). Also only one of the terms ($i = 0$) of the summation above can be nonzero, otherwise we get a violation to the adjunction inequality Theorem 8.5 (apply to the torus $\tau \times S^1$ which intersects T_0 at one point), so

$$SW_{X_{K_-}} = SW_{X_{K_+}} + SW_{X_{K_0}} \tag{13.48}$$

Similarly, when $K = K_+$ is a link of two components, then we get

$$SW_{X_{K_-}} = SW_{X_{K_+}} + SW_{X_{(S^3_K, C)}}$$

The last term is obtained from X_{K_0} by removing two disjoint copies of $T^2 \times D^2$ then identifying the boundary components with each other, hence by (13.43) (and by Remark 13.20) we have

$$SW_{X_{K_-}} = SW_{X_{K_+}} + SW_{X_{K_0}}(t^{-1} - t)^2 \tag{13.49}$$

We can combine (13.48) and (13.49) in one equation (13.50) by adjusting a definition:

$$SW^*_{X_K} = \begin{cases} SW_{X_K} & \text{if } K \text{ is a knot} \\ (t^{-1} - t)^{-1} SW_{X_K} & \text{if } K \text{ is a link} \end{cases}$$

$$SW^*_{X_{K_+}} = SW^*_{X_{K_-}} + (t - t^{-1}) SW^*_{X_{K_0}} \tag{13.50}$$

Now the result follows from the skein definition of the Alexander polynomial (2.2) □.

13.12 S–W for $S^1 \times Y^3$

One way to define the Seiberg–Witten invariant of a closed 3-manifold Y is to define as the Seiberg–Witten of the 4-manifold $X = Y \times S^1$. Following justifies this rationale: First pick $L \in Spin^c(X)$ which is pulled back from $Spin^c(Y)$, so $c_1^2(L) = 0$ and in particular this implies that the dimension of the moduli space $d(L)$ is zero, and vice versa (13.29).

Lemma 13.22. *(Witten's reduction) Solutions of Seiberg–Witten equations on X corresponding to $L \in Spin^c(X)$ are those ones pulled back from Y.*

Proof. Pick an S^1-invariant metric on $Y \times S^1$, then the first Seiberg–Witten equation (13.14) becomes equation (13.51). Square this equation and integrate over X. Integrating by parts (recall the Hermitian inner product is skew symmetric), we get (13.52)

$$\frac{\partial \psi}{\partial t} + \partial_A(\psi) = 0 \tag{13.51}$$

$$\int_X \langle \frac{\partial \psi}{\partial t} + \partial_A(\psi), \frac{\partial \psi}{\partial t} + \partial_A(\psi) \rangle = 0$$

$$||\frac{\partial \psi}{\partial t}||^2 + ||\partial_A(\psi)||^2 + \langle \psi\, ,\, \partial_A(\frac{\partial \psi}{\partial t}) - \frac{\partial}{\partial t}(\partial_A \psi) \rangle = 0 \tag{13.52}$$

The second quantity of the third term in (13.52) is contraction of the curvature with the vector field $\partial/\partial t$ in the S^1 direction, $\partial/\partial t \lrcorner F_A$. Write $\mathbf{F}_A = dt \wedge \alpha + *_3\beta$, hence the third term is $\langle \psi, \alpha.\psi \rangle = \langle \tau(\psi), \alpha \rangle$. Also from $\mathbf{F}_A^+ = (\mathbf{F}_A + *_4\mathbf{F}_A)/2$, we compute $\mathbf{F}_A^+ = [\, dt \wedge (\alpha + \beta) + *_3(\alpha + \beta)\,]/2$. The second Seiberg–Witten equation is $\mathbf{F}_A^+ = \sigma(\psi) = [dt \wedge \tau(\psi) + *_3\tau(\psi)]/2$ (Section 13.16) $\Rightarrow \tau(\psi) = \alpha + \beta$. So the equation (13.52) reads

$$||\frac{\partial \psi}{\partial t}||^2 + ||\partial_A(\psi)||^2 + ||\alpha||^2 + \langle \beta, \alpha \rangle = 0 \tag{13.53}$$

Also, by (13.29) $c_1^2(L) = 4d(L) \geq 0$. Since $c_1 = (i/2\pi)F_A$ and by convention $\mathbf{F}_A = iF_A$, then $4\pi^2 c_1^2(L) = \mathbf{F}_A \wedge \mathbf{F}_A = 2\langle\alpha,\beta\rangle volX$. Hence $\langle\alpha,\beta\rangle \geq 0$, thus we get no solutions when $d(L) > 0$, and when $d(L) = 0$ the solutions are pullback solutions of the following equations on Y, where $B \in \mathcal{A}_Y(L)$ (i.e. solutions on X are S^1 invariant):

$$\partial_B(\psi) = 0$$
$$*F_B = \tau(\psi)$$

For simplicity here we ignored the perturbation term of (13.22) in this discussion. □

When $b_1(Y) > 0$, Meng and Taubes defined the following quantity (cf. [HLe]):

$$\underline{SW}_Y = \sum_{L \in Spin^c(Y)} SW_Y(L) \,.\, c_1(L)/2$$

The convention of (13.33) is that the term $c_1(L)/2$ (equivalently $L/2$) is a formal coefficient. Recall coefficient L is denoted by t^L so as to satisfy the group ring property. Now recall the definition of the Milnor torsion $\nu(Y)$ from Section 2.7.

Theorem 13.23. *([MeT]) Let Y be a closed oriented 3-manifold with $b_1(Y) > 0$, then*

$$\underline{SW}_Y = \nu(Y)$$

For example, if $K \subset S^3$ a knot and $Y = S^3_K$, then this theorem and (2.5) gives:

$$SW_Y(t) := \sum_d SW_Y(d)\, t^{2d} = \underline{SW}_Y(t^2) = \nu(Y)(t^2) = \frac{\Delta_K(t^2)}{(t - t^{-1})^2} \tag{13.54}$$

Here $SW_Y(d)$ denotes the Seiberg–Witten invariant corresponding to the $Spin^c$ structure $L \to Y$ with $c_1(L) = 2d \in \mathbb{Z} \cong H^2(Y;\mathbb{Z})$ (these L are the characteristic classes of Y). From this we get another proof of Theorem 6.3 as follows: X_K is obtained by removing a copy of $T^2 \times D^2$ from X, and also from $Y_K \times S^1$, and then gluing them along their common T^3 boundaries. So applications of (13.38) and (13.40) along with (13.54) give:

$$SW_{X_K} = SW_X \,.\, (t^{-1} - t)^2 \,.\, SW_{S^3_K \times S^1} = SW_X \,.\, \Delta_K(t^2)$$

Remark 13.24. *This proof of Proposition 13.21 did not use the "T to be contained in a node neighborhood" assumption, as in the previous proof, but in any case to get the full conclusion of Theorem 6.3 we need the node neighborhood assumption. The reader can check from Figure 6.13 that when X is simply connected, the node neighborhood assumption (or the assumption of Exercise 13.2) implies the simply connectedness of X_K. Hence by Freedman X_K is homeomorphic to X.*

13.13 Moduli space near the reducible solution

Recall, at the reducible point $(A,0)$ the action of the gauge group $\mathcal{G}(L) = Map(X,S^1)$ on $\tilde{\mathcal{B}}(L) = \mathcal{A}(L) \times \Gamma(W^+)$ is not free (Definition 13.8). The stabilizer of $(A,0)$ are the constant gauge transformations. In this case quotienting $\tilde{\mathcal{B}}(L)$ with the gauge group can be performed in two steps. Let $\mathcal{G}_0(L)$ be the subgroup consisting of the kernel of the evaluation map, i.e. maps $s : X \to S^1$ with the property $s(x_0) = 1$, where x_0 is a base point of X. Then the fibration $\mathcal{G}_0(L) \to \mathcal{G}(L) \to S^1$ extends to the following fibration:

$$F := \mathcal{G}(L)/\mathcal{G}_0(L) \longrightarrow \tilde{\mathcal{M}}_\delta/\mathcal{G}_0(L) \longrightarrow \tilde{\mathcal{M}}_\delta/\mathcal{G}(L) = \mathcal{M}_\delta \tag{13.55}$$

The action of this so-called *based gauge group* $\mathcal{G}_0(L)$ on $\tilde{\mathcal{M}}_\delta$ is free, hence the intermediate quotient $\mathcal{M}^0_\delta = \tilde{\mathcal{M}}_\delta/\mathcal{G}_0(L)$ is smooth (for generic δ) even in the neighborhood of the reducible connection $\psi = 0$. This provides a useful tool for studying the neighborhood of the reducible solution (e.g. Section 14.1). Furthermore, when $H^1(X) = 0$ then $\mathcal{G}_0(L)$ is contractible, and so the fiber $F \cong S^1$, which can be identified by constant gauge transformations. As in (13.27) the derivative of the based gauge group action is still given by d. So by Proposition 13.12, in the neighborhood of a reducible solution, the Seiberg–Witten solution space $\mathcal{M}_\delta$ (A is generic) is given by $P^{-1}(0)/S^1$, where P is the S^1 equivariant surjection $P(a,\psi) = (d^+a, \not{D}_A\psi)$ (S^1 acting on the second factor),

$$P : ker(d^*) \oplus \Gamma(W^+) \longrightarrow \Omega^2_+(X) \oplus \Gamma(W^-) \tag{13.56}$$

One consequence of this is the Donaldson theorem, which gives restriction to the intersection form of a 4-manifold. Let X be a closed smooth oriented 4-manifold, and

$$q_X : H_2(X;\mathbb{Z}) \otimes H_2(X;\mathbb{Z}) \longrightarrow \mathbb{Z}$$

be the intersection form which we simply denote by $q_X(a,b) = a.b$. We say q_X positive definite, or negative definite if $a.a > 0$ or $a.a < 0$ respectively, for all a.

Theorem 13.25. *([D3]) If the intersection form q_X of a closed smooth simply connected 4-manifold X^4 is definite, then it must be diagonal.*

The proof uses the following algebraic lemma.

Lemma 13.26. *([El]) Let $q : V \times V \to \mathbb{Z}$ be any negative definite quadratic form, and $\mathcal{C}$ be the set of its characteristic elements of V (5.2), then the following holds:*

$$max\{q(a,a) \mid a \in \mathcal{C}\} + rank(q) \geq 0$$

Moreover, the inequality is an equality if and only if q is diagonal.

Proof. (of Theorem 13.25) We will prove this when $H^1(X) = 0$. Assume q_X is negative definite $H^2_+(X) = 0$, which implies $\sigma(X) = -b_2(X)$ and $\chi(X) = 2 + b_2(X)$. If q_X were not diagonalizable, then by Lemma 13.26 there must be a characteristic homology class (here we are representing it by a line bundle) $L \to X$, with $c_1(L)^2 + b_2(X) > 0$. Let $\mathcal{M}_\delta(L)$ be the corresponding Seiberg–Witten moduli space, which is compact by Proposition 13.9, and by Proposition 13.12 its dimension is given by

$$d = \frac{c_1(L)^2 + b_2}{4} - 1$$

And as a consequence of (5.3), d is a positive odd integer, say $2k-1$. Furthermore since $kerd^* \cap kerd^+ = H^1(X)$, (13.56) implies that in the neighborhood of $(A,0)$ the Seiberg–Witten solution space $\mathcal{M}_\delta$ is given by $Ker(\not{D}_A)/S^1 \cong \mathbb{C}^k/S^1 = cone(\mathbb{CP}^{k-1})$. Also it is clear that when $\psi = 0$, the Seiberg–Witten equations (Definition 13.11) have a unique solution up to gauge equivalence. So in particular $\mathcal{M}_\delta(L)$ is nonempty. By cutting out a small neighborhood of $(A,0)$ from $\mathcal{M}_\delta(L)$ we end up with a compact smooth manifold W with boundary $\partial W = \mathbb{CP}^{k-1}$. When $k=1$ this is an obvious contradiction. When $k>1$ it is still a contradiction for the following reason. By construction, W is the quotient of a compact smooth manifold $\tilde{W}$ with a free S^1 action, so $\tilde{W} \to W$ is the principal S^1 bundle; let $f : W \to BS^1 = \mathbb{CP}^\infty$ be its classifying map. By construction it is clear that f maps the fundamental class $[\partial W]$ nontrivially to $H_{2k-2}(\mathbb{CP}^\infty)$, which is a contradiction. □

Remark 13.27. *More generally, there is a relative version of the Seiberg–Witten invariants, defined for compact, oriented, smooth $Spin^c$-manifolds (X^4, L) with non-empty boundary, satisfying $b_2^+ \geq 1$. First of all, Seiberg–Witten equations are defined for oriented 3-manifolds with $Spin^c$ structures (Y, s) (this was briefly discussed in Sections 13.12 and 13.16), and they give certain homology groups $HM^\circ(Y,s)$ (which come in several flavors); they are called monopole–Floer homology groups. Then the relative Seiberg–Witten invariant $SW_X(L)$ is defined, and it takes values in $HM^\circ(Y,s)$, where $(Y,s) = \partial(X,L)$. Furthermore, any cobordism $(X.L)$ from one 3-manifold (Y_1, s_1) to another (Y_2, s_2), gives a homomorphism, $F(X,L) : HM^\circ(Y_1,s_1) \to HM^\circ(Y_2,s_2)$. Moreover (leaving technicalities aside), when a closed 4-manifold decomposes as a union of two such manifolds $M = X_1 \cup_Y X_2$, glued along the common boundary Y, then the invariant of M can be red off, by pairing the invariants (coming from each side) in $HM^\circ(Y)$ ([KM2]).*

13.14 Almost complex and symplectic structures

Now assume that X has an almost complex structure. This means that there is a principal $GL(2,\mathbb{C})$-bundle $Q \longrightarrow X$ such that, as complex bundles, there is an isomorphism

$$T(X) \cong Q \times_{GL(2,\mathbb{C})} \mathbb{C}^2$$

By choosing a Hermitian metric on $T(X)$ we can assume $Q \longrightarrow X$ is a $U(2)$ bundle, and the tangent frame bundle $P_{SO(4)}(TX)$ comes from Q by the reduction map

$$U(2) = (\ S^1 \times SU(2)\ )/\mathbb{Z}_2 \hookrightarrow (\ SU(2) \times SU(2)\ )/\mathbb{Z}_2 = SO(4)$$

Equivalently there is an endomorphism $I \in \Gamma(End(TX))$ with $I^2 = -Id$,

$$\begin{array}{ccc} T(X) & \xrightarrow{I} & T(X) \\ & \searrow \quad \swarrow & \\ & X & \end{array}$$

The $\pm i$ eigenspaces of I splits the complexified tangent space $T(X)_{\mathbb{C}}$,

$$T(X)_{\mathbb{C}} \cong T^{1,0}(X) \oplus T^{0,1}(X) = \Lambda^{1,0}(X) \oplus \Lambda^{0,1}(X)$$

This gives us a complex line bundle which is called the canonical line bundle:

$$K = K_X = \Lambda^{2,0}(X) = \Lambda^2(T^{1,0}) \longrightarrow X$$

Both K and K^{-1} are characteristic, i.e. $c_1(K^{\pm 1}) = w_2(X)$ mod 2. The bundle $K \to X$ defines a canonical $Spin_c(4)$ structure on X, given by the lifting of $f[\lambda, A] = ([\lambda, A], \lambda^2)$,

$$\begin{array}{ccc} & & Spin_c(4) \\ & F \nearrow & l \downarrow \\ U(2) & \xrightarrow{f} & SO(4) \times S^1 \end{array}$$

where l is the canonical projection and $F[\lambda, A] = [\lambda, A, \lambda]$. The transition function λ^2 gives the line bundle K, and the corresponding $\mathbb{C}^2$-bundles are given by

$$W^+ = \Lambda^{0,2}(X) \oplus \Lambda^{0,0}(X) = K^{-1} \oplus \mathbb{C} \qquad (13.57)$$

$$W^- = \Lambda^{0,1}(X) \qquad (13.58)$$

We can check this from the transition functions of W^+, say $x = z + jw \in \mathbb{H}$,

$$x \longmapsto \lambda x \lambda^{-1} = \lambda(z + jw)\bar{\lambda} = z + jw\bar{\lambda}\bar{\lambda} = z + jw\lambda^{-2}$$

Since we can identify $\bar{\Lambda}^{0,1}(X) \cong \Lambda^{1,0}(X)$, and $\Lambda^{0,2}(X)\otimes\Lambda^{1,0}(X) \cong \Lambda^{0,1}(X)$, we readily see the decomposition $T(X)_{\mathbb{C}} \cong W^+ \otimes \bar{W}^-$. As real bundles we have

$$\Lambda^+(X) \cong K \oplus \mathbb{R}$$

We can verify this by taking $\{e^1, e^2 = I(e^1), e^3, e^4 = I(e^3)\}$ to be a local orthonormal basis for $T^*(X)$, then

$$\Lambda^{1,0}(X) = < e^1 - ie^2, e^3 - ie^4 > \quad and \quad \Lambda^{0,1}(X) = < e^1 + ie^2, e^3 + ie^4 >$$

$$K = < f = (e^1 - ie^2) \wedge (e^3 - ie^4) >$$

$$\Lambda^+(X) = < \omega := f_1 = \frac{1}{2}(e^1 \wedge e^2 + e^3 \wedge e^4),\ f_2 = \frac{1}{2}(e^1 \wedge e^3 + e^4 \wedge e^2),\ f_3 = \frac{1}{2}(e^1 \wedge e^4 + e^2 \wedge e^3) >$$

ω is the global form $\omega(X,Y) = g(X, IY)$ where g is the Hermitian metric (which makes the basis $\{e^1, e^2, e^3, e^4\}$ orthogonal). Also since $f = 2(f_2 - if_3)$, we see as $\mathbb{R}^3$-bundles $\Lambda^+(X) \cong K \oplus \mathbb{R}(\omega)$. We can check

$$W^+ \otimes \bar{W}^+ \cong \mathbb{C} \oplus \mathbb{C} \oplus K \oplus \bar{K} = (K \oplus \mathbb{R})_{\mathbb{C}} \oplus \mathbb{C}$$

We see that ω, f_2, f_3 act as the Pauli matrices $\rho : \Lambda^+_{\mathbb{C}} \to sl(W^+)$:

$$\begin{aligned} \rho(\omega) &= \begin{pmatrix} i & 0 \\ 0 & -i \end{pmatrix} \\ \rho(f) &= 2\begin{pmatrix} 0 & -1 \\ 1 & 0 \end{pmatrix} - 2i\begin{pmatrix} 0 & -i \\ -i & 0 \end{pmatrix} = \begin{pmatrix} 0 & -4 \\ 0 & 0 \end{pmatrix} \\ \rho(\bar{f}) &= 2\begin{pmatrix} 0 & -1 \\ 1 & 0 \end{pmatrix} + 2i\begin{pmatrix} 0 & -i \\ -i & 0 \end{pmatrix} = \begin{pmatrix} 0 & 0 \\ 4 & 0 \end{pmatrix} \end{aligned} \tag{13.59}$$

If we write a section $\psi \in \Gamma(W^+) = \Gamma(\mathbb{C} \oplus K^{-1})$ by $\psi = (\alpha, \beta)$, then in terms of the basis $\{\omega, f, \bar{f}\}$, the formula (13.5) and the equality $\sigma(\psi) = \mathbf{F}_A$ is:

$$\begin{aligned} \sigma(\psi) \quad &= \quad (\frac{|\alpha|^2 - |\beta|^2}{2})\,\omega - \frac{i}{4}\,(\alpha\,\bar{\beta})\,f + \frac{i}{4}\,(\bar{\alpha}\,\beta)\,\bar{f} \\ \| \quad & \\ \mathbf{F}_A \quad &= \quad i\,F_A = i\,[F_A^{1,1} + F_A^{2,0} + F_A^{0,2}] \end{aligned}$$

From the second Seiberg–Witten equation, $\sigma(\psi) = \mathbf{F}_A = iF_A$, we get

$$F_A = -i\sigma(\psi) = (\frac{|\beta|^2 - |\alpha|^2}{2})\, i\,\omega - \frac{1}{4}\,(\alpha\,\bar{\beta})\,f + \frac{1}{4}\,(\bar{\alpha}\,\beta)\,\bar{f} \tag{13.60}$$

$$F_A^{2,0} = -\frac{1}{4}(\alpha\,\bar{\beta})\,f \tag{13.61}$$

$$F_A^{0,2} = \frac{1}{4}(\bar{\alpha}\,\beta)\,\bar{f} \tag{13.62}$$

$$F_A^{1,1} = (\frac{|\beta|^2 - |\alpha|^2}{2})\,i\,\omega \tag{13.63}$$

In the case where X is a Kähler surface the Dirac operator is given by (cf. [LaM])

$$\not{D}_A = \bar{\partial}_A^* + \bar{\partial}_A : \Gamma(W^+) \to \Gamma(W^-)$$

Hence from the Dirac part of the Seiberg–Witten equation we have

$$\bar{\partial}_A^*(\beta) + \bar{\partial}_A(\alpha) = 0 \Longrightarrow \bar{\partial}_A\bar{\partial}_A^*(\beta) + \bar{\partial}_A\bar{\partial}_A(\alpha) = 0 \tag{13.64}$$

Notice $\bar{\partial}_A\bar{\partial}_A(\alpha) = F_A^{0,2}\alpha = \frac{1}{4}|\alpha|^2\beta$. By taking the inner product of both sides the last equation by β and integrating over X we get the L^2 norms satisfying

$$||\alpha||^2||\beta||^2 + ||\bar{\partial}_A^*(\beta)||^2 = 0 \tag{13.65}$$

Hence both of the terms must be zero, in particular $\bar{\alpha}\beta = 0$ and $\bar{\partial}_A^*(\beta) = 0$. Then A is a holomorphic connection on K, and by (13.64) α is a holomorphic function, and β is an anti-holomorphic section of K^{-1}, hence it is a holomorphic 2-form values in $K \to X$. Also we must have that at any point one of α or β must be zero. Since the following quantity is constant, then either α or β must be zero:

$$c_1(K).\omega = \int_X c_1(K)\wedge\omega = \int_X \frac{i}{2\pi}F_A\wedge\omega = \frac{1}{4\pi}\int_X(|\beta|^2 - |\alpha|^2)dvol$$

This argument eventually gives the following result of Witten:

Theorem 13.28. *([Wi]) Let X be a minimal Kähler surface with $b_2^+(X) > 1$. Then $|SW_X(\pm K)| = 1$, where $K \to X$ is the canonical class. Furthermore, if $c_1^2(X) \geq 0$ then $SW_X(L) = 0$ for all other Spin^c structures L, hence $SW_X = t_K + t_K^{-1}$.*

Recall $E(2) = V_4$ (Section 12.2) is a Kähler surface with $b_2^+ = 3$ and $c_1 = 0$, hence $SW_{E(2)} = 1$. Rather than completing the proof of Theorem 13.28 we will review a stronger result of C.Taubes, which applies more generally to symplectic manifolds.

We call an almost complex manifold with Hermitian metric $\{X, I, g\}$ symplectic if $d\omega = 0$. Clearly a non-degenerate closed form ω and a Hermitian metric determine the almost complex structure I by $g(u,v) = \omega(u, Iv)$. Given ω, then I is called an almost complex structure taming the symplectic form ω.

By the discussion preceding Proposition 13.6, there is a unique connection A_0 of $K \to X$ such that the induced Dirac operator D_{A_0} on W^+ restricted to the trivial summand $\underline{\mathbb{C}} \to X$ is the exterior derivative d. Let $\mathbf{u}_0$ be a section of $W^+ = \mathbb{C} \oplus K^{-1}$, with constant $\mathbb{C}$ component and $\|\mathbf{u}_0\| = 1$. Taubes's first fundamental observation is

$$\not{D}_A(\mathbf{u}_0) = 0 \quad \text{if and only if} \quad d\omega = 0$$

This can be seen by applying the Dirac operator to both sides of $i\mathbf{u}_0 = \rho(\omega).\mathbf{u}_0$, and observing that by the choice of $\mathbf{u}_0$ the term $\nabla_{A_0}(\mathbf{u}_0)$ lies entirely in the K^{-1} component:

$$\begin{aligned} i\not{D}_{A_0}(\mathbf{u}_0) &= \sum \rho(e^i)\nabla_i\,(\rho(\omega)\,\mathbf{u}_0) \\ &= \sum \rho(e^i)\,[\,\nabla_i\,(\rho(\omega))\,\mathbf{u}_0 + \rho(\omega)\,\nabla_i\,(\mathbf{u}_0)\,] \\ &= \sum \rho(e^i)\nabla_i\,(\rho(\omega))\,\mathbf{u}_0 \\ 2i\,\not{D}_{A_0}(\mathbf{u}_0) &= \rho((d+d^*)\omega)\,\mathbf{u}_0 = \rho((d - *d)\omega)\,\mathbf{u}_0 \end{aligned}$$

The last equality holds since $\omega \in \Lambda^+(X)_{\mathbb{C}} \oplus \mathbb{C}$, and by naturality, the Dirac operator on $\Lambda^*(X)_{\mathbb{C}}$ is $d + d^*$, and since $d = - * d*$ on 2 forms and ω is self dual,

$$2i\,\not{D}_{A_0}(\mathbf{u}_0) = \rho(- * d\omega \oplus d\omega)\mathbf{u}_0 \qquad \square$$

Theorem 13.29. *(Taubes) Let (X,ω) be a closed symplectic manifold with $b_2(X)^+ \geq 2$ and $L \to X$ be the canonical line bundle, then $SW_X(L) = \pm 1$.*

Proof. Let $\psi = \alpha\mathbf{u}_0 + \boldsymbol{\beta} = (\alpha, \beta) \in \Gamma(W^+) = \Gamma(\mathbb{C} \oplus K^{-1})$, where $\mathbf{u}_0$ is a unit section with constant component in $\mathbb{C}$ and $\alpha : X \to \mathbb{C}$. Consider the perturbed Seiberg–Witten equations (where r is constant):

$$\not{D}_A(\psi) = 0 \tag{13.66}$$
$$F_A^+ = F_{A_0}^+ + r\,[\,-i\sigma(\psi) + i\,\omega\,] \tag{13.67}$$

where $(A, \psi) \in \mathcal{A}(L) \times \Gamma(W^+)$. By (13.60) the second equation is

$$F_A^+ - F_{A_0}^+ = r\,[\,(\frac{|\beta|^2 - |\alpha|^2}{2} + 1)\,i\omega - \frac{1}{4}(\alpha\bar{\beta})f + \frac{1}{4}(\bar{\alpha}\beta)\bar{f}\,] \tag{13.68}$$

$(A,\psi) = (A_0, \mathbf{u}_0)$ and $r = 0$ is the trivial solution. We will show that for $r >> 1$, up to gauge equivalence there is a unique solution to these equations. Write $A = A_0 + ia$, after a gauge transformation we can assume that a is co-closed, $d^*(a) = 0$. It suffices to show that for $r \longmapsto \infty$ these equations admit only (A_0, u_0) as a solution. From Proposition 13.6, Weitzenbock formula (13.11), and abbreviating $\nabla_{A_0}(\mathbf{u}_0) = b$, we get

$$I\!D^2_A(\psi) = I\!D^2_A(\boldsymbol{\beta}) + (\nabla^*_a \nabla_a \alpha)\mathbf{u}_0 - 2 < \nabla_a \alpha, b > + \frac{1}{2}\rho(F^+_A - F^+_{A_0})\, \alpha \mathbf{u}_0 \qquad (13.69)$$

$$I\!D^2_A(\boldsymbol{\beta}) = (\nabla^*_A \nabla_A\, \boldsymbol{\beta}) + \frac{s}{4}\,\boldsymbol{\beta} + \frac{1}{2}\rho(F^+_{A_0})\,\boldsymbol{\beta} + \frac{1}{2}\rho(F^+_A - F^+_{A_0})\,\boldsymbol{\beta} \qquad (13.70)$$

By applying ρ to (13.68), using (13.59) we get

$$\frac{1}{2}\rho(F^+_A - F^+_{A_0}) = \frac{r}{4}\begin{pmatrix} (|\alpha|^2 - |\beta|^2 - 2) & 2\alpha\bar{\beta} \\ 2\bar{\alpha}\beta & -(|\alpha|^2 - |\beta|^2 - 2) \end{pmatrix} \qquad (13.71)$$

Here we are denoting $\alpha\mathbf{u}_0 = (\alpha, 1)$, $\boldsymbol{\beta} = (0, \beta)$, then

$$\frac{1}{2}\,\rho(F^+_A - F^+_{A_0})\alpha\,\mathbf{u}_0 = \frac{r}{4}\,(|\alpha|^2 - |\beta|^2 - 2)\alpha\, u_0 + \frac{r}{2}\,|\alpha|^2\beta \qquad (13.72)$$

$$\frac{1}{2}\,\rho(F^+_A - F^+_{A_0})\,\boldsymbol{\beta} = \frac{r}{2}\alpha|\beta|^2\, u_0 - \frac{r}{4}\,(|\alpha|^2 - |\beta|^2 - 2)\beta \qquad (13.73)$$

By substituting (13.72) in (13.69) we get

$$\begin{aligned} I\!D^2_A(\psi - \boldsymbol{\beta}) &= [\, \nabla^*_a \nabla_a \alpha + \frac{r}{4}\,(|\alpha|^2 - |\beta|^2 - 2)\,\alpha\,]\,\mathbf{u}_0 \\ &\quad -2 < \nabla_a \alpha, b > + \frac{r}{2}\,|\alpha\,|^2\boldsymbol{\beta} \end{aligned} \qquad (13.74)$$

Substituting (13.73) in (13.70), then substituting this new (13.70) into (13.74), we obtain

$$\begin{aligned} 0 = I\!D^2_A(\psi) &= [\, \nabla^*_a \nabla_a \alpha + \frac{r}{4}\alpha\,(|\alpha|^2 - |\beta|^2 - 2)\,]\,\mathbf{u}_0 - 2 < \nabla_a \alpha, b > \\ &\quad + [\, \nabla^*_A \nabla_A + \frac{s}{4} + \frac{1}{2}\,\rho(F^+_{A_0}) + \frac{r}{4}(|\alpha|^2 + |\beta|^2 + 2)\,]\,\boldsymbol{\beta} \end{aligned} \qquad (13.75)$$

By recalling that $\mathbf{u}_0$ and $\boldsymbol{\beta}$ are orthogonal sections of W^+, we take the inner product of both sides of (13.75) with $\boldsymbol{\beta}$ and integrate over X, and obtain

$$\int_X (\, |\nabla_A \boldsymbol{\beta}\,|^2 + \frac{r}{4}\,|\beta|^4 + \frac{r}{2}|\beta|^2 + \frac{r}{4}|\alpha|^2|\beta|^2\,) =$$

$$2\int_X (<< \nabla_a \alpha\,,\, b >\,,\, \boldsymbol{\beta} > - \frac{s}{4}\,|\beta|^2 - \frac{1}{2} < \rho(F^+_{A_0})\boldsymbol{\beta}\,,\, \boldsymbol{\beta} >$$

$$\text{Hence,}\quad \int_X |\nabla_A \boldsymbol{\beta}|^2 + \frac{r}{4}|\beta|^4 + \frac{r}{2}|\beta|^2 + \frac{r}{4}|\alpha|^2|\beta|^2 \le \int_X c_1|\beta|^2 + c_2|\beta||\nabla_a\alpha| \tag{13.76}$$

where c_1 and c_2 are positive constants depending on the Riemanian metric and the base connection A_0. Choose $r \gg 1$, by calling $c_2 = 2c_3$, we get

$$\int_X (\,|\nabla_A \boldsymbol{\beta}|^2 + \frac{r}{4}|\beta|^4 + \frac{r}{4}|\beta|^2 + \frac{r}{4}|\alpha|^2|\beta|^2\,) \le \int_X (c_1 - \frac{r}{4})\,|\beta|^2 + 2c_3|\beta||\nabla_a\alpha| =$$

$$-\int_X \left[\,(r/4 - c_1)^{1/2}|\beta| - c_3(r/4-c_1)^{-1/2}|\nabla_a\alpha|\,\right]^2 + \frac{c_3^2}{(r/4-c_1)}|\nabla_a\alpha|^2 \le \int_X \frac{C}{r}|\nabla_a\alpha|^2$$

for some C, depending on the metric and A_0. In particular we have

$$\begin{aligned} \int_X r\,|\beta|^2 - \frac{4C}{r}|\nabla_a\alpha|^2 &\le 0 \\ \int_X 4c_2\,|\beta|\,|\nabla_a\alpha| - \frac{4C}{r}|\nabla_a\alpha|^2 &\le \int_X (r - 4c_1)\,|\beta|^2 \end{aligned}$$

$$\text{Hence,}\quad \int_X c_2\,|\beta|\,|\nabla_a\alpha| - \frac{2C}{r}|\nabla_a\alpha|^2 \le 0 \tag{13.77}$$

Now by self adjointness of the Dirac operator, and by $\alpha u_0 = \psi - \boldsymbol{\beta}$, we get

$$\begin{aligned} <\not{D}_A^2(\psi)\,,\,\alpha\mathbf{u}_0> &= <\not{D}_A^2(\psi - \boldsymbol{\beta})\,,\,\alpha\mathbf{u}_0> + <\not{D}_A^2\,(\boldsymbol{\beta})\,,\,\alpha\mathbf{u}_0> \\ &= <\not{D}_A^2(\psi - \boldsymbol{\beta})\,,\,\alpha\mathbf{u}_0> + <\boldsymbol{\beta}\,,\,\not{D}_A^2(\psi - \boldsymbol{\beta})> \end{aligned} \tag{13.78}$$

We calculate (13.78) by using (13.74) and get (recall $\mathbf{u}_0$ component of $\nabla_a\alpha$ is zero)

$$\begin{aligned} 0 =<\not{D}_A^2(\psi)\,,\,\alpha\mathbf{u}_0> &= |\nabla_a\alpha|^2 + \frac{r}{4}|\alpha|^4 - \frac{r}{4}|\alpha|^2|\beta|^2 - \frac{r}{2}|\alpha|^2 \\ &\quad + \frac{r}{2}|\alpha|^2|\beta|^2 - 2 << \nabla_a\alpha, b>, \boldsymbol{\beta}> \end{aligned}$$

$$\begin{aligned} \int_X |\nabla_a\alpha|^2 + \frac{r}{4}|\alpha|^4 - \frac{r}{2}|\alpha|^2 &\le \int_X 2 << \nabla_a\alpha, b>, \boldsymbol{\beta}> - \frac{r}{4}|\alpha|^2|\beta|^2 \\ &\le \int_X 2 << \nabla_a\alpha, b>, \boldsymbol{\beta}> \le \int_X c_2|\nabla_a\alpha|\,|\beta| \end{aligned}$$

By choosing $c_4 = 1 - 2C/r$ and by using (13.77), we see

$$\int_X c_4 \, |\nabla_a \alpha|^2 + \frac{r}{4}|\alpha|^4 - \frac{r}{2}|\alpha|^2 \;\leq\; 0 \tag{13.79}$$

Now since for a connection A on $K \to X$ the form $(i/2\pi)F_A$ represents the Chern class $c_1(K)$, and also since ω is a self dual 2-form, we can write

$$\int_X F_A \wedge \omega = -2\pi \, i \, c_1(K).\,\omega \;\;\Rightarrow\;\; \int_X F_A \wedge \omega = \int_X F_A^+ \wedge \omega$$

$$\int_X (F_A^+ - F_{A_0}^+) \wedge \omega = 0$$

By (13.68) this implies:

$$\frac{r}{2}\int_X (2 - |\alpha|^2 + |\beta|^2) = 0 \tag{13.80}$$

By adding (13.80) to (13.79) we get

$$\int_X c_4 \, |\nabla_a \alpha|^2 + \frac{r}{2}|\beta|^2 + r(1 - \frac{1}{2}|\alpha|^2)^2 \leq 0 \tag{13.81}$$

Assume $r \gg 1$, then $c_4 \geq 0$ and hence $\nabla_a \alpha = 0$ and $\beta = 0$ and $|\alpha| = \sqrt{2}$, hence,

$$\beta = 0 \;\text{ and }\; \alpha = \sqrt{2}e^{i\,\theta} \quad \text{and} \;\; \nabla_a(e^{i\,\theta}) = d(e^{i\,\theta}) + i\,a\,e^{i\,\theta} = 0$$

Hence, $a = d(-\theta)$, recall that we also have $d^*(a) = 0$, which gives

$$0 =< d^* d(\theta), \theta >=< d(\theta), d(\theta >) >= ||d(\theta)||^2$$

Hence, $a = 0$ and $\theta =$ Constant. So up to a gauge equivalence $(A, \psi) = (A_0, u_0)$.

□

Remark 13.30. *The more general version of Theorem 13.29 has an additional conclusion that if $SW_X^-(k) \neq 0$, then $|k.\omega| \leq L.\omega$, with the equality if and only if $k = \pm L$ ([Ta3], [Ta4]). In particular if X admits a symplectic structure then we must have $L.\omega \geq 0$.*

Remark 13.31. *In case $b_2^+(X) = 1$, Theorem 13.29 goes through except that we have to use the chamber dependent Seiberg–Witten invariant $SW_X^- = SW_{X,H}^- = \pm 1$ (Section 13.6).*

13.15 Antiholomorphic quotients

It is an old problem as to whether the quotient of a simply connected smooth complex surface by an antiholomorphic involution $\sigma : \tilde{X} \to \tilde{X}$ (an involution which anticommutes with the almost complex homomorphism $\sigma_* J = -J\sigma_*$) is a "standard" manifold (i.e. connected sums of $S^2 \times S^2$ and $\pm\mathbb{CP}^2$). A common example of an antiholomorphic involution is the complex conjugation on a complex projective algebraic surface with real coefficients. It is known that the quotient of $\mathbb{CP}^2$ by complex conjugation is S^4 (Arnold, Massey, Kuiper); and for every d there is a curve of degree d in $\mathbb{CP}^2$ whose 2-fold branched cover has a standard quotient ([A17]). This problem makes sense only if the antiholomorphic involution has a fixed point, otherwise the quotient has fundamental group $\mathbb{Z}_2$ and hence it cannot be standard. By "connected sum" theorem, Seiberg-Witten invariants of "standard" manifolds vanish, so it is natural question to ask whether Seiberg–Witten invariants of the quotients vanish. The following is the answer.

Theorem 13.32 ([Wg]). *Let $\tilde{X}$ be a minimal Kähler surface of general type, with a free antiholomorphic involution $\sigma : \tilde{X} \to \tilde{X}$, with the quotient $X = \tilde{X}/\sigma$, then $SW_X = 0$.*

Proof. Let h be the Kähler metric on $\tilde{X}$ with a Kähler form $\omega(X,Y) = h(X, JY)$. Then $\tilde{g} = h + \sigma^* h$ is an invariant metric on $\tilde{X}$ with the Kähler form $\tilde{\omega} = \omega - \sigma^*\omega$. Let g be the "push-down" metric on X. Now we claim that all $SW_X(L) = 0$ for all $L \to X$, because if $L \to (X, g)$ is the characteristic line bundle supporting any solution (A, ψ), then the pull-back pair $(\tilde{A}, \tilde{\psi})$ is a solution for the pull-back line bundle $\tilde{L} \to \tilde{X}$ with the pull-back $Spin^c$ structure, hence

$$0 \le dim\mathcal{M}_{\tilde{L}}(\tilde{X}) = \frac{1}{4}c_1^2(\tilde{L}) - \frac{1}{4}(3\sigma(\tilde{X}) + 2\chi(\tilde{X}))$$

But $\tilde{X}$ being a minimal Kähler suface of general type $3\sigma(\tilde{X}) + 2\chi(\tilde{X}) = K^2_{\tilde{X}} > 0$, hence $c_1^2(\tilde{L}) > 0$. This implies that $(\tilde{A}, \tilde{\psi})$ must be an irreducible solution (i.e. $\psi \neq 0$), otherwise $F^+_{\tilde{A}} = 0$ would imply $c_1^2(\tilde{L}) < 0$. By (13.65) the nonzero solution $\psi = \alpha u_0 + \beta$ must have either one of α or β zero (so the other one is nonzero), and since $\tilde{\omega} \wedge \tilde{\omega}$ is the volume element we get (13.82). We also know that $\sigma^*(\tilde{\omega}) = -\tilde{\omega}$,

$$\tilde{\omega}.c_1(\tilde{L}) = \frac{i}{2\pi}\int \tilde{\omega} \wedge F^+_{\tilde{A}} = \frac{i}{2\pi}\int \tilde{\omega} \wedge (\frac{|\beta|^2 - |\alpha|^2}{2})\, i\,\tilde{\omega} \neq 0 \tag{13.82}$$

But since $\sigma^* c_1(\tilde{L}) = c_1(\tilde{L})$, and σ is an orientation preserving map, we get a contradiction

$$\tilde{\omega}.c_1(\tilde{L}) = \sigma^*(\tilde{\omega}.c_1(\tilde{L})) = -\tilde{\omega}.c_1(\tilde{L}) \qquad \square$$

Remark 13.33. *There are some formulas relating the Seiberg-Witten invariant of branched covers $\tilde{X} \to X$ to that of X, which this theorem is an example of, e.g. [Pa], [RW].*

13.16 S–W equations on $\mathbb{R} \times Y^3$

Let $X^4 = \mathbb{R} \times Y^3$; if a *Spin*^{c} structure on X (an integral lifting of $w_2(X) \in H^2(X;\mathbb{Z}_2)$ which is represented by a complex line bundle $L \to X$) comes from a *Spin*^{c} structure on Y, we can identify both $\mathbb{C}^2$-bundles $W^{\pm} \to \mathbb{R} \times Y^3$ with the pull back of a $\mathbb{C}^2$-bundle $W \to Y$ by the projection $\mathbb{R} \times Y \to Y$. Clifford multiplication with dt identifies W^+ with W^-. More specifically, *Spin*^{c} structures on $\mathbb{R} \times Y$ are induced from the imbedding

$$Spin^c(3) = (SU(2) \times S^1)/\mathbb{Z}_2 \hookrightarrow (SU(2) \times SU(2) \times S^1)/\mathbb{Z}_2 = Spin^c(4)$$

given by the map $[q,\lambda] \mapsto [q, q^{-1}, \lambda]$. In the notation of Section 13.1 the isomorphism $W^+ \mapsto W^-$ is induced by $[q,\lambda] \mapsto [q^{-1}, \lambda]$. Also we have the following identification:

$$\Lambda^1(Y) \cong \Lambda^2_+(X) \text{ given by the map } a \mapsto \frac{1}{2}[\, dt \wedge a + *_3\, a\,]$$

For example, if $a \in \Omega^1(X)$, we identify $d^+(a) \leftrightarrow da/dt + *_3\, d_Y(a) \ \in \Omega^1(Y)$ since

$$\begin{aligned} 2d^+(a) = da + *_4\, da &= dt \wedge \frac{da}{dt} + d_Y(a) + *_4[\, dt \wedge \frac{da}{dt} + d_Y(a)\,] \\ &= dt \wedge [\frac{da}{dt} + *_3\, d_Y(a)] + *_3\, [\frac{da}{dt} + *_3\, d_Y(a)] \end{aligned}$$

The Clifford multiplication $T^*Y \otimes W \to W$ identifies the trace zero endomorphisms of W with the complexified 1-forms $(W \otimes \bar{W})_0 \cong \Lambda^1(Y)_{\mathbb{C}}$, so we can define a map $\tau : W \to \Lambda^1_{\mathbb{C}}$ by sending ψ to the image of $\psi \otimes \bar{\psi}$ in $\Lambda^1_{\mathbb{C}}$, then in the above identification $\tau(\psi)$ corresponds to $\sigma(\psi)$, i.e. $2\sigma(\psi) = dt \wedge \tau(\psi) + *_3\tau(\psi)$.

By viewing Seiberg–Witten equations (13.14), (13.15) as equations in the quotient space $\mathcal{B}(L) = \tilde{\mathcal{B}}(L)/\mathcal{G}(L)$, we can apply gauge transformations (13.16) to change the dt component of the connection parameter $A = A_0 + a \in \{A_0\} + \Omega^1(X)$ by any $f(x,t)dt$, hence make it zero. Such a connection is called a connection in *temporal gauge*. So $A = A(t,x)$ can be viewed as a family of connections on Y parametrized by t; hence the Dirac operator on X is $D\!\!\!/_A(\psi) = d\psi/dt + \partial_A(\psi)$, where ∂_A is the Dirac operator on Y, which is expressed by $\partial_A\psi = \partial_{A_0}\psi + a.\psi$. For simplicity here we are abbreviating $\rho(a)\psi = a\psi$. Then the Seiberg–Witten equations on $\mathbb{R} \times Y$ give flow equations on the configuration space $\tilde{\mathcal{B}}(L) = \Omega^1(Y^3) \oplus \Gamma(W)$ of the Seiberg–Witten equations of the 3-manifold Y:

$$da/dt = -*_3\, F_A - i\tau(\psi) \qquad (13.83)$$

$$d\psi/dt = -\partial_A(\psi) \qquad (13.84)$$

These are the downward gradient flow equations of the function $C/2 : \tilde{\mathcal{B}}(L) \to \mathbb{R}$,

$$C(A,\psi) = \int_Y (A - A_0) \wedge (F_A + F_{A_0}) + \langle \psi, \partial_A(\psi) \rangle$$

To see this take a variation, $A = A_0 + a(t) \Rightarrow \dot{A} = \dot{a}$, and use $F_A = F_{A_0} + da$, $F_{\dot{A}} = d\dot{a}$, the Stokes theorem and the self adjointness of the Dirac operator:

$$\begin{aligned} dC_{(A,\psi)}(\dot{A},\dot{\psi}) &= \int_Y \dot{a} \wedge (F_A + F_{A_0}) + a \wedge F_{\dot{A}} + \langle \dot{\psi}, \partial_A \psi \rangle + \langle \psi, \partial_A \dot{\psi} \rangle + \langle \psi, \dot{a}.\psi \rangle \\ &= \int_Y 2[\dot{a} \wedge da + \dot{a} \wedge F_{A_0} + \langle \dot{\psi}, \partial_A \psi \rangle] + \langle \dot{a}.\psi, \psi \rangle \\ &= \int_Y 2[\dot{a} \wedge F_A + \langle \dot{\psi}, \partial_A \psi \rangle] + 2i\langle \dot{a}, \tau(\psi) \rangle \qquad \text{from (13.7)} \\ &= 2\int_Y [\langle \dot{a} \,, *_3 F_A \rangle + \langle \dot{a} \,, i\tau(\psi) \rangle + \langle \dot{\psi} \,, \partial_A \psi \rangle] \\ &= 2 < (\dot{a} \,, \dot{\psi}), (*_3 F_A + i\tau(\psi), \partial_A \psi) > \end{aligned}$$

where $< .,. >$ is the metric on $\tilde{\mathcal{B}}(L)$. So $*_3 F_A + i\tau(\psi)$ is the gradient vector field of the functional $\frac{1}{2}C$. In particular at the gradient flow solutions we have (compare [KM2])

$$dC(\dot{a},\dot{\psi}) = 2 < (\dot{a},\dot{\psi}), (-\dot{a},-\dot{\psi}) >= -2(|\dot{a}|^2 + |\dot{\psi}|^2) \tag{13.85}$$

Remark 13.34. *A gauge transformation $s : X \to S^1$ (13.16) alters C by*

$$C(s^*(A,\psi)) - C(A,\psi) = \int_Y s^{-1}ds \wedge da = \int_Y s^{-1}ds \wedge F_A = \langle 2\pi i c_1(L) \cup h, [Y] \rangle$$

where $s^{-1}ds = s^*(d\theta) \in \Omega^1(Y)$ is a closed 1-form representing a class $h \in H^1(Y;\mathbb{Z})$.

13.17 Adjunction inequality

Theorem 13.35. *([KM], [MST]) Let X be a closed smooth 4-manifold with $b_2^+(X) > 1$ with nonzero Seiberg–Witten invariant corresponding to a line bundle $L \to X$, and $\Sigma \subset X$ be a smooth closed genus $g(\Sigma) > 1$ oriented surface with $[\Sigma] \neq 0$ and $\Sigma.\Sigma \geq 0$. Then*

$$2g(\Sigma) - 2 \geq \Sigma.\Sigma + |\langle c_1(L).\Sigma \rangle| \tag{13.86}$$

Proof. ([KM]) We prove this when $\Sigma.\Sigma = 0$. Let $Y = S^1 \times \Sigma \subset \partial(D^2 \times \Sigma) \subset X$. We first stretch X near Y metrically, i.e. we pick a metric ds^2_R on X which is the product metric $dt^2 + ds^2_Y$ on $[-R, R] \times S^1 \times \Sigma \subset X$. Call this metrically stretched Riemannian manifold X_R (Figure 13.7). Over $[-R, R] \times Y$ we identify $Spin^c(4)$ bundles $W^{\pm} \to [-R, R] \times Y$ with the pull back of the $Spin^c(3)$ bundle $W \to Y$. Choose a base connection A^0_R on X_R restricting fixed connections $B_{\pm}$ on the complement of the product region $X_{\pm}$, independent of R. By Remark 13.34, $Cs^*(A, \psi) - C(A, \psi) = 2\pi i \langle c_1(L) \cup h, \partial[X_+]\rangle$, and by construction both $c_1(L)$ and h are restrictions of cohomology classes of X_+. Hence $Cs^*(A, \psi) = C(A, \psi)$. This implies that the function C is defined on the quotient $\mathcal{B}(L)$ not just on $\tilde{\mathcal{B}}(L)$,

$$C : \mathcal{B}(L) \to \mathbb{R}$$

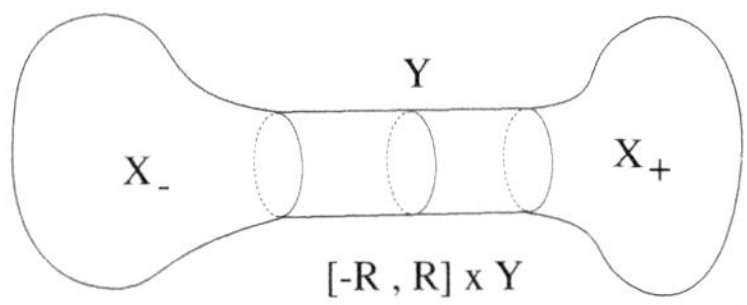

Figure 13.7: Deforming the metric of X to get X_R

Recall by Proposition 13.9, the Seiberg–Witten solution space in $\mathcal{B}(L)$ is compact. Now we pick a solution $(A_R, \psi_R) = (A^0_R + a_R, \psi_R)$ on X_R for R sufficiently large, such that A_R is in temporal on the product region. Let $(A_R(t), \psi_R(t))$ denote the restrictions of (A_R, ψ_R) to $\{t\} \times Y$. Now we claim that the following quantity is negative:

$$l(R) = C(A_R(R), \psi_R(R)) - C(A_R(-R), \psi_R(-R))$$

and satisfies an uniform bound C_0 independent of R. This is because by Proposition 13.9 and Remark 13.10 we can choose gauge transformations $s_{\pm}$ on $X_{\pm}$ so that $s^*_{\pm}(A_R) - B_{\pm}$ and its first derivatives are uniformly bounded, which gives uniform bounds to quantities

$$C(s^*_{\pm} A_R(\pm R), s^*_{\pm} \psi_R(\pm R)) = C(A_R(\pm R), \psi_R(\pm R))$$

This implies that there is an interval $[-1, 1] \times Y$ on which the change $\Delta C(1)$ is at most C_0/R, so by taking the limit of the solutions as $R \mapsto \infty$, we get

$$0 = \Delta C(1) = \int_0^1 \frac{dC}{dt} dt = \int_0^1 dC(\dot{a}, \dot{\psi}) dt = -2 \int_0^1 (|\dot{a}|^2 + |\dot{\psi}|^2) dt$$

Hence $(A(t), \psi(t))$ is constant, giving a translation invariant solution (A, ψ) on $\mathbb{R}\times Y$ in temporal gauge. Here Σ is a 2-manifold with constant scalar curvature s; we chose the metric so that Σ has the unit area such that its scalar curvature (twice the Gaussian curvature) is $-2\pi(4g-4)$. Also by (13.15), (13.7), (13.8) and (13.17), we calculate

$$\sqrt{2}\,|F_A^+| = \sqrt{2}\,|\sigma(\psi)| = |\rho(\sigma(\psi))| = |\boldsymbol{\sigma}(\psi)| = \frac{1}{2}|\psi|^2 \le 2\pi(2g-2) \;\Rightarrow\; |F_A^+| \le \sqrt{2}\pi(2g-2)$$

Because A is translation-invariant in a temporal gauge, the curvature $F_A = F_A^+ + F_A^-$ is a 2-form pulled back from $Y \Rightarrow F_A \wedge F_A = 0$, which implies $|F_A^+| = |F_A^-|$, therefore we have

$$\langle c_1(L), [\Sigma]\rangle = \frac{i}{2\pi}\int_\Sigma \mathbf{F}_A \le \frac{1}{2\pi}(\sup|F_A|)\ \mathrm{Area}(\Sigma) \le 2g-2 \qquad \square$$

Exercise 13.3. *(Blowup formula 13.7) If X is a closed smooth 4-manifold with $b_2^+(X) > 1$ with a nonzero Seiberg–Witten invariant corresponding to a line bundle $L \to X$, then the blownup manifold $\tilde{X} = X\#k\bar{\mathbb{CP}}^2 = X\#\bar{\mathbb{CP}}^2\# .. \#\bar{\mathbb{CP}}^2$ has nonzero Seiberg–Witten invariant corresponding to the line bundle $\tilde{L}$ with $c_1(\tilde{L}) = c_1(L) + [\bar{\mathbb{CP}}^1] + .. + [\bar{\mathbb{CP}}^1]$ [FS3].*

Exercise 13.4. *Prove Theorem 13.35 under the condition $\Sigma.\Sigma = k > 0$ (Hint blow up $X \mapsto X\#k\bar{\mathbb{CP}}^2$ until the condition $\Sigma.\Sigma = 0$ holds, and use Exercise 13.3).*

The proof of the last theorem leads to a solution of the so-called Thom conjecture.

Theorem 13.36. *([KM]) Let $\Sigma \subset \mathbb{CP}^2$ be a smooth closed oriented surface of genus g, representing the homology class $d[\mathbb{CP}^1] \in H_2(\mathbb{CP}^2; \mathbb{Z})$ with $d \ge 3$, then $g \ge (d-1)(d-2)/2$.*

Proof. Let $X = \mathbb{CP}^2\#k\bar{\mathbb{CP}}^2$ with $k = d^2$, and let $H, E_1, \ldots, E_k$ denote the $\mathbb{CP}^1$'s in each factor of X. Now let $\Sigma \subset X$ be the imbedding induced from the first factor, and $\tilde{\Sigma}$ be the surface $\Sigma\#E_1\#\ldots\#E_k$. Let $L \to X$ be the characteristic line bundle corresponding to $3H - (E_1 + \ldots + E_k)$, then by Proposition 13.12 the dimension of the solutions of Seiberg–Witten equations is $d(L) = 0$. Since $b_2^+(X) = 1$, Theorem 13.35 does not apply directly, but one can use the metric dependent $SW^{\pm}_{X,g,\delta}(L)$ (Section 13.6) if we know it is nonzero. Then the proof of Theorem 13.35 gives the adjunction inequality for $\tilde{\Sigma}$, which then implies the required result. By [Hit] $X = \mathbb{CP}^2\#k\bar{\mathbb{CP}}^2$ admits Kähler metric g_+ with positive scalar curvature, and the corresponding Kähler form ω_+ has the property $L.\omega_+ > 0$, so in this chamber the Seiberg–Witten moduli space is empty. Hence by (13.35) it is enough to produce a metric g_- on $X = \mathbb{CP}^2\#k\bar{\mathbb{CP}}^2$ such that $L.\omega_- < 0$. This is done by taking a sequence of metrics g_n on the manifolds X_n with long necks (Figure 13.7), such that the corresponding ω_n are normalized $H \cup \omega_n = 1$ and $||\omega_n||_{L^2} \le 1$.

Then as $n \mapsto \infty$ there is a subsequence converging to a harmonic L^2 form ω on compact subsets of the two disjoint union of manifolds $Z := X_- \cup ([0,\infty)\times Y) \smile ((-\infty,0]\times Y)\cup X_+$ with cylindrical ends (here we are identifying compact subsets of Z and X_R for $R >> 1$),

$$\begin{aligned}[\omega_n].L &= [\omega_n].[\tilde{\Sigma}] - (d-3)[\omega_n].H \\ &= [\omega_n].[\tilde{\Sigma}] - (d-3) \longrightarrow 3-d\end{aligned}$$

The last implication follows from the fact that as $n \mapsto \infty$ the harmonic forms ω_n of X_n limit to a cohomology class of $X_\infty = X_- \cup ([0,\infty)\times Y)$ which lies in the image of the compactly supported cohomology $H^2_c(X_\infty)$ and hence is zero (check $||\omega||^2$ is constant), hence $[\omega_n][\tilde{\Sigma}] \longrightarrow 0$. Hence when $g \geq 3$ we can find a metric g_- satisfying $L.\omega_- < 0$. □

Exercise 13.5. *If X is a closed smooth 4-manifold with $b_2^+(X) > 1$ with nonzero Seiberg–Witten invariant, show that there is no imbedded 2-sphere $S \subset X$ with $[S].[S] \geq 1$.*

Remark 13.37. *Similar to the Seiberg–Witten theory, there is another gauge theory defined for smooth 4-manifolds with Spin^c structure (X, L), called Heegaard–Floer theory. Heegaard–Floer theory was introduced by Ozsvath and Szabo ([OS3], [OS4]). Similar to Remark 13.27, the Heegaard–Floer invariant of (X, L) takes values in so-called Heegaard–Floer homology groups $HF^\circ(Y, s)$ of its boundary $(Y, s) = \partial(X, L)$ (they come in several flavors). In [KLT] it was shown that Heegaard–Floer and monopole–Floer homology groups are isomorphic to each other.*

Remark 13.38. *In Chapters 9 and 10, we constructed exotic copies of some smooth 4-manifolds M, by twisting them along corks (W, f). Our proofs were indirect, in the sense that we started with W, and then built M around it. So, a natural question is, given a decomposition $M^4 = N \smile_\partial W$ as in (10.1), when the cork twisting M along W changes its Seiberg–Witten (or Heegard–Floer) invariant? For this, one has to study the action of map $f : \partial W \to \partial W$ on its Floer homology. In some cases, knowing that the cork-twisted manifold M is exotic, enables us to conclude that the induced map on the Floer homology $f_* : HF^\circ(\partial W) \to HF^\circ(\partial W)$ is nontrivial ([AD], [AKa1], [AKa2]).*

Chapter 14

Some applications

14.1 10/8 theorem

By a theorem of J.H.C.Whitehead, the homotopy type of a simply connected closed smooth 4-manifold is determined by its intersection form,

$$q_X : H_2(X;\mathbb{Z}) \otimes H_2(X;\mathbb{Z}) \longrightarrow \mathbb{Z}$$

Furthermore by C.T.C Wall, q_X determines the h-cobordism class of X [Wa2]. S. Donaldson (cf. [DK]) showed that if q_X is definite then it is diagonalizable (Theorem 13.25):

$$q_X = <1> \oplus <1> \oplus \ldots <1>$$

We call q_X, even if q(a,a) is even for all a, otherwise we call q_X odd. Since integral liftings c of the second Stiefel–Whitney class w_2 of X are characterized by $c.a = a.a$ for all $a \in H_2(X;\mathbb{Z})$, the condition of q_X being even is equivalent to X being spin. From the classification of even unimodular integral quadratic forms (e.g. [MH]) and the Rohlin's Theorem 5.3 it follows that the intersection form of a closed smooth spin manifold is

$$q_X = 2kE_8 \oplus lH \tag{14.1}$$

where E_8 is the 8×8 intersection matrix given by the following diagram:

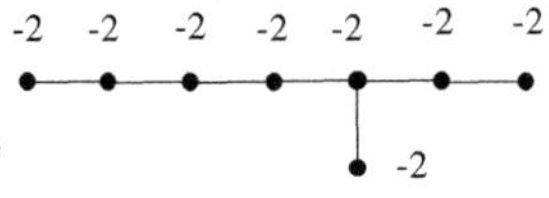

and H is the form $H = \begin{pmatrix} 0 & 1 \\ 1 & 0 \end{pmatrix}$. The intersection form of $S^2 \times S^2$ realizes the form H, and the K3 surface (quadric in $\mathbb{CP}^3$) realizes $2E_8 \oplus 3H$ (Proposition 12.2). Donaldson had shown that if $k = 1$, then $l \geq 3$ ([D2]). Clearly, connected sums of the K3 surfaces realize $2kE_8 \oplus 3kH$. In general it is conjecture that in (14.1) we must necessarily have $l \geq 3k$. This is called the 11/8 conjecture. M. Furuta has shown the following:

Theorem 14.1. *([Fu]) Let X be a smooth oriented closed spin 4-manifold with the intersection form $q_X = 2kE_8 \oplus lH$, then $l \geq 2k + 1$.*

Proof: We will only sketch the proof of $l \geq 2k$. We pick $L \longrightarrow X$ to be the trivial bundle (it is characteristic since X is spin). Notice that the spinor bundles

$$V^{\pm} = P \times_{\rho_{\pm}} \mathbb{C}^2 \longrightarrow X$$

$\rho_{\pm} : x \longmapsto q_{\pm} x$, are quaternionic vector bundles. That is, there is an action $j : V^{\pm} \to V^{\pm}$ defined by $[p, x] \to [p, xj]$, which is clearly well defined. This action commutes with

$$\partial\!\!\!/ : \Gamma(V^+) \longrightarrow \Gamma(V^-)$$

Let A_0 be the trivial connection, and write $\pm A = A_0 \pm i\, a \in \mathcal{A}(L)$. Recall by Section 13.1, the bundle $W^+ = V^+ \otimes L^{-1/2} \to X$ is obtained from the action $x \mapsto qx\lambda^{-1}$,

$$\begin{aligned} \partial\!\!\!/_A(\psi j) &= \sum \rho(e^k)\,[\, \nabla_k + i\, a \,]\,(\psi j) = \sum \rho(e^k)\,[\, \nabla_k(\psi) j + (\psi j)(-ia) \,] \\ &= \sum \rho(e^k)\,[\, \nabla_k(\psi) j + \psi(ia) j \,] = \sum \rho(e^k)\,[\, \nabla_k - ia \,](\psi) j = \partial\!\!\!/_{-A}(\psi) j \end{aligned}$$

$\mathbb{Z}_4$ action $(A, \psi) \longmapsto (-A, \psi j)$ on $\Omega^1(X) \times \Gamma(W^+)$ preserves the compact set

$$\mathcal{M}_0 = \tilde{\mathcal{M}} \cap ker(d^*) \oplus \Gamma(W^+)$$

where $\tilde{\mathcal{M}} = \{(a, \psi) \in \Omega^1(X) \oplus \Gamma(W^+) \mid \partial\!\!\!/_A(\psi) = 0 \,,\; F_A^+ = \sigma(\psi)\}$. For example, from the local description of σ in Section 13.2 we can check that

$$\sigma(\psi j) = \sigma(z + jw) j = \sigma(-\bar{w} + j\bar{z}) = -\sigma(\psi) = -\,F_A^+ = F_{-A}^+$$

This $\mathbb{Z}_4$ in fact extends to an action of the subgroup $G = Pin(2)$ of $SU(2)$ which is generated by $< S^1 \,,\; j >$, where S^1 acts trivially on Ω^* and by complex multiplication on $\Gamma(W^+)$, and j acts by -1 on Ω^*, and by quaternionic multiplication on $\Gamma(W^+)$. In particular we get a G-equivariant map

$$\varphi : \mathbb{V} = ker(d^*) \oplus \Gamma(V^+) \longrightarrow \Omega^2_+ \oplus \Gamma(V^-) = \mathbb{W}$$

$$L = \begin{pmatrix} d^+ & 0 \\ 0 & \partial\!\!\!/ \end{pmatrix} \quad \text{and} \quad \theta(a,\psi) = (\,\sigma(\psi)\,,\, a\psi\,)$$

with $\varphi^{-1}(0) = \mathcal{M}_0$ and $\varphi(v) = L(v) + \theta(v)$ with L linear Fredholm and θ quadratic. We apply the "usual" Kuranishi technique (cf. [La]) to obtain a finite-dimensional local model $V \longmapsto W$ for φ as follows:

We let $\mathbb{V} = \oplus V_\lambda$ and $\mathbb{W} = \oplus W_\lambda$, where V_λ and W_λ are λ eigenspaces of $L^*L : \mathbb{V} \to \mathbb{V}$ and $LL^* : \mathbb{W} \to \mathbb{W}$ respectively. Since L^*L is a multiplication by λ on V_λ, for $\lambda > 0$ we have isomorphisms $L : V_\lambda \xrightarrow{\cong} W_\lambda$. Now pick $\Lambda > 0$ and consider projections:

$$\oplus_{\lambda \le \Lambda} W_\lambda \overset{p_\Lambda}{\longleftarrow} \mathbb{W} \overset{1-p_\Lambda}{\longrightarrow} \oplus_{\lambda > \Lambda} W_\lambda$$

Consider the local diffeomorphism (recall implicit function thm) $f_\Lambda : \mathbb{V} \longrightarrow \mathbb{V}$ given by:

$$u = f_\Lambda(v) = v + L^{-1}(1 - p_\Lambda)\theta(v) \iff L(u) = L(v) + (1 - p_\Lambda)\theta(v)$$

The condition $\varphi(v) = 0$ is equivalent to $p_\Lambda\, \varphi(v) = 0$ and $(1 - p_\Lambda)\, \varphi(v) = 0$, but

$$(1 - p_\Lambda)\, \varphi(v) = 0 \iff (1 - p_\Lambda)\, L(v) + (1 - p_\Lambda)\, \theta(v) = 0 \iff$$
$$(1 - p_\Lambda)\, L(v) + L(u) - L(v) = 0 \iff L(u) = p_\Lambda\, L(v) \iff u \in \oplus_{\lambda \le \Lambda} V_\lambda$$

Hence, $\varphi(v) = 0 \iff p_\Lambda\, \varphi(v) = 0$ and $u \in \oplus_{\lambda \le \Lambda} V_\lambda$, let

$$\varphi_\Lambda : V = \oplus_{\lambda \le \Lambda} V_\lambda \longrightarrow W = \oplus_{\lambda \le \Lambda} W_\lambda \quad \text{where} \quad \varphi_\Lambda(u) = p_\Lambda\, \varphi\, f_\Lambda^{-1}(u)$$

Hence in the local diffeomorphism $f_\Lambda : \mathcal{O} \xrightarrow{\approx} \mathcal{O} \subset \mathbb{V}$ takes the piece of the compact set $f_\Lambda(\mathcal{O} \cap \mathcal{M}_0)$ into the finite-dimensional subspace $V \subset \mathbb{V}$, where $\mathcal{O}$ is a neighborhood of $(0,0)$. As a side fact, note that near $(0,0)$ we have

$$\mathcal{M}(L) \approx \mathcal{M}_0(L)/S^1$$

We claim that for $\Lambda \gg 1$, the local diffeomorphism $f_\Lambda : \mathcal{O} \cap V \xrightarrow{\approx} \mathcal{O} \cap V$ extends to a ball $B_R \subset V$ of large radius R containing the compact set $\mathcal{M}_0(L)$, i.e. we can make the zero set $\varphi_\Lambda^{-1}(0)$ a compact set. We see this by applying the Banach contraction principle. For example, for a given $u \in B_R$, showing that there is $v \in V$ with $f_\Lambda(v) = u$ is equivalent to showing that the map $T_u(v) = u - L^{-1}(1 - p_\Lambda)\theta(v)$ has a fixed point. Since $L^{-1}(1 - p_\Lambda)$ has eigenvalues $1/\lambda$ on each W_λ in an appropriate Sobolev norm we can write

$$||T_u(v_1) - T_u(v_2)|| \le \frac{C}{\Lambda}||\theta(v_1) - \theta(v_2)|| \le \frac{C}{\Lambda}||v_1 - v_2||$$

Vector subspaces V_λ and W_λ are either quaternionic or real depending on whether they are subspaces of $\Gamma(V^{\pm})$ or $\Omega^*(X)$. For a generic metric we can make the cokernel of $\partial\!\!\!/$ zero, hence the dimension of the kernel (as a complex vector space) is $ind(\partial\!\!\!/) = -\sigma/8 = 2k$, and since $H^1(X) = 0$ the dimension of the cokernel of d^+ (as a real vector space) is $b^+ = l$. Hence φ_Λ gives a G-equivariant map

$$\varphi : \mathbb{H}^{k+y} \oplus \mathbb{R}^x \longrightarrow \mathbb{H}^y \oplus \mathbb{R}^{l+x}$$

with compact zero set. From this Furuta shows that $l \geq 2l + 1$. Here we show an easier argument of D. Freed which gives a slightly weaker result of $l \geq 2k$. Let E_0 and E_1 be the complexifications of the domain and the range of φ; consider E_0 and E_1 as bundles over a point x_0 with projections $\pi_i : E_i \to x_0$, and with 0-sections $s_i : x_0 \to E_i$, $i = 0, 1$. Recall $K_G(x_0) = R(G)$, and we have Bott isomorphisms $\beta(\rho) = \pi_i^*(\rho)\, \lambda_{E_i}$, for $i = 0, 1$, where λ_{E_i} are the Bott classes. By compactness we get an induced map φ^* on the corresponding equivariant K-groups (e.g. [Bt], [At]):

$$\begin{array}{ccc} K_G(B(E_1), S(E_1)) & \xrightarrow{\varphi^*} & K_G(B(E_0), S(E_0)) \\ \approx \uparrow \beta & & \approx \uparrow \beta \\ R(G) & & R(G) \end{array}$$

Consider $s_i^*(\lambda_{E_i}) = \sum (-1)^k \Lambda^k(E_i) = \Lambda_{-1}(E_i) \in R(G)$, then by some ρ we have

$$\Lambda_{-1}(E_1) = s_1^*(\lambda_{E_1}) = s_0^* \varphi^*(\lambda_{E_1}) = s_0^*(\pi_0^*(\rho)\, \lambda_{E_0}) = \rho\, \Lambda_{-1}(E_0)$$

So in particular $tr_j(\Lambda_{-1}(E_0))$ divides $tr_j(\Lambda_{-1}(E_1))$. By recalling $j : E_i \to E_i$

$$tr_j(\Lambda_{-1}(E_i)) = det(I - j) \quad \text{for } i = 0, 1$$

Since $(z, w)j = (z + jw)j = -\bar{w} + j\bar{z} = (-\bar{w}, \bar{z})$ j acts on $\mathbb{H} \otimes \mathbb{C}$ by matrix

$$A = \begin{pmatrix} 0 & 0 & 1 & 0 \\ 0 & 0 & 0 & -1 \\ -1 & 0 & 0 & 0 \\ 0 & 1 & 0 & 0 \end{pmatrix}$$

so $det(I - A) = 4$, and j acts on $\mathbb{R} \otimes \mathbb{C}$ by $j(x) = -x$ so $det(I - (-I)) = 2$. Hence $4^{k+y}\, 2^x$ divides $4^y\, 2^{l+x}$ which implies $l \geq 2k$. □

14.2 Cappell–Shaneson homotopy spheres

In [CS] Cappell and Shaneson constructed a family of homotopy spheres, Σ_m, $m = 0, 1, \ldots$, some of them covering exotic $\mathbb{RP}^4$'s. They asked whether they are diffeomorphic to S^4. Σ_m is obtained by first taking the mapping torus $M(f)$ of the punctured 3-torus T_0^3 with a diffeomorphism $f : T_0^3 \to T_0^3$ induced by an integral matrix $C_m \in SL(3, \mathbb{Z})$,

$$C_m = \begin{pmatrix} 0 & 1 & 0 \\ 0 & 1 & 1 \\ 0 & 1 & m+1 \end{pmatrix} \tag{14.2}$$

and then gluing it to a $S^2 \times B^2$ with the nontrivial diffeomorphism of $S^2 \times S^1$ along their common boundaries. By [ARu] C_0 is conjugate to the matrix B of 4.2. In [AK1] it was shown that Σ_0 is obtained from S^4 by Gluck twisting (Section 6.1). Already in Section 4.5 a handlebody of the mapping torus $M(f)$ has been constructed. So Figure 14.1 describes Σ_0, which is just the Figure 4.22 drawn in circle-with-dot notation. The extra 1-framed 2-handle linking 1-handle A describes the $S^2 \times B^2$ part of $\Sigma_0 = M(f) \smile S^2 \times B^2$.

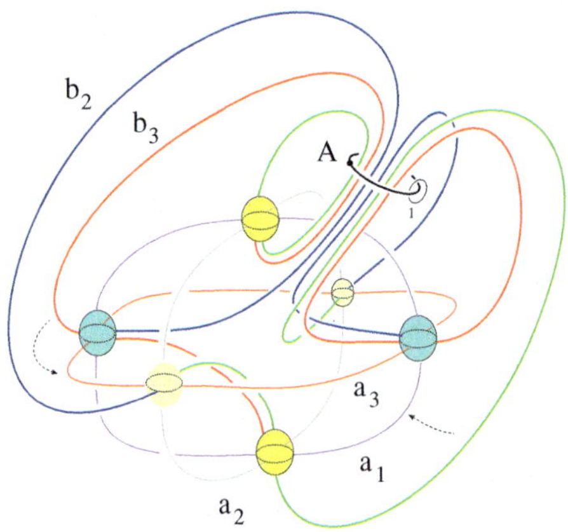

Figure 14.1: Σ_0

It turns out that all Σ_m are diffeomorphic to S^4 ($m = 0$ case in [G3], and the other cases in [A5]). Both proofs are based on the basic Proposition 14.2, proved by techniques developed earlier. Let us go through this. First let $M = \Sigma_0 - B^4$ be the punctured homotopy sphere Σ_0, clearly proving $M = B^4$ is equivalent to proving $\Sigma_0 = S^4$.

Proposition 14.2. *([AK4]) Figure 14.2, which consists of two* 1*-handles and two* 2*-handles, describes a handlebody of the homotopy ball* M. *Furthermore,* M *smoothly imbeds into* S^4. *So by the Schoenflies theorem, [Bn], [Maz], it is homeomorphic to* B^4.

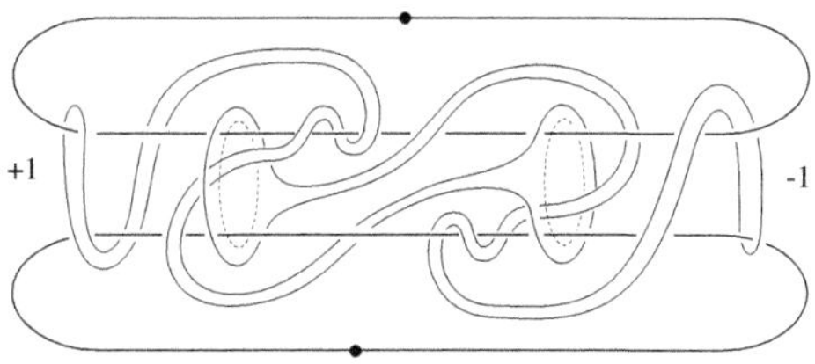

Figure 14.2: M

Proof. First notice that two of the three 2-handles a_1, a_2, a_3 (coming from T_0^3) of the Figure 14.1 are canceled by 3-handles, i.e. by sliding them over each other we can make them parallel, so after further sliding we get two unknotted zero-framed circles away from the main figure. This is because, if we erase 1-handle A from the figure, what is left collapses and becomes B^4; if we follow the 1-handle circle of A during this isotopy we get Figure 14.3. Hence if we wish we can erase a_2 and a_3 from Figure 14.1.

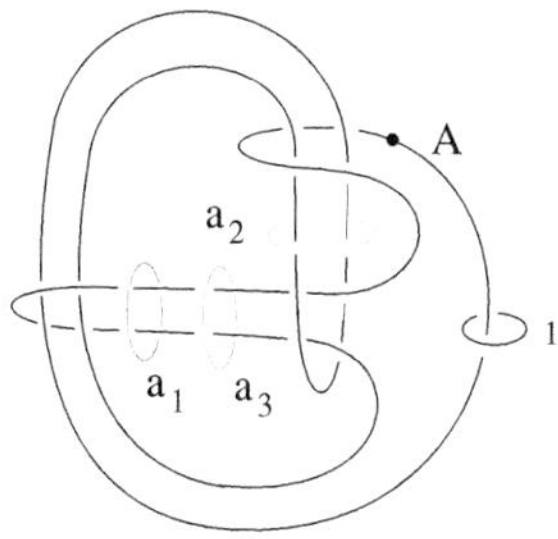

Figure 14.3

Next, we redraw Figure 14.1 in all circle-with-dot notation. For this we ned to identify the framings of the 2-handles. It is easy to see that 2-handle a_1 has zero framing, but we must be extra careful identifying the framings of handles b_1, b_2, b_3 in this transition. For this, we draw the figure without handle a_1, as in Figure 14.4, and carefully check the writhe of each of the b_1, b_2, b_3 circles, and see that the framings are $b_1 = -1, b_2 = 2, b_3 = -1$.

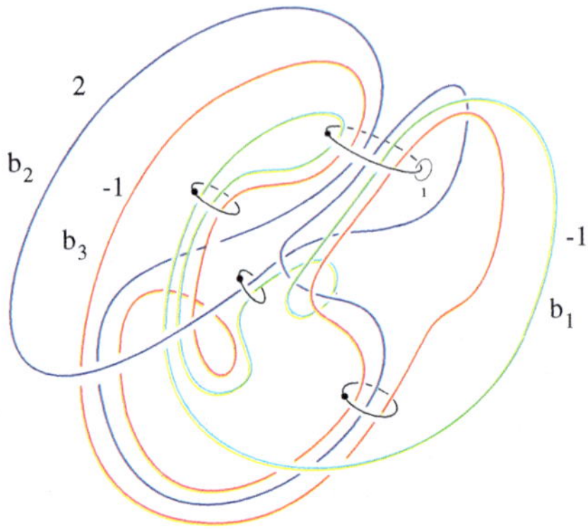

Figure 14.4

Having identified the framings, we go back to Figure 14.1 and obtain Figure 14.5 by first sliding b_1 over a_1, and b_3 over a_3 (thereby changing their framings) and then drawing the resulting figure in the circle-with-dot notation.

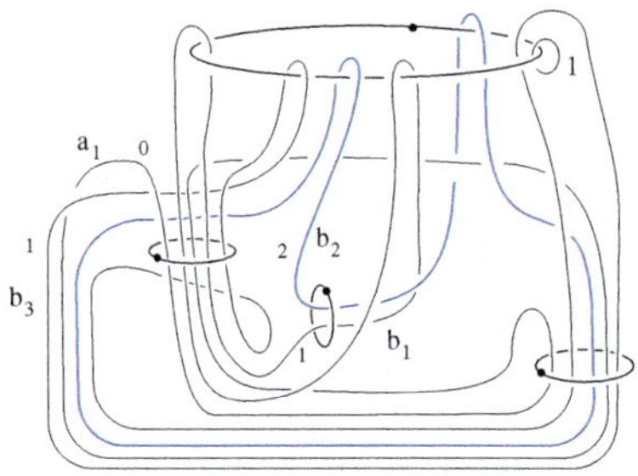

Figure 14.5

Then by canceling the 2-handle b_2 with the middle 1-handle in Figure 14.5 we get Figure 14.6, and in that figure by canceling the 2-handle b_3 with the left 1-handle we get Figure 14.7, and then by an isotopy we get Figure 14.8. Finally, in Figure 14.8, by canceling 2-handle b_1 with the 1-handle next to it we obtain Figure 14.9. Then by further isotopies we arrive at the first picture of Figure 14.10. The dotted lines in Figures 14.9 and 14.10 signify the "ribbon move" of the ribbon 1-handle.

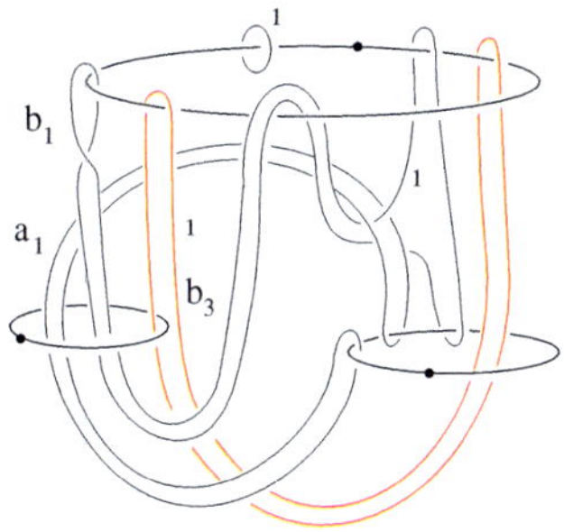

Figure 14.6

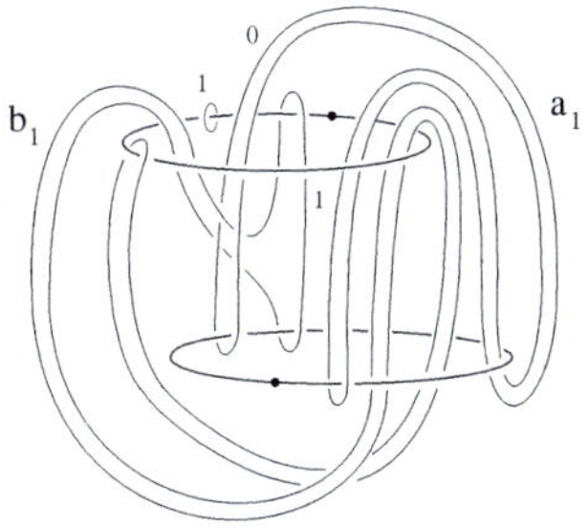

Figure 14.7

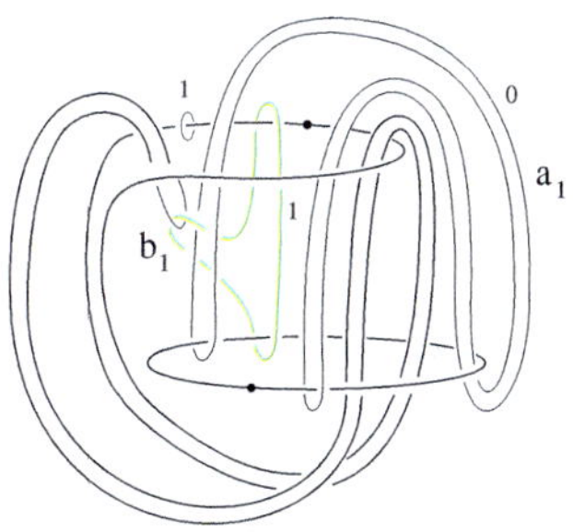

Figure 14.8

Similarly to Figure 1.20 (and Exercises 1.10), Figure 14.10 describes an imbedding $S^2 \hookrightarrow S^4$ consisting of the union of two ribbon disks (one is in the upper hemisphere the other in the lower hemisphere) meeting along the common boundary ribbon knot K. The small 1-framed linking circle in the figures shows that this 2-sphere is being Gluck twisted (Section 6.1). Now we need to Gluck twist S^4 along this imbedding $S^2 \hookrightarrow S^4$.

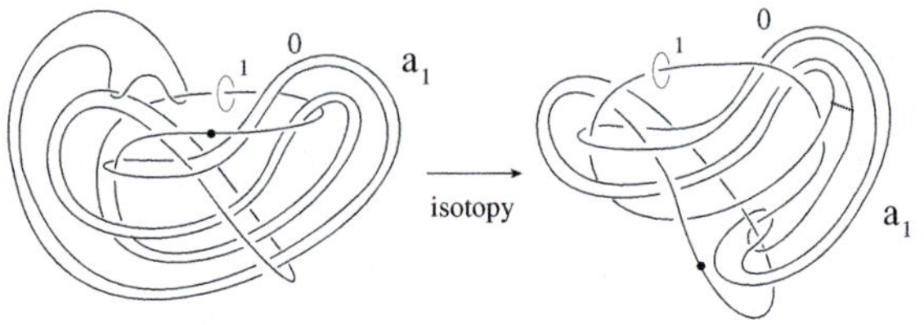

Figure 14.9

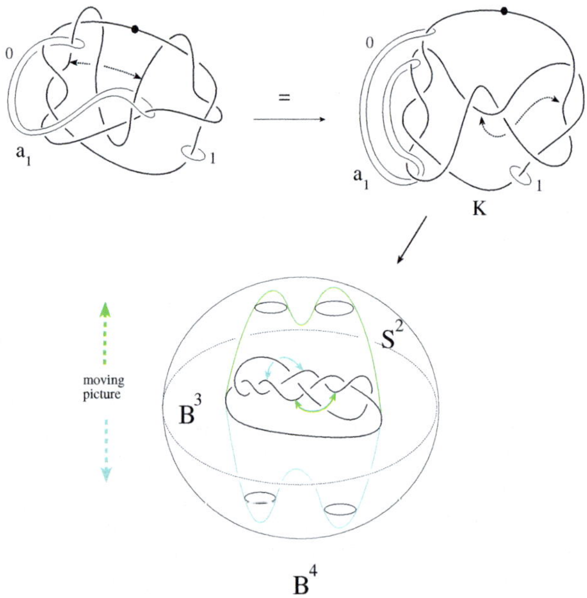

Figure 14.10: $S^2 \hookrightarrow S^4$

The first two pictures of Figure 14.11 show the Gluck twisting of S^4 along this 2-sphere. In the third picture we introduce a canceling 2/3 handle pair, for future convenience (check that the 2-handle of this pair is attached to the unknot).

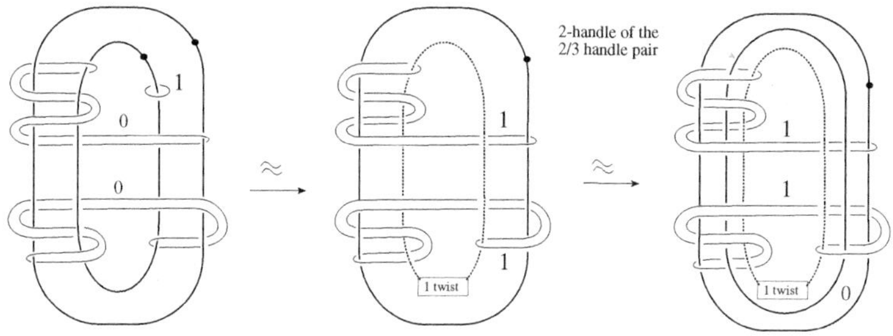

Figure 14.11

By sliding over the new 2-handle, we can unlink the clasp of one of the 1-framed 2-handles from the rest (this changes its framing to -1), as shown in the first picture of Figure 14.12. Now clearly the -1 framed 2-handle is free to cancel the 1-handle. So the first picture of Figure 14.12 can be viewed as a handlebody consisting of a pair of 2-handles and a pair of 3-handles. Now we immediately turn this handlebody upside down (by using the method of Section 3.1) and obtain a handlebody of M consisting of a pair of 1-handles and a pair of 2-handles.

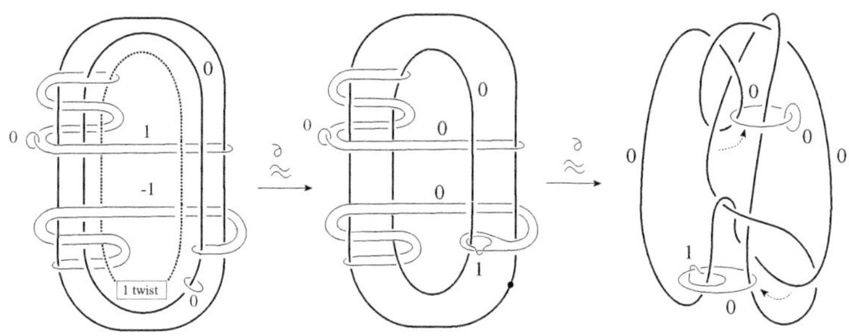

Figure 14.12

For this, we take the dual circles of the 2-handles (as indicated in the first picture of Figure 14.12) and find any diffeomorphism from the boundary to the boundary of $(S^1 \times B^3)\#(S^1 \times B^3)$ and attach 2-handles to $(S^1 \times B^3)\#(S^1 \times B^3)$ along the images of these dual circles. The second and third pictures of Figure 14.12 are first obtained by blowing up, sliding, blowing down operations, and then by a handle slide. Using similar steps we obtain the first second and third pictures of Figure 14.13. Then finally, in the last picture, we blow down ± 1 framed circles to arrive at Figure 14.2, which was our goal. The proof of the last part of this proposition is in the following exercise. □

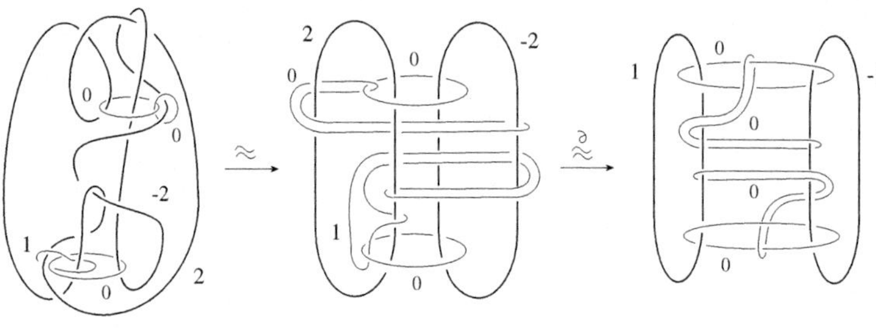

Figure 14.13

Exercise 14.1. *(a) Show that M imbeds into S^4. (Hint attached two 2-handles to M, as indicated in Figure14.14, to show that the handlebody collapses to become S^4.)*
(b) Show that as a 2-complex M has the fundamental group presentation
$\pi_1(M) = \{x, y \mid xyx = yxy, x^4 = y^3\}$ *(a presentation of the trivial group).*

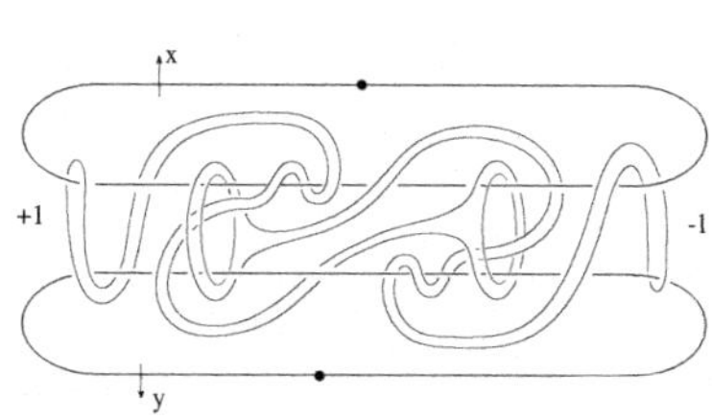

Figure 14.14: M

Remark 14.3. *By using Proposition 14.2 one can prove that M is in fact diffeomorphic to B^4 not just homeomorphic to it ([G3]). In Section 14.3 a recent direct proof of this ([A21]) will be explained in a sequence of exercises.*

In the general case of $m > 0$, imitating the above steps gives Figure 14.15, which is the handlebody of $M_m = \Sigma_m - B^4$, where $\Sigma_m = M(f) \cup S^2 \times B^2$ and f is the diffeomorphism corresponding to the matrix C_m (see [G1]).

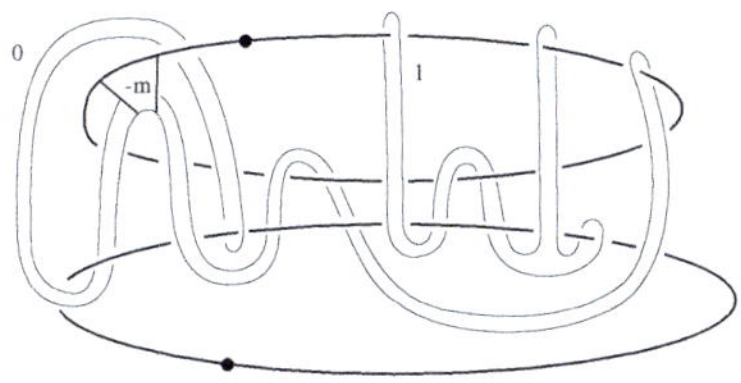

Figure 14.15: $M_m = \Sigma_m - B^4$

Theorem 14.4. *When $m > 0$, Σ_m is diffeomorphic to S^4 ([A5])*

Proof. We first find a specific boundary diffeomorphism, $\partial M_m \approx S^3$ (Figure 14.16).

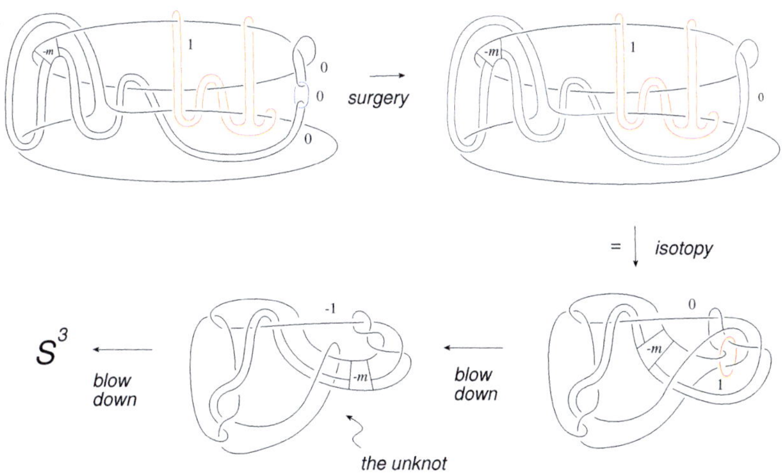

Figure 14.16

This diffeomorphism above shows that the α and the β circles on the boundary of Figure 14.17 are isotopic to each other, in fact each is a 1-framed unknot in S^3.

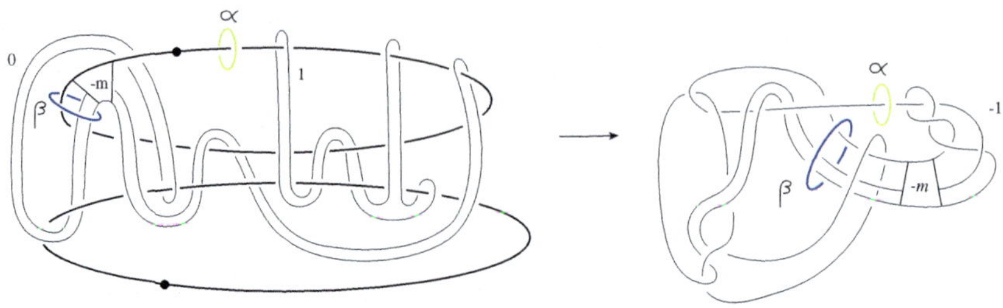

Figure 14.17

Hence by attaching a -1 framed 2-handle to either α or β gives $S^1 \times S^2$, which we can cancel immediately with a 3-handle, i.e. we get diffeomorphisms of the handlebodies (the second and the third handlebodies have 3-handles):

$$\Sigma_m \approx \Sigma_m + \alpha^{-1} \approx \Sigma_m + \beta^{-1}$$

Now observe that by attaching a -1 framed 2-handle to Σ_m along β, and then by sliding the other 2-handle going through it, as shown in Figure 14.18, we get Σ_{m-1} with a 2-handle attached to α with -1 framing, so we have

$$\Sigma_m + \beta^{-1} \approx \Sigma_{m-1} + \alpha^{-1}$$

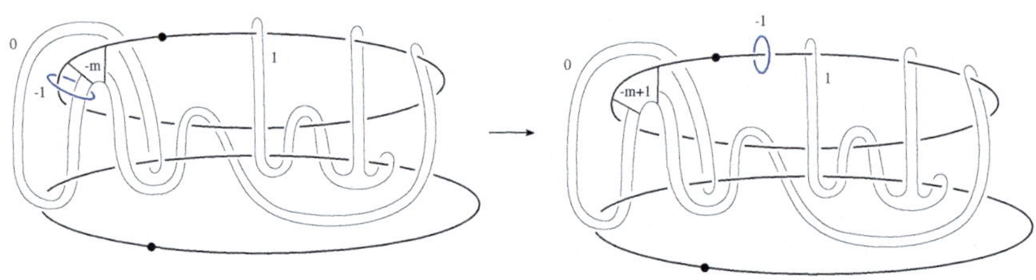

Figure 14.18: $\Sigma_m + \beta^{-1} \approx \Sigma_{m-1} + \alpha^{-1}$

Hence we have diffeomorphisms $\Sigma_m \approx \Sigma_{m-1} \approx \ldots \approx \Sigma_0 \approx S^4$. □

14.3 Flexible contractible 4-manifolds

A flexible contractible manifold W is a smooth contractible 4-manifold which has only one 0- handle and 1- and 2-handles, such that the 2-handles are represented by 0-framed unknots (i.e. erasing circles with dots results in a 0-framed unlink). We call the manifold W^* (Figure 14.19) obtained by zero and dot exchanges of the handles of W the *twin* of the handlebody W (i.e. applying the operation of $S^2 \times B^2 \leftrightarrow B^3 \times S^1$ to the handles of W). It is clear from the description that, this gives a decomposition of the 4-sphere, $S^4 = W \cup -W^*$ (Figure 14.20), which is obtained by carving the upper and lower hemispheres of the standard handlebody of S^4, from its equator (similar to the process described in Section 8.4). Note that if a smooth homotopy ball W with $\partial W = S^3$ admits a handlebody structure, which is flexible with twin $W^* = B^4$, then this implies that $W = B^4$. Here we will use this to verify Remark 14.3 ([A21]).

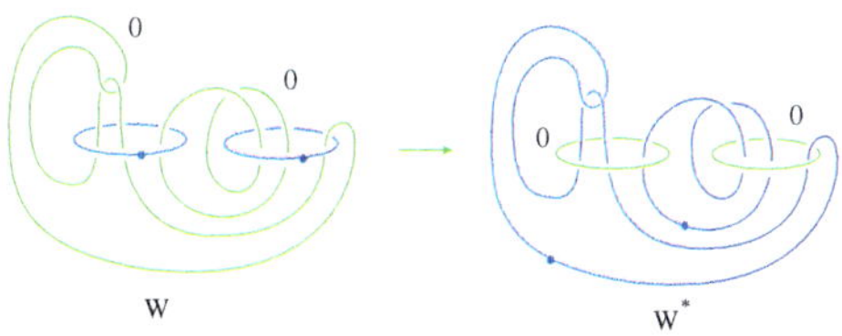

Figure 14.19: $W \to W^*$

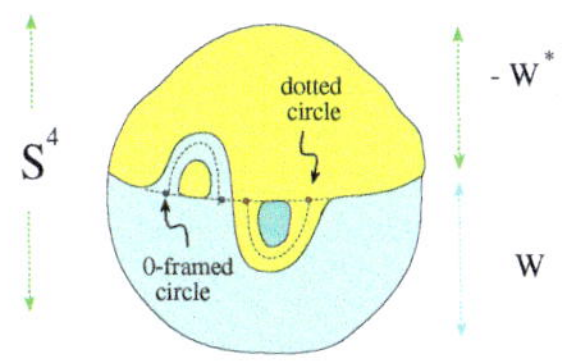

Figure 14.20: $S^4 = W \cup -W^*$

Exercise 14.2. *Show that applying the diffeomorphism of Section 2.3 to Figure 14.14 gives Figure 14.21, which is a flexible handlebody description of M (where ±1 marked dotted circle means every strand that goes through it gets twisted by ±1 full twist).*

Figure 14.21: M

Exercise 14.3. *Show that Figure 14.22 is the twin of Figure 14.21, and that (a) applying the indicated handle slides to Figure 14.22 (along the small arrows) gives Figure 14.23; (b) the handlebody structure of Figure 14.23 gives the fundamental group presentation* $\pi_1(M) = \{x, y \mid xyx = yxy, x^3 = y^2\}$ *(another presentation of the trivial group).*

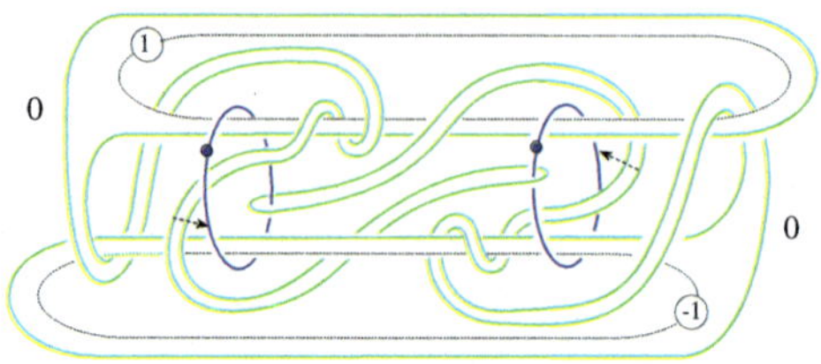

Figure 14.22: M^*

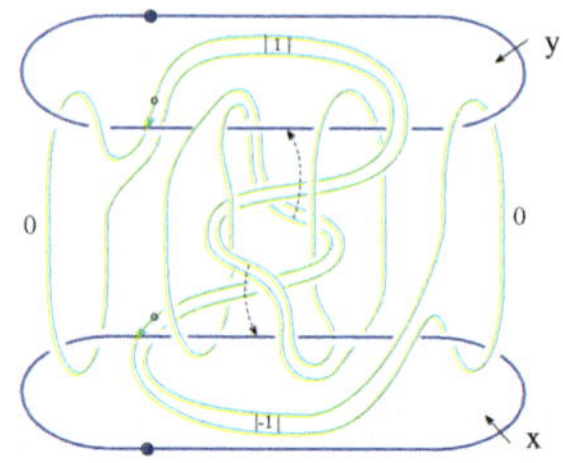

Figure 14.23: M^*

Exercise 14.4. *Show that sliding* 2*-handles of Figure 14.23 over its* 1*-handles (as indicated by the arrows of the figure) gives Figure 14.24. Show that by applying the diffeomorphism of Section 2.3 to Figure 14.24 we can get Figure 14.25; and after the indicated handle slides we get Figure 14.26.*

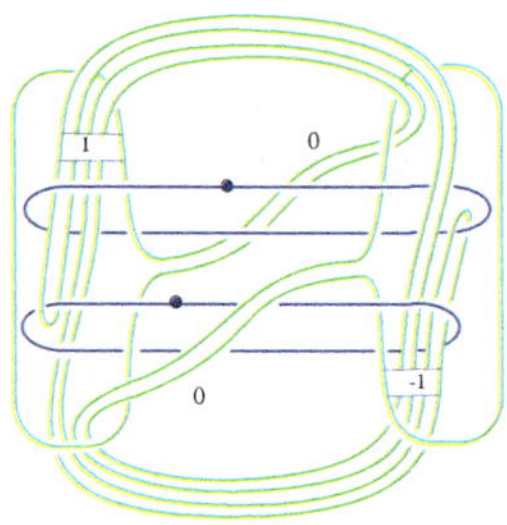

Figure 14.24: M^*

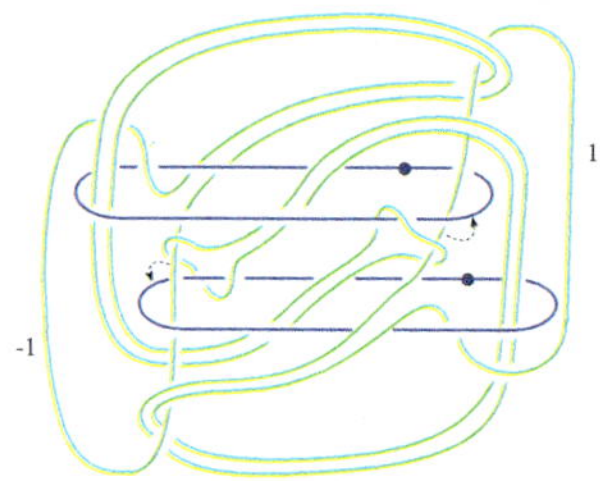

Figure 14.25: M^*

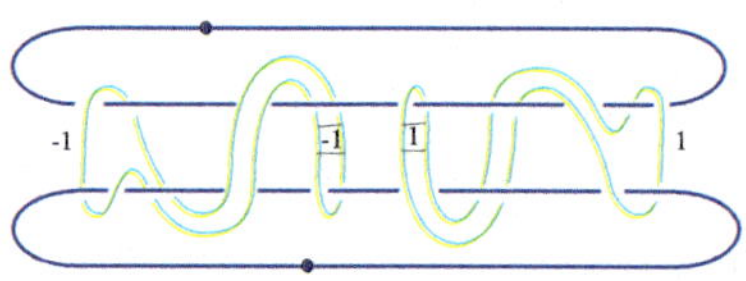

Figure 14.26: M^*

Exercise 14.5. *First notice that Figure 14.27 is an isotopic copy of Figure 14.26, then show that after the indicated handle slides it turns into Figure 14.28, then Figure 14.29.*

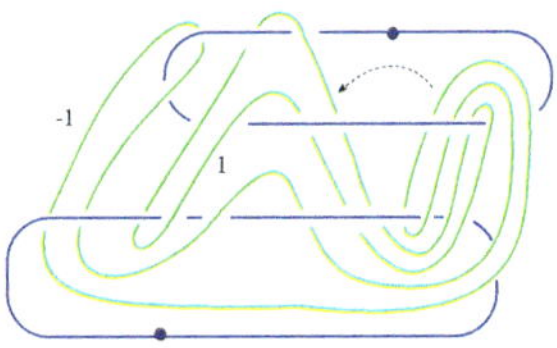

Figure 14.27: M^*

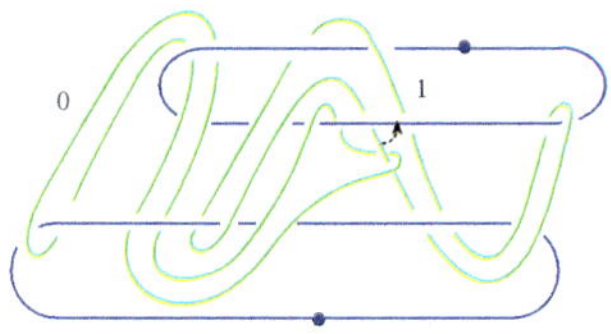

Figure 14.28: M^*

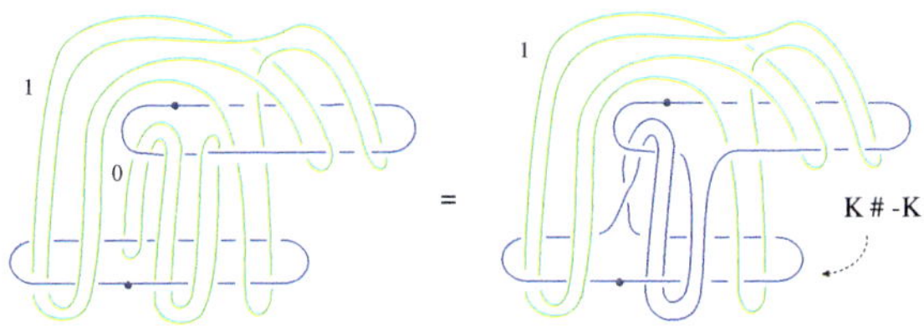

Figure 14.29: M^*

Exercise 14.6. *The first picture of Figure 14.30 is a shorthand for Figure 14.29 (the dotted line indicates the ribbon move, which gives the ribbon* 1*-handle of Figure 14.29). Show that the indicated handle slide applied to Figure 14.30 gives a handlebody of* M^* *which we will call* C_2*, since it is the special case of the handlebody* C_n *of Figure 14.31.*

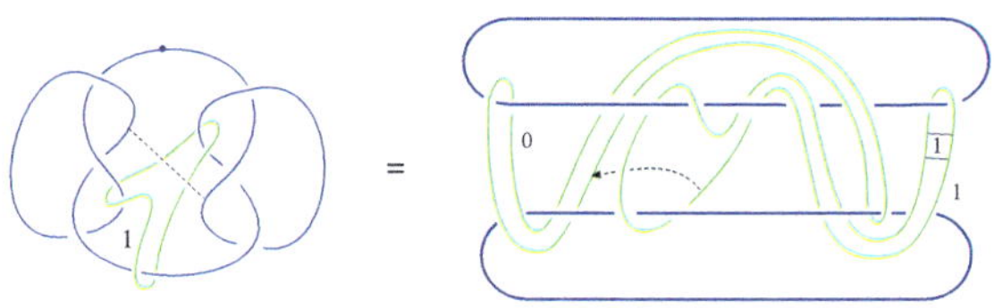

Figure 14.30: M^*

Exercise 14.7. *First show that the two picures of Figure 14.31 are related to each other by an isotopy (more precisely by sliding the* 1*-framed* 2*-handle over the* 1*-handles). Then by induction prove that* $C_n \approx C_{n-1} \approx .. \approx C_0 \approx B^4$*. (Hint introduce a canceling* 2/3*- handle pair, along* γ*, to Figure 14.31, then perform the handle slides indicated in Figure 14.32.)*

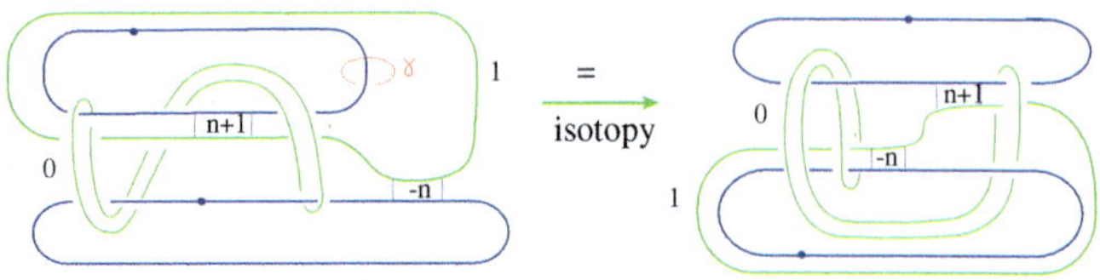

Figure 14.31: C_n

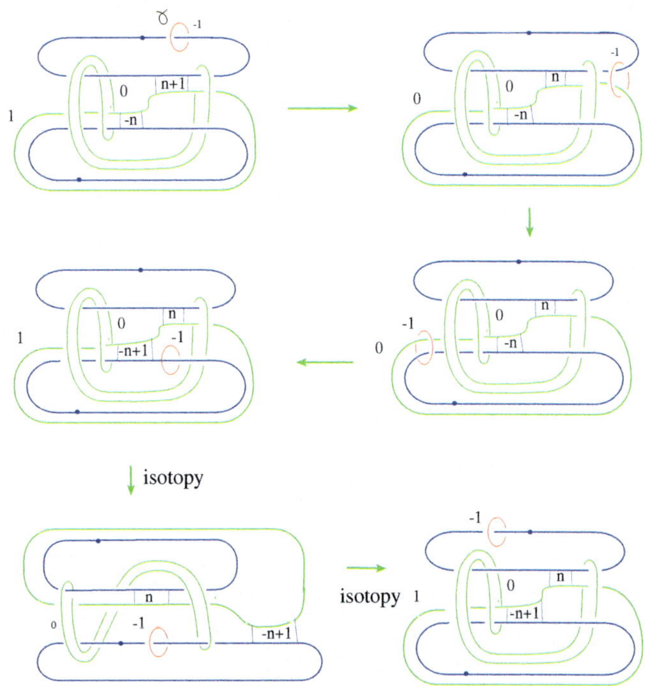

Figure 14.32: Proving $C_n \approx C_{n-1}$

14.4 Some small closed exotic manifolds

Here by techniques developed in earlier chapters, we will construct handlebodies of some closed symplectic 4-manifolds ([A9], [A10]) built by A. Akhmedov and D. Park in [AP1], [AP2]. They are closed smooth 4-manifolds M and N, which are exotic copies of the standard manifolds $\mathbb{CP}^2\#3\bar{\mathbb{CP}}^2$ and $\mathbb{CP}^2\#2\bar{\mathbb{CP}}^2$, respectively. They are symplectic since they are obtained by a symplectic gluing process from basic symplectic pieces, and they are exotic because their Kodaira dimensions (an invariant defined for symplectic manifolds in [L]) are different from that of the standard manifolds [HL], [LZ]. Alternative ways of building exotic $\mathbb{CP}^2\#3\bar{\mathbb{CP}}^2$'s have been given in [FS4] and [BK]. At the end we will discuss the related *reverse engineering technique* of R. Fintushel and R. Stern [FS5].

14.4.1 An exotic $\mathbb{CP}^2\#3\bar{\mathbb{CP}}^2$

M is obtained from two codimension zero pieces, glued along their common bondaries:

$$M = \tilde{E}_0 \smile_\partial \tilde{E}_2$$

To construct the pieces, we start with the product of a genus 2 surface and the torus $E = \Sigma_2 \times T^2$, and let $E_0 = \Sigma_2 \times T^2$ be the punctured E. Figure 4.10 shows E_0 and the adiffeomorphism $f : \partial E_0 \xrightarrow{\approx} \Sigma_2 \times S^1$. As explained in Section 3.2, to describe the diffeomorphism f we are indicating where the "basic" arcs of the figure are thrown by f. Now let $< a_1, b_1, a_2, b_2 >$ and $< C, D >$ be the standard circles generating the first homology of Σ_2 and T^2 respectively, i.e. the cores of the 1-handles (as indicated at the top of Figure 14.33). Let $\tilde{E}_0$ be the manifold obtained from E_0 by doing 1-log transforms to the four subtori $(a_1 \times C, a_1)$, $(b_1 \times C, b_1)$, $(a_2 \times C, C)$ and $(a_2 \times D, D)$ (see Section 6.3).

Lemma 14.5. *The handlebody of $\tilde{E}_0$ and the diffeomorphism $f : \partial\tilde{E}_0 \xrightarrow{\approx} \Sigma_2 \times S^1$ is as shown in Figure 14.33.*

Proof. By performing the indicated handle slide to E_0 (indicated by dotted arrow) in Figure 14.34, we obtain a second equivalent picture of E_0. By performing 1-log transforms to Figure 14.34 along $(a_1{\times}C, a_1)$, $(b_1{\times}C, b_1)$ (as described in Section 6.3) we obtain the first picture of Figure 14.35, and then by handle slides we obtain the second picture. First by an isotopy then a handle slide to Figure 14.35 we obtain the first and second pictures in Figure 14.36. By a further isotopy we obtain the first picture of Figure 14.37, and then by 1-log transforms to $(a_2 \times C, C)$, $(a_2 \times D, D)$ we obtain the second picture in Figure 14.37. By introducing canceling handle pairs we express this last picture by a simpler looking first picture of Figure 14.38. Then by indicated handle slides we obtain the second picture of Figure 14.38, which is $\tilde{E}_0$ as drawn in Figure 14.33. □

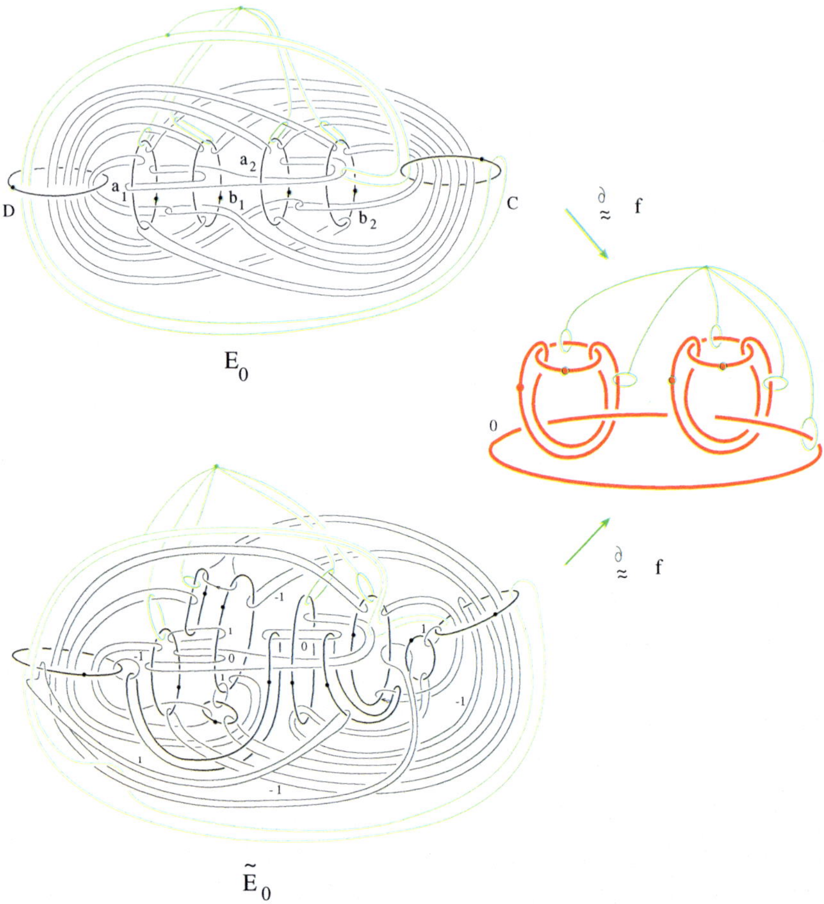

Figure 14.33: $f : \partial E_0 \xrightarrow{\approx} \Sigma_2 \times S^1$ and $f : \partial \tilde{E}_0 \xrightarrow{\approx} \Sigma_2 \times S^1$

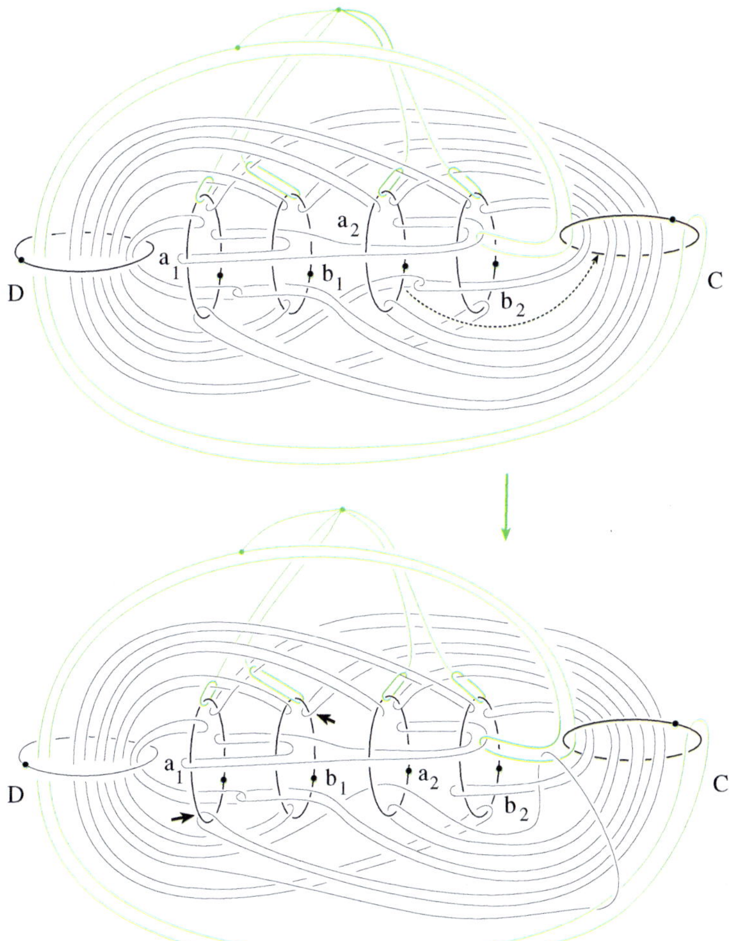

Figure 14.34: Handle slide (the pair of thick arrows in the second picture indicate where we will perform log transforms next)

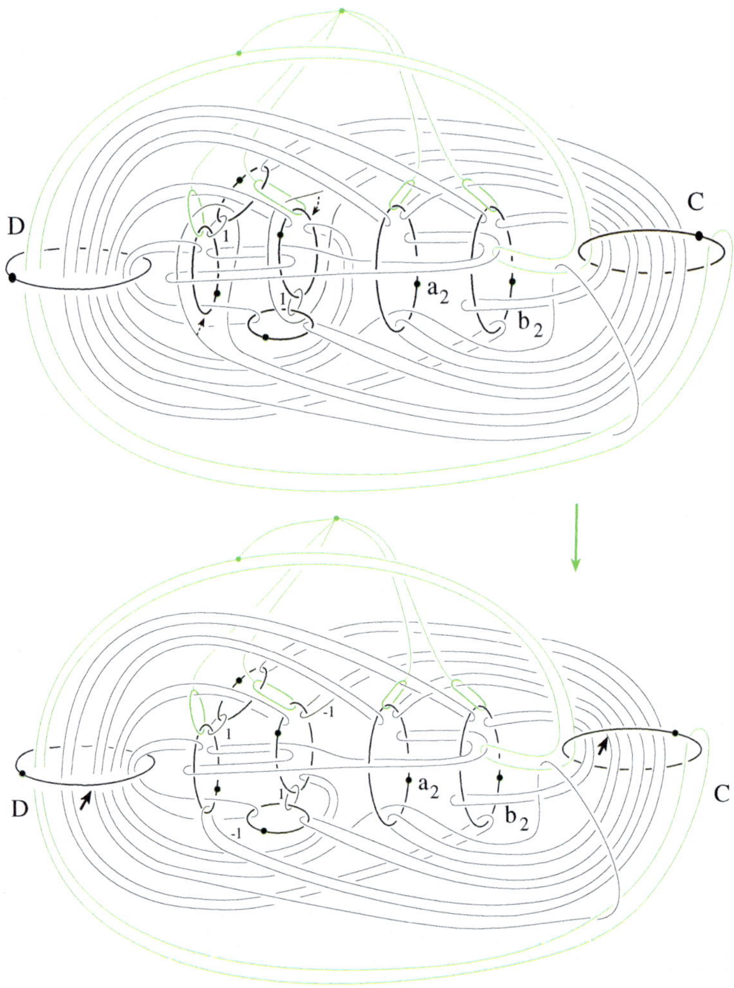

Figure 14.35: First we perform log transforms along $(a_1 \times C, a_1), (b_1 \times C, b_1)$, then handle slides (thick arrows indicate where we will perform log transforms next)

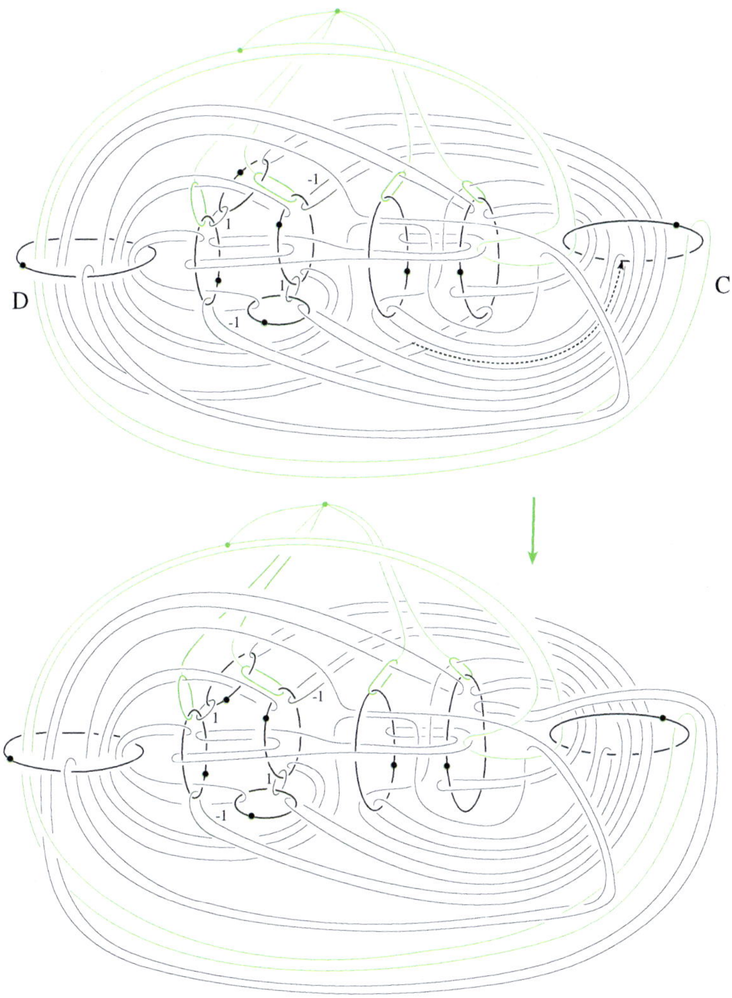

Figure 14.36: Isotopy and handle slides

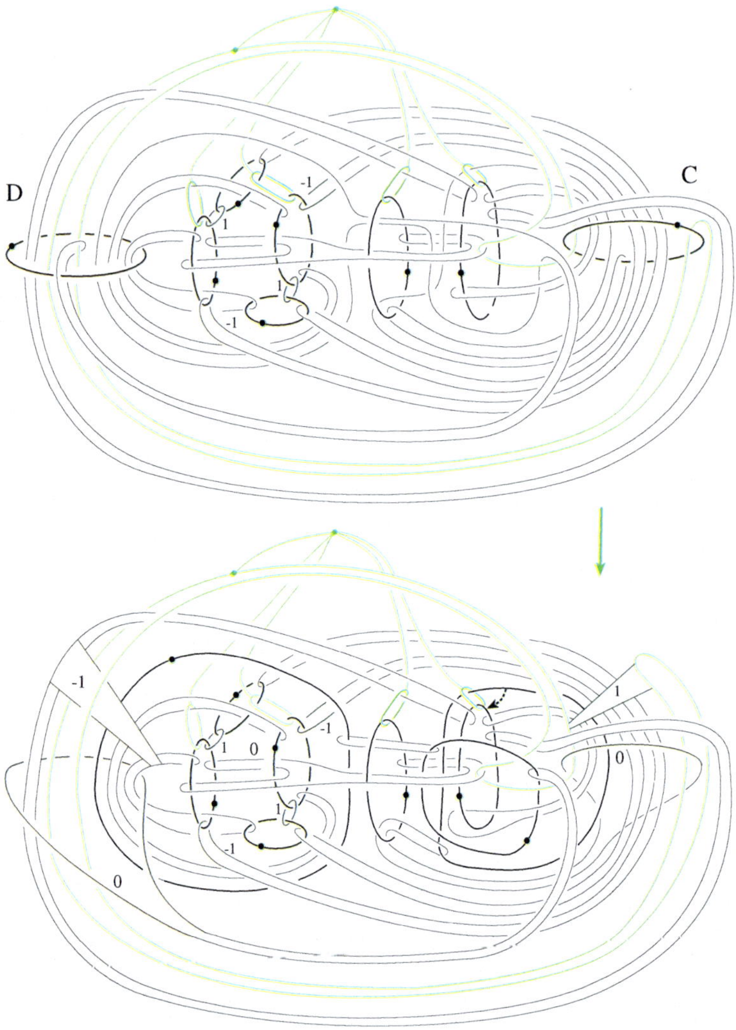

Figure 14.37: Log transforms along $(a_2 \times C, C), (a_2 \times D, D)$

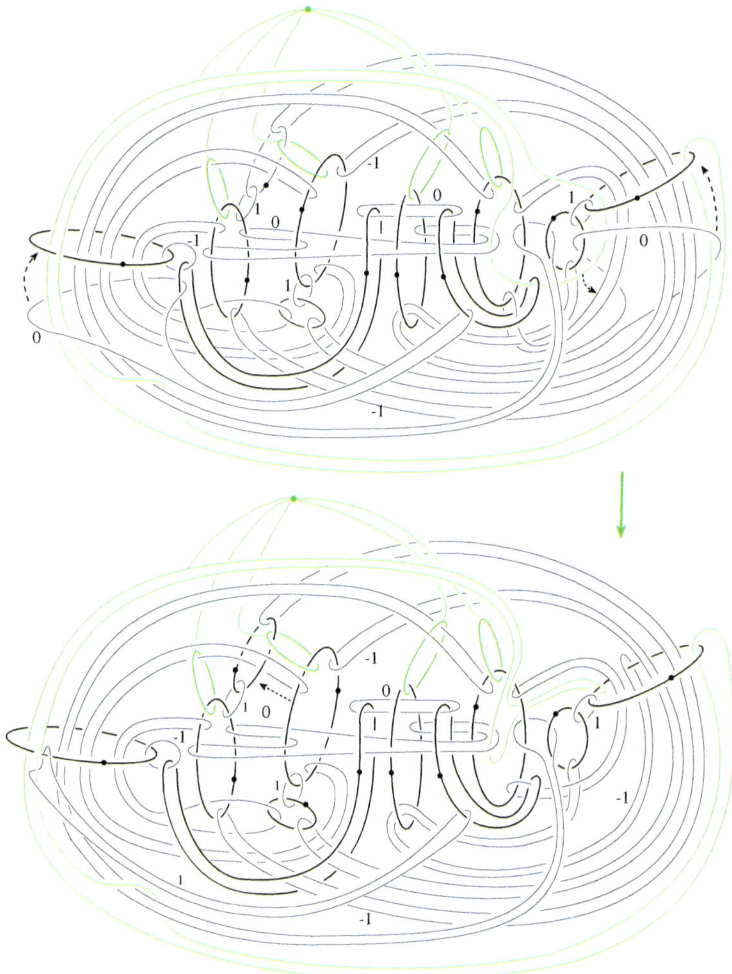

Figure 14.38: More isotopies and handle slides, getting $\tilde{E}_0$

To built the other piece, let $K \subset S^3$ be the trefoil knot, and $S^3_0(K)$ be the 3-manifold obtained by doing 0-surgery to S^3 along K. Being a fibered knot, K induces a fibration $T^2 \to S^3_0(K) \to S^1$ and the fibration

$$T^2 \to S^3_0(K) \times S^1 \to T^2 \tag{14.3}$$

Let T^2_1 and T^2_2 be the vertical (fiber) and the horizontal (section) tori of this fibration, intersecting at one point p . Smoothing $T^2_1 \cup T^2_2$ at p gives an imbedded genus 2 surface with self intersection 2, so by blowing up the total space twice (at points on this surface) we get a genus 2 surface $\Sigma_2 \subset (S^3_0(K) \times S^1)\#2\bar{\mathbb{CP}}^2$, with trivial normal bundle. Define:

$$\tilde{E}_2 = (S^3_0(K) \times S^1)\#2\bar{\mathbb{CP}}^2 - \Sigma_2 \times D^2$$

Now to see $\tilde{E}_2$ we construct a handlebody of $(S^3_0(K) \times S^1)\#2\bar{\mathbb{CP}}^2$ so that $\Sigma_2 \times D^2$ is clearly visible inside. This process is explained in the steps of Figure 14.39:

(1) The first picture is $S^3_0(K) \times S^1$ (Section 6.5). In this picture the horizontal $T_2 \times D^2$ is visible, but not the vertical torus T^2_1 (which consists of the Seifert surface of K capped off by the 2-handle given by the 0-framed trefoil). We redraw the first picture so that both vertical and horizontal tori are visible. (The reader can check this by canceling 1- and 2-handle pairs from the second picture, and obtaining the first picture).

(2) In $(S^3_0(K) \times S^1)\#2\bar{\mathbb{CP}}^2$, by sliding the 2-handle of T_1 over the 2-handle of T_2 (and by sliding over the two $\bar{\mathbb{CP}}^1$'s of the $\bar{\mathbb{CP}}^2$'s) we obtain a picture of the imbedding $\Sigma_2 \times D^2 \subset (S^3_0(K) \times S^1)\#2\bar{\mathbb{CP}}^2$.

(3) The rest of the steps, (4) and (5), are isotopies and handle slides (indicated by dotted arrows) to obtain the last picture $(S^3_0(K) \times S^1)\#2\bar{\mathbb{CP}}^2$, where $\Sigma_2 \times D^2$ is clearly visible inside.

Figure 14.40 is a combination of the last picture of Figure 14.39 and Figure 14.33, except that in this figure, by using Exercise 1.10, we have replaced the ribbon complement by a pair of 1-handles and a 2-handle. We now want to remove this $\Sigma_2 \times D^2$ from the handlebody of $(S^3_0(K) \times S^1)\#2\bar{\mathbb{CP}}^2$ and replace it with $\tilde{E}_0$ by the boundary identification. The arcs in Figure 14.33 and Figure 14.40 describe the diffeomorphism f, which guides us how to do the boundary identification, resulting in the handlebody of M in Figure 14.41.

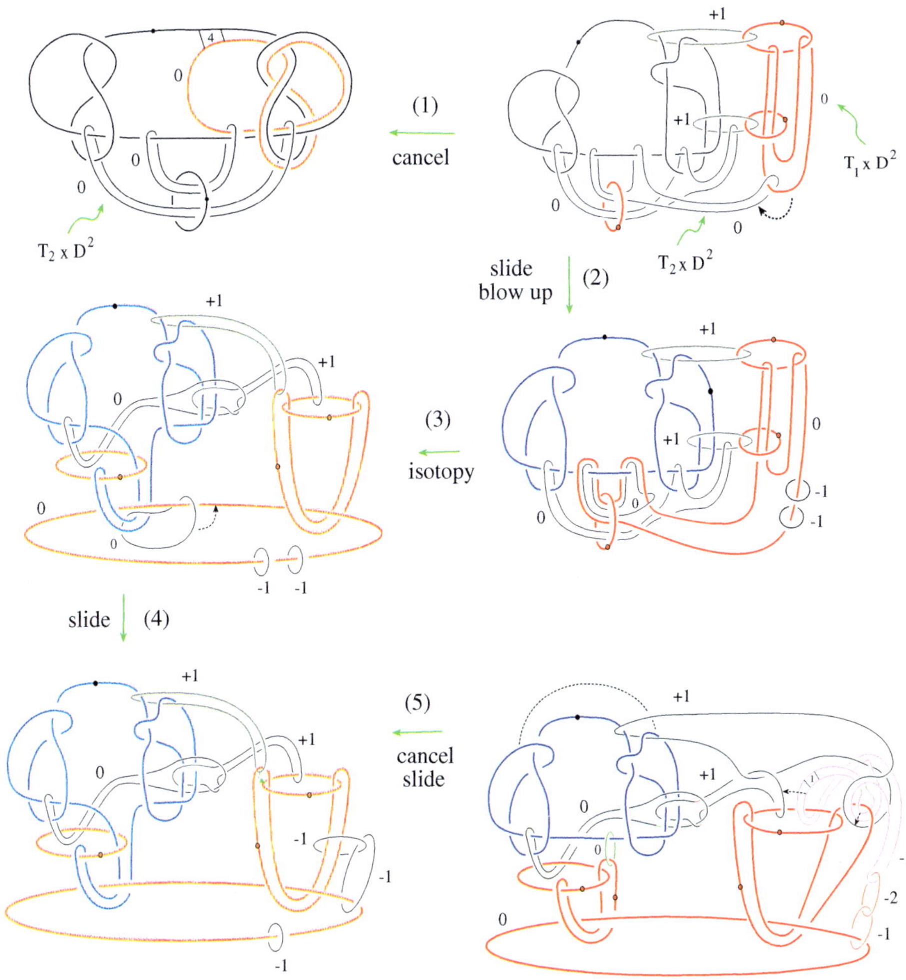

Figure 14.39: $(S^3_0(K) \times S^1)\#2\overline{\mathbb{CP}}^2$

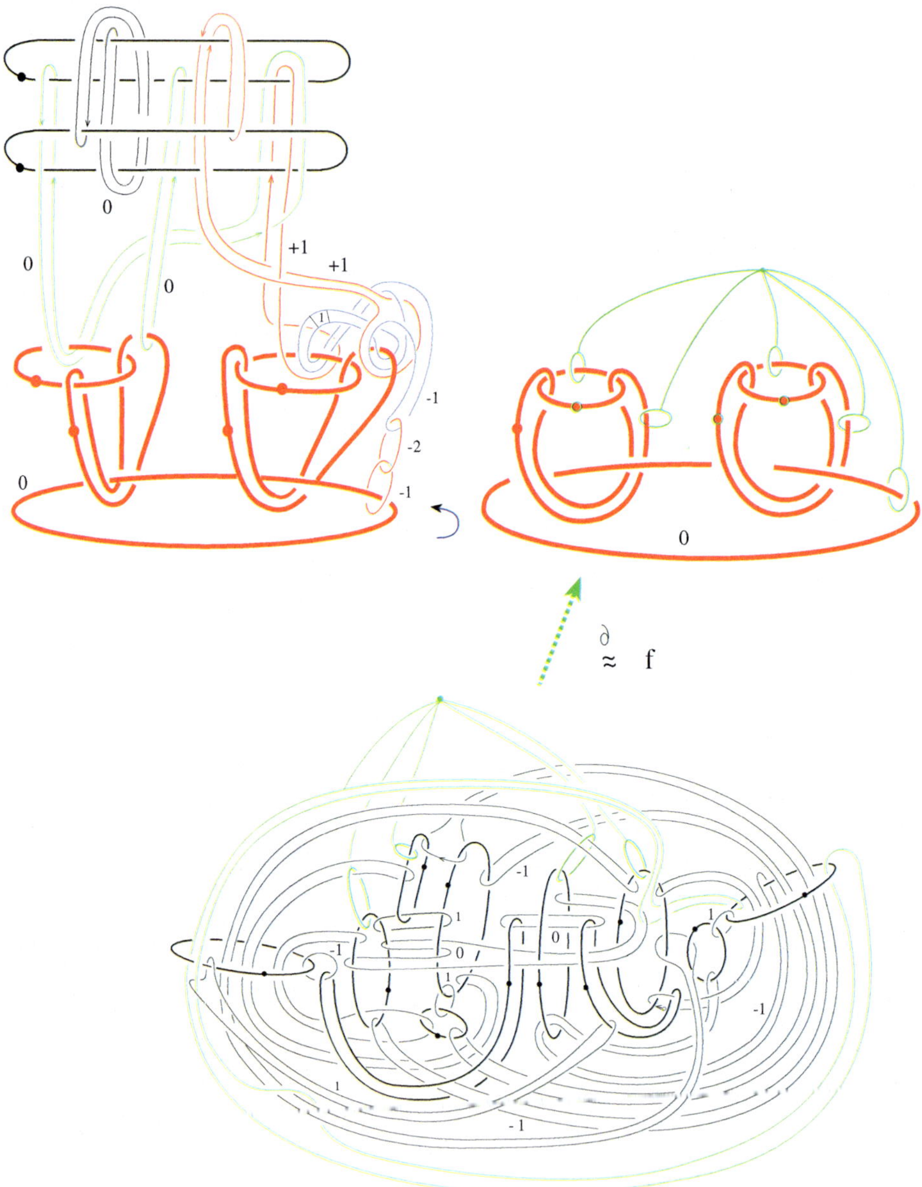

Figure 14.40: Getting ready to replace $\Sigma_2 \times D^2$ with $\tilde{E}_0$

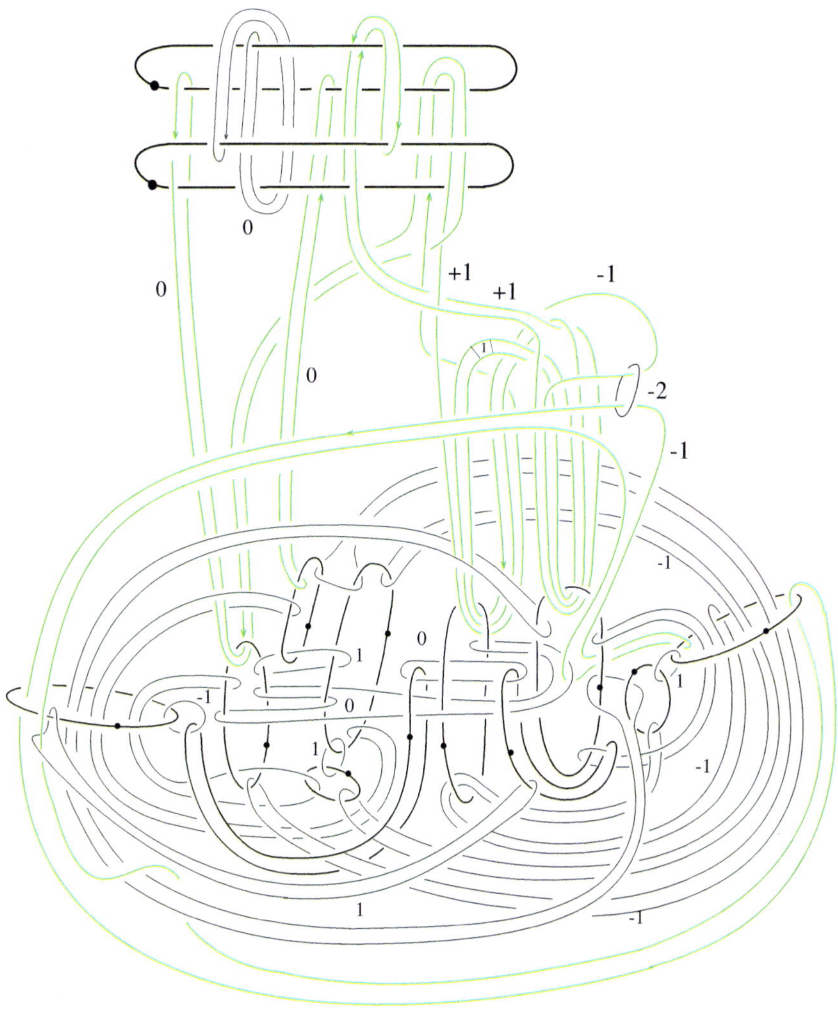

Figure 14.41: M

14.4.2 An exotic $\mathbb{CP}^2\#2\bar{\mathbb{CP}}^2$

Akhmedov and Park's other manifold N consists of two codimension zero pieces glued along the common boundaries: $\tilde{E}_0$ (which is defined in the previous section) and $\tilde{E}_1$,

$$N = \tilde{E}_0 \smile_\partial \tilde{E}_1$$

$\tilde{E}_1$ is defined as follows. As before, let $S^3_0(K)$ be the 3-manifold obtained by 0-surgering S^3 along the trefoil knot K. Consider the induced fibration $T^2 \to S^3_0(K) \times S^1 \to T^2$ of (14.3). Let T^2_1 and T^2_2 be the vertical (fiber) and the horizontal (section) tori of this fibration, transversally intersecting one point p. The degree 2 map $S^1 \to S^1$ induces an imbedding $S^1 \times D^3 \hookrightarrow S^1 \times D^3$, and by using one of the circle factors of T^2 this induces an imbedding $\rho : T^2 \times D^2 \hookrightarrow T^2_1 \times D^2$. The torus $T^2_3 := \rho(T^2 \times 0) \subset T^2_1 \times D^2$ intersects a normal disk D^2 of T^2_1 transversally at two points $\{p_1, p_2\}$, hence it intersects T^2_2 transversally at $\{p_1, p_2\}$. Therefore by blowing up the total space at p_1, we get a pair of imbedded pair tori $T^2_3 \cup T^2_2 \subset (S^3_0(K) \times S^1)\#\bar{\mathbb{CP}}^2$, each with self intersection -1, and transversally intersecting each other at one point p_2. Smoothing $T^2_2 \cup T^3_2$ at p_2 gives a genus 2 surface, $\Sigma_2 \subset (S^3_0(K) \times S^1)\#\bar{\mathbb{CP}}^2$, with trivial normal bundle $\Sigma_2 \times D^2$, then let:

$$\tilde{E}_1 = (S^3_0(K) \times S^1)\#\bar{\mathbb{CP}}^2 - \Sigma_2 \times D^2$$

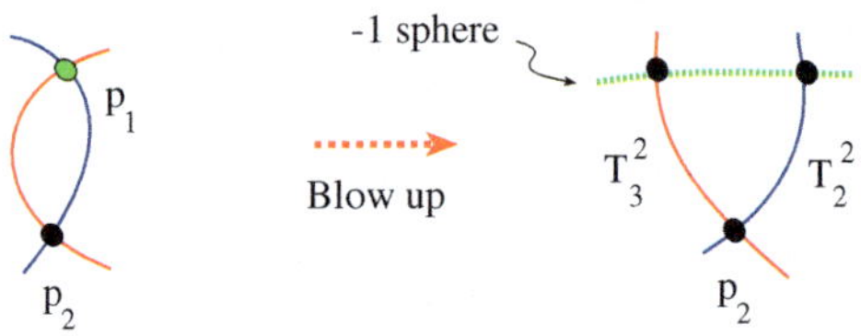

Figure 14.42

The first picture of Figure 14.43 is $S^3_0(K) \times S^1$ (Figure 6.12), where in this picture the horizontal $T_2 \times D^2$ is visible, but not the vertical torus T^2_1 (which consists of the Seifert surface of K capped off by the 2-handle bounded by the trefoil knot). In the second picture of Figure 14.43, as before, we redraw this handlebody so that both the vertical and the horizontal tori are visible. By an isotopy and creating a canceling 1- and 2 handle pair we obtain the third and the fourth pictures of Figure 14.43.

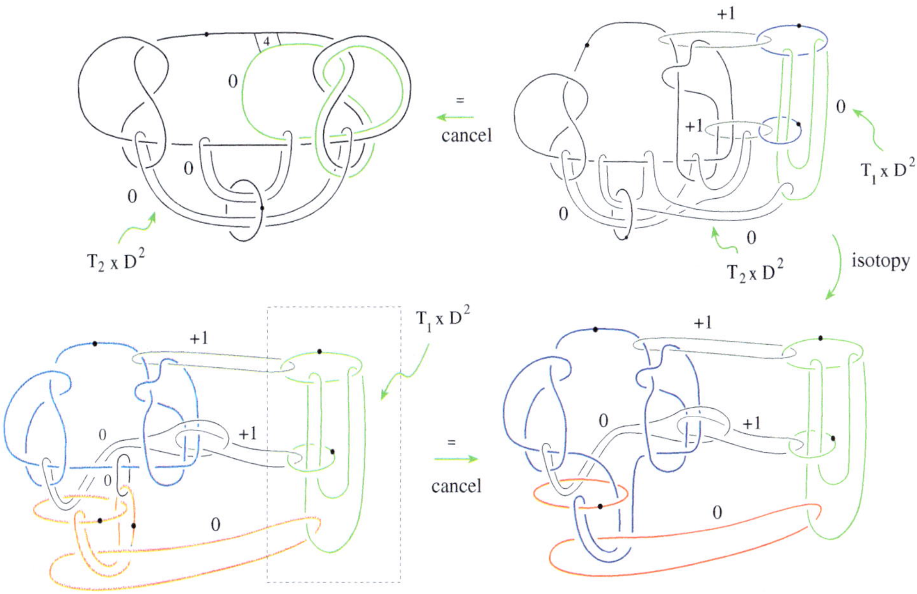

Figure 14.43: $S^3_0(K) \times S^1$

Now we want to find an imbedded copy of $T^2 \times D^2$ inside the neighborhood $T_1^2 \times D^2$ of the vertical torus (indicated by a dotted rectangle) imbedded by the degree 2 map, i.e. we want to locate T_3^2. For this, we first start with $S^1 \times D^2$ inside of another $S^1 \times D^2$ imbedded by a degree two map. This the first picture of Figure 14.44 (Heegaard picture). Here the 1-handle of the sub $S^1 \times D^2$ is B and the 1-handle of the ambient $S^1 \times D^2$ is A. Crossing this picture with S^1 we get a handle description of $T^2 \times D^2$ inside another $T^2 \times D^2$ imbedded by the degree 2 map, which is the second picture of Figure 14.44. By changing 1-handle notations to circle-with-dot notation, we get the last picture of Figure 14.44, where the degree 2 imbedding of $T^2 \times D^2$ inside $T^2 \times D^2$ is clearly visible. This picture is just $T^2 \times D^2$, drawn in a nonstandard way so that the degree 2 imbedded copy of the sub $T^2 \times D^2$ is clearly visible inside. Next we want to install this picture inside the dotted rectangle (i.e. inside of $T_1^2 \times D^2$) in Figure 14.43.

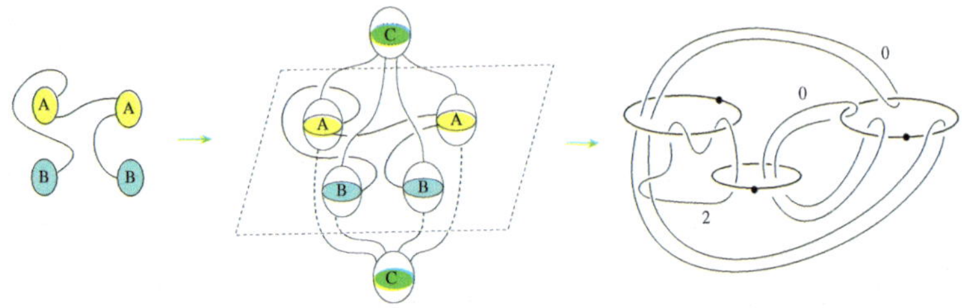

Figure 14.44: Degree 2 imbedding $S^1 \times (S^1 \times D^2) \hookrightarrow S^1 \times (S^1 \times D^2)$

Figure 14.45, describes a concrete diffeomorphism h (in terms of handle slides) from the nonstandard copy of $T^2 \times D^2$ in Figure 14.44 to the standard copy of $T^2 \times D^2$. Then we apply the inverse h^{-1} of this diffeomorphism to the region enclosed by the dotted rectangle in Figure 14.43. This gives the second picture of Figure 14.46. Think of Figure 14.46 as a piece of $S^3_0(K) \times S^1$ (i.e. the region enclosed by the dotted rectangle in Figure 14.43). Now we want to work only in this region (i.e. manipulate it by diffeomorphisms), then at the end install it back into Figure 14.43. This will help to simplify our pictures, otherwise we have to carry all of the rest of Figure 14.43 every step of the way, even though we are only working inside $T^2_1 \times D^2$.

After blowing up the second picture of Figure 14.46 at the indicated point we get the first picture of Figure 14.47, i.e. we have performed the operation

$$S^3_0(K) \times S^1 \mapsto (S^3_0(K) \times S^1)\#\bar{\mathbb{CP}}^2$$

(actually the figure shows only part of it). The thick arrow in Figure 14.46 indicates where we have performed the blowing up operation. Then by isotopies, handle slides (indicated by dotted arrows), and handle cancelations, we go from the first picture of Figure 14.47 to the last picture of Figure 14.48. By installing this last picture inside the dotted rectangle in Figure 14.43 we obtain the whole picture of $(S^3_0(K) \times S^1)\#\bar{\mathbb{CP}}^2$, which is Figure 14.49 (the two pictures are the same, in the second one we draw the slice 1-handle as a pair standard 1-handles and a 2-handle, as discussed in Exercise 1.10), where we see clearly the imbedded copy $\Sigma_2 \times D^2 \subset (S^3_0(K) \times S^1)\#\bar{\mathbb{CP}}^2$ discussed in the introduction. Now finally, to obtain N we need to remove this copy of $\Sigma_2 \times D^2$ from inside $(S^3_0(K) \times S^1)\#\bar{\mathbb{CP}}^2$ and replace it with $\tilde{E}_0$. The arcs in Figure 14.50 (describing diffeomorphism f) show us how to do this, resulting in the handlebody picture of N in Figure 14.51. Here we have used the convention that, when not indicated, the framings of the 2-handles are zero.

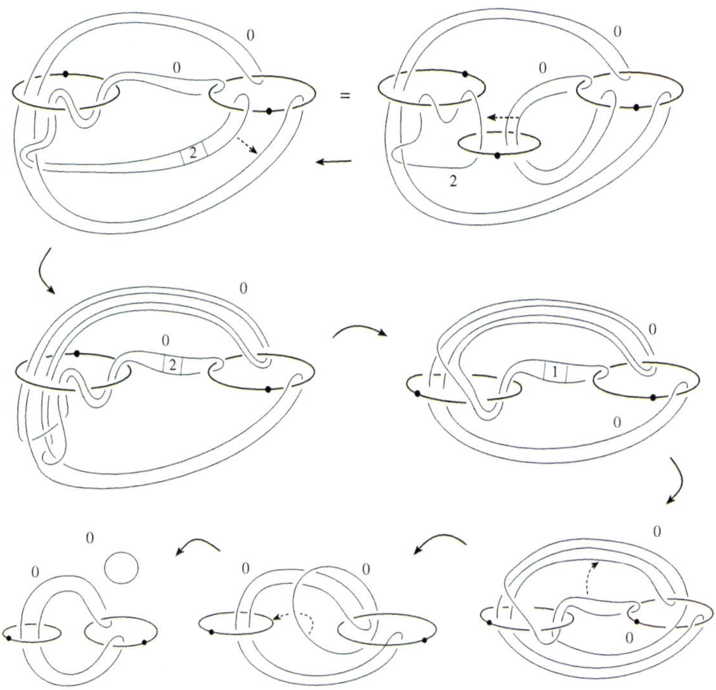

Figure 14.45: Checking the boundary diffeomorphism

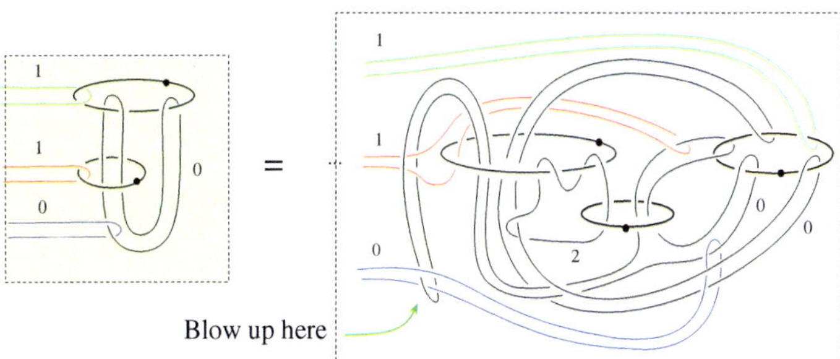

Figure 14.46: $T^2 \times D^2$ containing a degree 2 imbedded copy of itself, with some reference arcs drawn to describe the imbedding $T^2 \times D^2 \subset S^3_0(K) \times S^1$

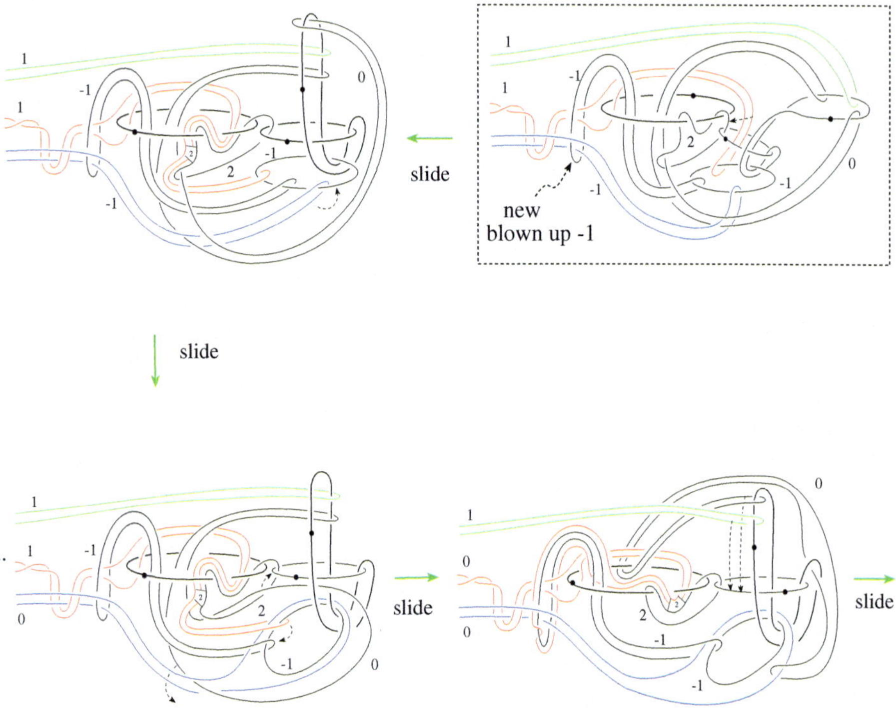

Figure 14.47: Blowup $(T^2 \times D^2) \to (T^2 \times D^2)\#\overline{\mathbb{CP}}^2$, then simplify.

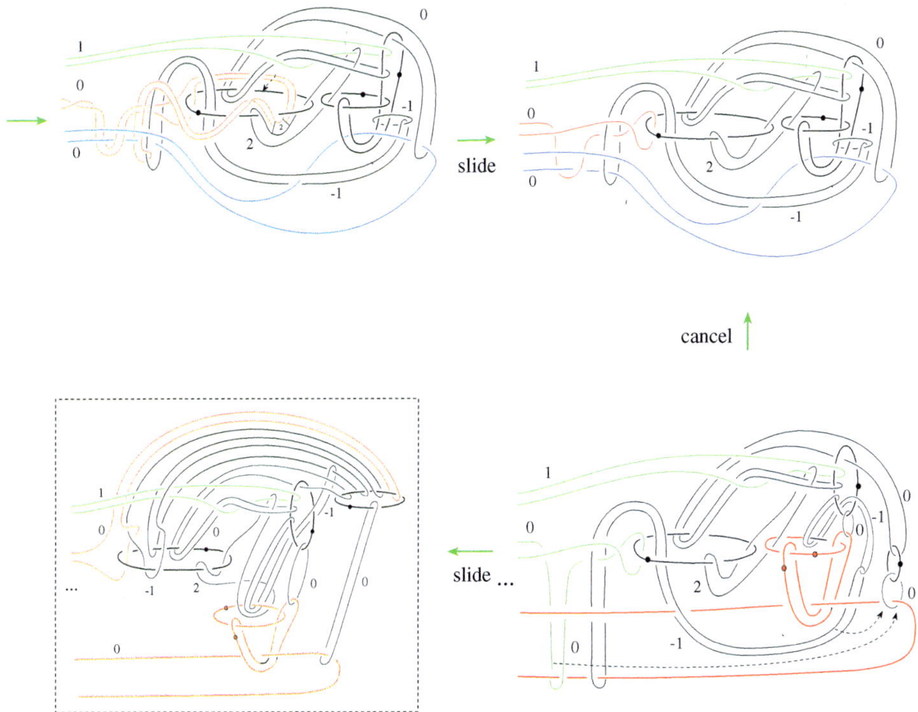

Figure 14.48: Figure 14.47 continued

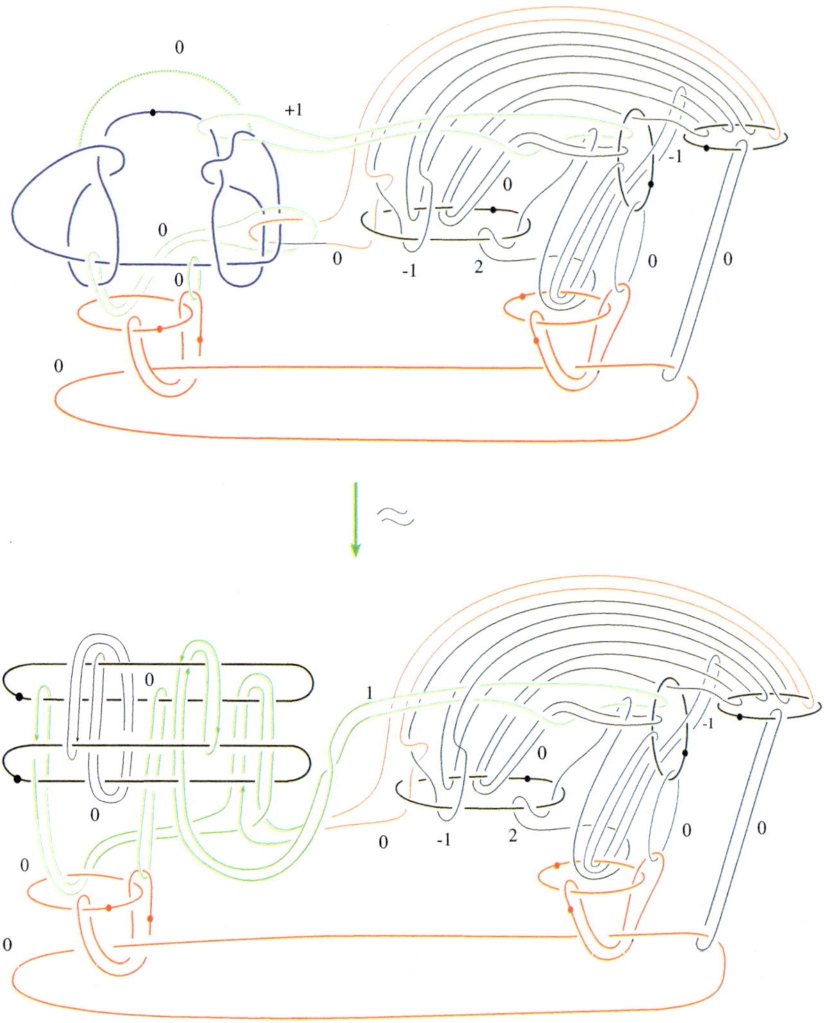

Figure 14.49: Installing $(T^2 \times D^2)\#\overline{\mathbb{CP}}^2$ (Figure 14.34) into Figure 14.30

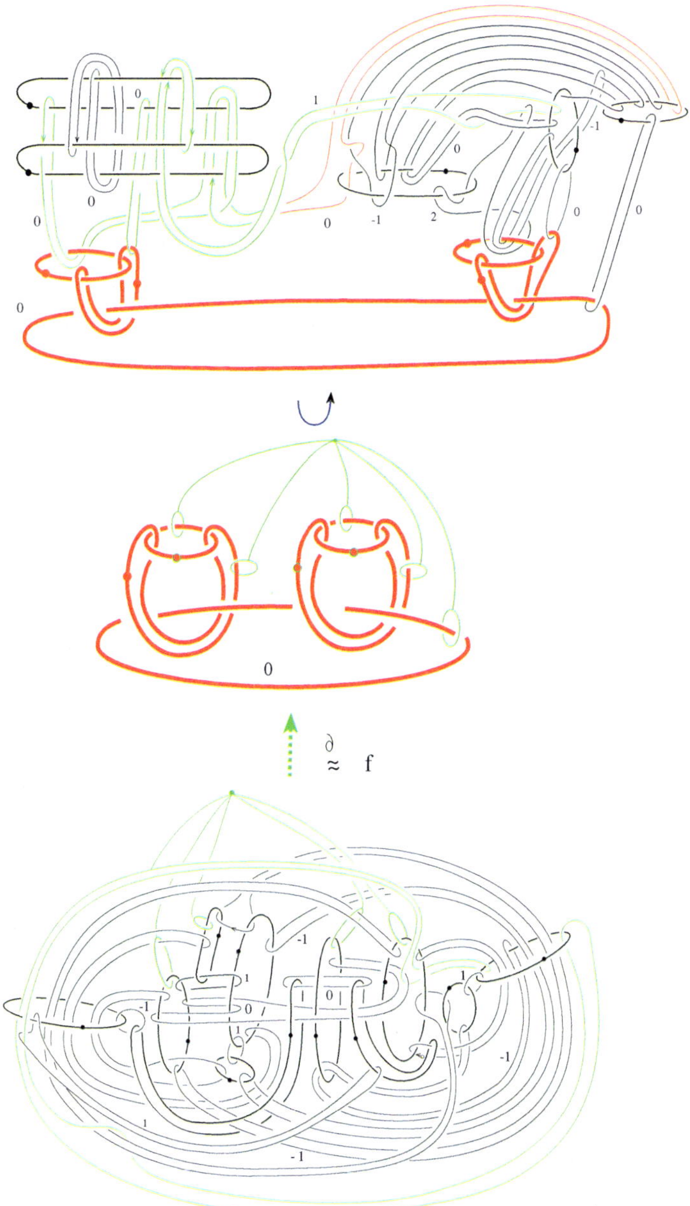

Figure 14.50: Getting ready to replace $\Sigma_2 \times D^2$ with $\tilde{E}_1$

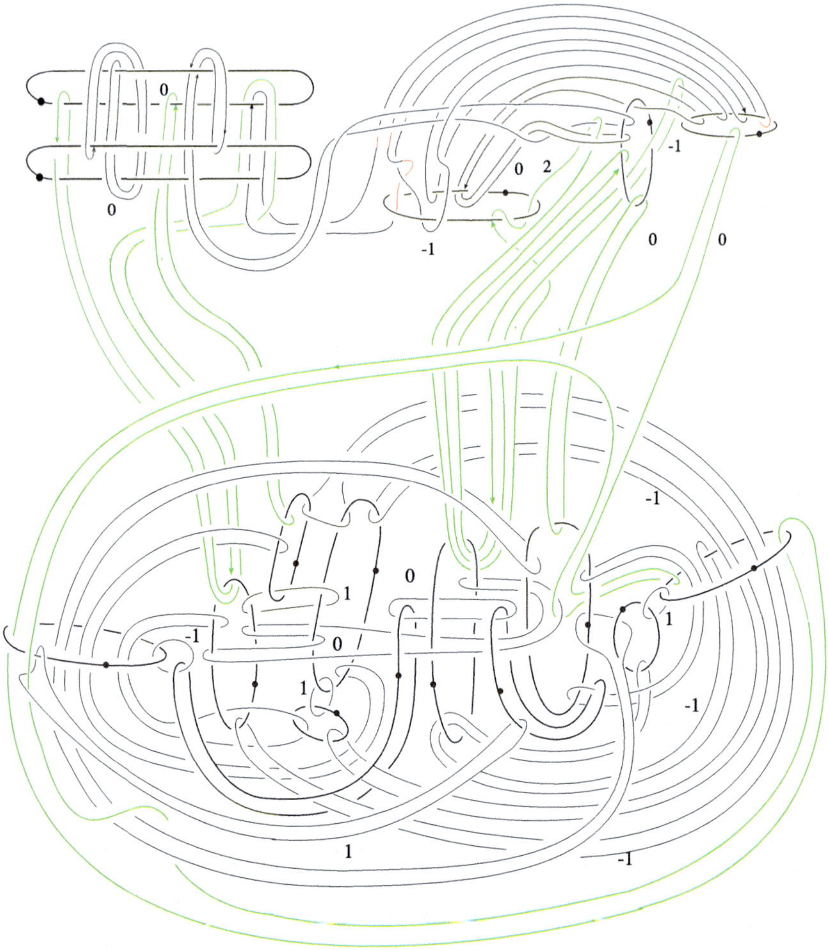

Figure 14.51: N

14.4.3 Fintushel–Stern reverse engineering

Let T be a homologically essential torus in a closed smooth X_0^4, imbedded with trivial normal bundle $N = T^2 \times B^2 \subset X_0$, also assume that the homology class of $[T^2]$ is primitive, hence there is a surface in X_0 intersecting T at one point, so the loop b of Figure 14.52 (top picture) bounds a surface in the complement of N. Also assume that the loop c represents a nontrivial element in $H_1(X - N; \mathbb{R})$. Let X be the manifold obtained by a 0-log transform operation from X_0 along T. Now perform Luttinger surgery (Section 6.4) $\varphi_{o,p} : X \mapsto X_{0,p}$. This is the operation of cutting out N from X and re-gluing by a diffeomorphism taking the loops $\{a, b, c\}$ to $\{a', b', c'\}$, as in Figure 14.52. It is clear from the pictures that $X_{0,p}$ can also be obtained by the p-log transformation from X_0. Furthermore, the core tori of N inside X and $X_{0,p}$ are homologically trivial, because their dual loops c and c' represent a nontrivial element in $H_1(X - N; \mathbb{R})$ by assumption. In the notation of Theorem 13.18, $M = X - N$ and $M_\gamma = X_{0,p}$, where $\gamma = p\mathbf{b} + \mathbf{c}$, hence

$$SW^0_{X_{0,p}} = SW^0_{M_{\mathbf{c}}} + pSW^0_{M_{\mathbf{b}}} \tag{14.4}$$

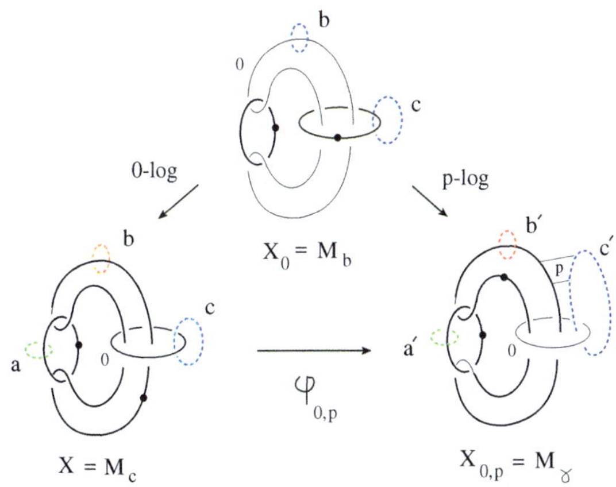

Figure 14.52: Luttinger surgery $X \mapsto X_{0,p}$

Clearly, $M_c = X$ and $M_b = X_0$. And since the inclusion $H_2(N) \to H_2(X)$ is the zero map, the map $H_2(X) \to H_2(X, N) \cong H_2(X - N, \partial N)$ is monic, hence there is a unique $Spin^c$ structure k on X extending a given one on M. By the same argument applied to $X_{0,p}$ there is a unique $Spin^c$ structure k_p on $X_{0,p}$ extending a given one on M. Notice that under the above hypotheses we have $b_2^+(X_{0,p}) = b_2^+(X_0) - 1$ and $b_1(X_{0,p}) = b_1(X_0) - 1$.

Therefore the formula (14.4) becomes ([FS4])

$$SW_{X_{0,p}}(k_p) = SW_X(k) + p\sum_i SW_{X_0}(k_0 + i[T_0]) \tag{14.5}$$

In particular, if there is an imbedded torus T_d of self intersection 0 in X_0 with $T_d.T = 1$ the adjunction inequality implies that the sum of the above formula can have at most one nonzero term, and we get

$$SW_{X_{0,p}} = SW_X + pSW_{X_0} \tag{14.6}$$

So if $SW_{X_0} \neq 0$, there are infinitely many smoothly different manifolds distinguished by their Seiberg–Witten invariants.

Reverse engineering is a method suggested in [FS4] for producing exotic copies of a given simply connected model manifold R with $b_2^+(R) \geq 1$. For this, we first find a symplectic model manifold M with $\chi(M) = \chi(R)$ and $\sigma(M) = \sigma(R)$, containing $n = b_1(M)$ pairwise disjoint Lagrangian tori $T_i \subset M$, each containing a non-separating loop c_i, such that these loops $\{c_i\}$ generate $H_1(M;\mathbb{R})$. Note that torus surgeries of a 4-manifold do not change its Euler characteristic χ and signature σ (this M will play the role of X_0 in the diagram of Figure 14.52). Now by performing consecutively 1-log transformations to M along the Lagrangian tori $\{T_i\}$, we obtain symplectic manifolds $M_1, M_2, \dots, M_n$, with properties $b_1(M_i) = b_1(M) - i$ and $b_2^+(M_i) = b_2^+(M) - i$. In particular $b_1(M_n) = 0 = b_1(R)$ and $b_2^+(M_n) = b^+(R)$. So if M_n is simply connected, in formula (14.5) we can use M_n for $X_{0,1}$ and M_{n-1} for X_0. In [FPS] this program was carried out for $R = \mathbb{CP}^2 \# 3\bar{\mathbb{CP}}^2$ and $M = Sym^2(\Sigma_3)$, where Σ_3 is a closed surface of genus 3.

Exercise 14.8. *(a) Explain why the manifolds M and N, which are built in the previous sections, are homotopy equivalent to (hence homeomorphic to) $\mathbb{CP}^2 \# 3\bar{\mathbb{CP}}^2$ and $\mathbb{CP}^2 \# 2\bar{\mathbb{CP}}^2$, respectively. (Hint: They are built from the manifolds with the same signature and Euler characteristic, and Luttinger surgery doesn't change those quantities). (b) From the handlebody pictures of M and N (Figure 14.41 and Figure 14.51) check that they are simply connected (this is done in [A9] by hand calculation, and in [A10] by the aid of group theory software GAP).*

References

[A1] S. Akbulut, *On 2-dimensional homology classes of 4-manifolds*, Math. Proc. Camb. Phil. Soc. 82 (1977) 99–106.

[A2] S. Akbulut, *A solution to a conjecture of Zeeman*, Topology 30(3) (1991) 513–515.

[A3] S. Akbulut, *A fake 4-manifold*, Contemp. Math. 35 (1984) 75–141.

[A4] S. Akbulut, *Knots and exotic structures on 4-manifolds*, J. Knot Theor. Ramif. 2(1) (1993) 1–10.

[A5] S. Akbulut, *Cappell-Shaneson homotopy spheres are standard*, Ann. Math. 171 (2010) 2171–2175.

[A6] S. Akbulut, *A fake cusp and a fishtail*, Turkish J. Math. 1 (1999) 19–31.

[A7] S. Akbulut, *The Dolgachev surface (disproving Harer-Kas-Kirby conjecture)*, Comment. Math. Helv. 87(1) (2012) 187–241.

[A8] S. Akbulut, *Lectures on Seiberg-Witten invariants*, Turk. J. Math. 20 (1996) 95–118.

[A9] S. Akbulut, *The Akhmedov-Park exotic* $\mathbb{CP}^2\#3\bar{\mathbb{CP}}^2$, Adv. Math. 274 (2015) 928–947.

[A10] S. Akbulut, *The Akhmedov-Park exotic* $\mathbb{CP}^2\#2\bar{\mathbb{CP}}^2$, Adv. Math. 274 (2015) 948–967.

[A11] S. Akbulut, *A Fake compact contractible 4-manifold*, J. Differ. Geom. 33 (1991) 335–356.

[A12] S. Akbulut, *An exotic 4-manifold*, J. Differ. Geom. 33 (1991) 357–361.

[A13] S. Akbulut, *Topology of multiple log transforms of* 4*-manifolds*, Int. J. Math. 24 (2013) 1350052.

[A14] S. Akbulut, *Variations on Fintushel-Stern knot surgery on 4-manifolds*, Turk. J. Math. (2001) 81–92.

[A15] S. Akbulut, *Constructing a fake* 4*-manifold by Gluck construction to a standard* 4*-manifold*, Topology 27(2) (1988) 239–243.

[A16] S. Akbulut, *Twisting 4-manifolds along* RP^2, Proc. of GGT/2009, Int. Press (2010) 137–141.

[A17] S. Akbulut, *Quotients of complex surfaces under complex conjugation*, J. Reine Angew. Math. 447 (1994) 83–90.

[A18] S. Akbulut, *On representing homology classes of 4-manifolds*, Invent. Math. 49 (1978) 193–198.

[A19] S. Akbulut, *Scharlemanns manifold is standard*, Ann. Math.149 (1999) 497–510.

[A20] S. Akbulut, *Isotoping* 2*-spheres in* 4*-manifolds*, Proc. GGT (2014) 267–269. http://arxiv.org/abs/1406.5654v1

[A21] S. Akbulut, *Cork twisting Schoenflies problem*, J. GGT 8 (2014) 35–43.

[A22] S. Akbulut, *Notes on geometric topology* `http://www.math.msu.edu/~akbulut/papers/notes.pdf`

[A23] S. Akbulut, *Nash homotopy spheres are standard*, Proc. GGT/2010, Int. Press (2011) 135–144.

[AD] S. Akbulut and S. Durusoy, *An involution acting non-trivially on Heegaard-Floer homology*, Fields Inst. Commun. 47 (2005) 1–9.

[AKa1] S. Akbulut and C. Karakurt, *Action of the cork twist on Floer homology*, Proc. of GGT/2011, Int. Press (2012) 42–52.

[AKa2] S. Akbulut and C. Karakurt, *Heegaard Floer homology of Mazur type manifolds*, P. Am. Math. Soc.142 (2014) 4001–4013.

[AKa3] S. Akbulut and C. Karakurt, *Every* 4*-manifold is BLF*, J. GGT 2 (2008) 83–106.

[AKi] S. Akbulut and H.C. King, *Topology of Real Algebraic Sets*, MSRI book series no. 25, Springer-Verlag (1992) ISBN-10 1461397413.

[AK1] S. Akbulut and R. Kirby, *An exotic involution on* S^4, Topology 18 (1979) 75–81.

[AK2] S. Akbulut and R. Kirby, *Mazur manifolds*, Mich. Math. J. 26 (1979) 259–284.

[AK3] S. Akbulut and R. Kirby, *Branched covers of surfaces in 4-manifolds*, Math. Ann. 252 (1980) 111–131.

[AK4] S. Akbulut and R. Kirby, *A potential smooth counterexample in dimension 4 to the Poincare conjecture, The Schoenflies conjecture, and the Andrews-curtis conjecture*, Topology 24(4) (1985) 375–390.

[AM1] S. Akbulut and R. Matveyev, *Exotic structures and adjunction inequality*, Turk. J. Math. 21 (1997) 47–53.

[AM2] S. Akbulut and R. Matveyev, *A note on contact structures*, Pac. J. Math. 182 (1998) 201–204.

[AM3] S. Akbulut and R. Matveyev, *A convex decomposition theorem for 4-manifolds*, Int. Math. Res. Notices, no. 7 (1998) 371–381.

[AMc] S. Akbulut and J.D. McCarthy *Casson's Invariant for Oriented Homology 3-Spheres -an Exposition*, Math. Notes 36, Princeton Univ. Press (1990) ISBN 0-691-08563-3.

[AO1] S. Akbulut and B. Ozbağci, *Lefschetz fibrations on compact stein surfaces*, Geom. Topol. 5 (2001) 319–334.

[AO2] S. Akbulut and B. Ozbağci, *On the topology of compact Stein surfaces*, Int. Math. Res. Notices 15 (2002) 769–782.

[AR] S. Akbulut and D. Ruberman, *An absolutely exotic contractible 4-manifold*, (to appear in Comment Math. Helv.) http://arxiv.org/pdf/1410.1461v1.pdf.

[AY1] S. Akbulut and K. Yasui, *Gluck twisting 4-manifolds with odd intersection form*, (to appear in Math. Res. Lett.) arXiv:1205.6038v1.

[AY2] S. Akbulut and K. Yasui, *Corks, plugs and exotic structures*, J. Gokova Geom. Topol. 2 (2008) 40–82.

[AY3] S. Akbulut and K. Yasui, *Knotting corks.* J. Topol. 2(4) (2009) 823–839.

[AY4] S. Akbulut and K. Yasui, *Small exotic Stein manifolds*, Comment Math. Helv. 85 (2010) 705–721.

[AY5] S. Akbulut and K. Yasui, *Infinitely many small exotic Stein fillings*, J. Symplect. Geom. 12(4) (2014) 673–684.

[AY6] S. Akbulut and K. Yasui, *Stein manifolds and corks*, J. GGT 6 (2012) 58–79.

[AY7] S. Akbulut and K. Yasui, *Cork twisting exotic Stein 4-manifolds*, J. Differ. Geom. 93(1) (2013) 1–36.

[ARu] I. Aitchison and J. Rubinstein, *Fibered knots and involutions on homotopy spheres*, Four manifold theory, AMS Contemp. Math. 35 (1984) 1–74.

[AEMS] A. Akhmedov, J.B. Etnyre, T.E. Mark and I. Smith, *A note on Stein fillings of contact manifolds*, Math. Res. Lett. 15(6) (2008) 1127–1132.

[AP1] A. Akhmedov and B.D. Park, *Exotic 4-manifolds with small Euler characteristic*, Invent. Math. 173 (2008) 209–223.

[AP2] A. Akhmedov and B.D. Park, *Exotic smooth structures on small 4-manifolds with odd signatures*, Invent. Math. 181 (2010) 209–223.

[Ar] M.F. Arikan, *On support genus of a contact structure*, J. Gokova Geol. Topol. 1 (2007) 96–115.

[Aro] N, Aronszajn, *A unique continuation theorem for elliptic differential equations or inequalities of the second order*, J. Math. Pur. Appl. 36 (1957) 235–239.

[As] D. Asimov, *Round handles and non-singular Morse-Smale flows*, Ann. Math. second series 102(1) (1975) 41–54.

[At] M. Atiyah, *K-theory*, W.A. Benjamin, Inc. (1967) 67–31266.

[Au] D. Auckly, *Families of four dimensional manifolds that become mutually diffeomorphic after one stabilization,* Topol. Appl. 127 (2003) 277–298.

[ADK1] D. Auroux, S.K. Donaldson and L. Katzarkov, *Luttinger surgery along Lagrangian tori and non-isotopy for singular symplectic plane curves*, Math. Ann. 326 (2003) 185–203.

[ADK2] D. Auroux, S.K. Donaldson and L. Katzarkov, *Singular Lefschetz pencils*, Geom. Topol. 9 (2005) 10431114.

[BK] S. Baldridge and P. Kirk, *A symplectic manifold homeomorphic but not diffeomorphic to* $\mathbb{CP}^2\#3\overline{\mathbb{CP}}^2$, Geom. Topol. 12 (2008), 919–940.

[Ba1] I. Baykur, *Topology of broken Lefschetz fibrations and near-symplectic 4-manifolds*, Pac. J. Math. 240(2) (2009) 201–230.

[Ba2] I. Baykur, *Kähler decomposition of 4-manifolds*, Algebr. Geom. Topol. 6 (2006) 1239–1265.

[Ba3] I. Baykur, *Existence of broken Lefschetz fibrations*, Int. Math. Res. Notices (2008) 1–15.

[Bl] F. van der Blij, *An invariant of quadratic forms mod 8*, Indagat. Math. 21 (1959) 291–293.

[Bt] R. Bott, *Lectures on $K(X)$*, W.A. Benjamin, Inc. (1969) 8053-1051-7.

[Br] W. Browder, *Surgery on simply-connected manifolds*, Springer-Verlag (1972) ISBN-13: 978-3642500220.

[Bn] M. Brown, *A proof of the generalized Schoenflies theorem*, B. Am. Math. Soc. 66 (1960) 74–76.

[Bo] S. Boyer, *Simply-connected 4-manifolds with a given boundary*, T. Am. Math. Soc. 298(1) (1986) 331–357.

[B] R. Brusse, The Kähler class and the C^∞ properties of Kähler surfaces, NY J. Math. 2 (1996) 103–146.

[CS] S. Cappell and J. Shaneson, *Some new four-manifolds*, Ann. Math. 104 (1976) 61–72.

[CH] A. Casson and J. Harer, *Some homology lens spaces which bound rational homology balls,* Pac. J. Math. 96 (1981) 23–36.

[CCM] F. Catanese, C. Ciliberto and M. Mendes Lopes, *On the classification of irregular surfaces of general type with non birational bicanonical map*, T. Am. Math. Soc. 350 (1998) 275–308.

[C] J. Cerf, *La stratification naturelle des espaces de fonctions diffrentiables reeles et la theoreme de la pseudoisotopie*, Publ. Math. I.H.E.S. 39 (1970).

[CE] K. Cieliebak and Y. Eliashberg, *From Stein to Weinstein and Back. Symplectic geometry of affine complex manifolds*, AMS Colloquium Publications, vol. 59, AMS, ISBN 978- 0-8218-8533-8.

[CGH] V. Colin, E. Giroux and K. Honda, *Finitude homotopique et isotopique des structures de contact tendues* [Homotopy and isotopy finiteness of tight contact structures] Publ. Math. Inst. I.H.E.S. 109 (2009) 245-293.

[Co] M.M. Cohen, *A Course in Simple Homotopy Theory*, GTM, vol.10, Springer (1973), ISBN-10: 0387900551.

[CFHS] C.L. Curtis, M.H. Freedman, W.C. Hsiang and R. Stong, *A decomposition theorem for h-cobordant smooth simply-connected compact 4-manifolds*, Invent. Math. 123(2) (1996) 343–348.

[D1] S.K. Donaldson, *Irrationality and the h-cobordism conjecture*, J. Differ. Geom. 26 (1987) 141–168.

[D2] S.K. Donaldson, *Connections, cohomology and the intersection forms of 4-manifolds*, J. Differ. Geom. 24 (1986) 275–341.

[D3] S.K. Donaldson, *An application of gauge theory to four-dimensional topology*, J. Differ. Geom. 18(2) (1983) 279-315.

[D5] S.K. Donaldson, *Lefschetz fibrations in symplectic geometry*, Doc. Math. J. DMV, Extra Volume ICMII (1998), 309-314.

[DK] S.K. Donaldson and P. Kronheimer, *The Geometry of 4-manifolds*, Oxford Math. Monographs (1990) ISBN-13: 978-0198502692.

[E1] Y. Eliashberg, *Few remarks about symplectic filling*, Geom. Topol. 8 (2004) 277–293.

[E2] Y.Eliashberg, *Topological characterization of Stein manifolds of dimension* > 2, Int. J. Math. 1(1) (1990) 29–46.

[E3] Y. Eliashberg, *Classification of overtwisted contact structures on 3-manifolds*, Invent. Math. 98 (1989) 623–637.

[E4] Y. Eliashberg, *On symplectic manifolds with some contact properties*, J. Differ. Geom. 33 (1991) 233–238.

[EP] Y. Eliashberg and L. Polterovich *New applications of Luttinger's surgery*, Comm. Math. Helvetici (1994) 512–522.

[El] N. D. Elkies, *A characterization of $\mathbb{Z}^n$ lattice*, Math Res. Lett. 2 no.3 (1995) 321–326.

[Et1] J. Etnyre, *On symplectic fillings*, Algebr. Geom. Topol. 4 (2004) 73-80.

[Et2] J. Etnyre, *Lectures on open book decompositions and contact structures*, arXiv:math.SG/0409402 v3 (2005).

[EH] J. Etnyre and K. Honda, *On symplectic cobordisms*, Math. Ann. 323 (2002) 31–39.

[F] M. Freedman, *The topology of 4-dimensional manifolds*, J. Differ. Geom. 17 (1982) 357–453.

[FQ] M. Freedman and F. Quinn, *Topology of 4-manifolds*, Princeton Univ. Press, PMS-39, (1990) ISBN: 9780691602899.

[FPS] R. Fintushel, B.D. Park and R. J. Stern, *Reverse engineering small 4-manifolds*, Algebr. Geom. Topol. 8 (2007) 2103–2116.

[FS1] R. Fintushel and R. Stern, *Knots, links, and 4-manifolds,* Invent. Math. 134(2) (1998) 363–400.

[FS2] R. Fintushel and R.J. Stern, *Rational blowdowns of smooth 4-manifolds*, J. Differ. Geom. 46(2) (1997) 181–235.

[FS3] R. Fintushel and R.J. Stern, *Immersed spheres in* 4-*manifolds and the immersed Thom Conjecture*, Turk. J. Math.19 (1995) 145–157, also in Proc. GGT 3 (1994) 27–39.

[FS4] R. Fintushel and R.J. Stern, *Surgery on nullhomologous tori*, Geom. Topol. Monogr. 18 (2012) 61–81.

[FS5] R. Fintushel and R.J. Stern, *Pinwheels and nullhomologous surgery on* 4-*manifolds with* $b^+ = 1$, Algebr. Geom. Topol. 11 (2011) 1649–1699.

[FS6] R. Fintushel and R.J. Stern, *Six lectures on four 4-manifolds*, Low dimensional topology, IAS/Park City Math. Ser., 15, Amer. Math. Soc., Providence, RI, 2009 (265315) MR2503498.

[Fu] M. Furuta, *Monopole equation and the 11/8-conjecture*, Math. Res. Lett. 8(3) (2001) 279–291.

[Ga] D. Gay, *Explicit concave fillings of 3-manifolds*, Math. Proc. Cambridge Phil. Soc. 133 (2002) 431–441.

[Ga2] D. Gay, *Four-dimensional symplectic cobordisms containing three-handles*, Geom. Topol. 10 (2006), 1749–1759.

[GK] D. Gay and R. Kirby, *Constructing Lefschetz-type fibrations on four-manifolds*, Geom. Topol. 11 (2007), 2075–2115.

[Ge] H. Geiges, *An Introduction to Contact Topology*, Cambridge Studies in Advanced Math 109, ISBN 9780521865852.

[Gi] E. Giroux, *Contact geometry: from dimension three to higher dimensions*, Proceedings of the International Congress of Mathematicians (Beijing 2002) 405–414.

[Gi2] E. Giroux, *Convexité en topologie de contact*, Comment. Math. Helvetici, 66 (1991) 637–677.

[G1] R. Gompf, *On Cappell-Shaneson 4-spheres*, Topol. Appl. 38(2) (1991) 123–136.

[G2] R. Gompf, *Handlebody contruction of Stein surfaces*, Ann. Math. 148 (1998) 619–693.

[G3] R. Gompf, *Killing the Akbulut-Kirby 4 -sphere, with relevance to the Andrews-Curtis and Schoenflies problems,* Topology 30 (1991) 97–115.

[G4] R. Gompf, *The topology of symplectic manifolds,* Turk. J. Math. 25 (2001) 43–59.

[G5] R. Gompf, *Minimal genera of open 4-manifolds,* arXiv:1309.0466.

[GS] R. Gompf and A. Stipsicz, *4-Manifolds and Kirby Calculus,* Graduate Studies in Mathematics, 20. AMS (1999) ISBN 0-8218-0994-6.

[Go] C. McA. Gordon, *Dehn surgery and satellite knots*, T. Am. Math. Soc. 275(2) (1983).

[Ham] M.J.D. Hamilton, *Homology classes of negative square and embedded surfaces in 4-manifolds*, Bull. London Math. Soc., 45 (2013) 1221–1226.

[Ha] J. Harer, *Pencils of curves on 4-manifolds*, PhD thesis, UC Berkeley, 1974.

[H] M.W. Hirsch, *Differential Topology*, Grad texts in math 33, Springer-Verlag, ISBN 0-387-90148-5.

[Hi] F. Hirzebruch, *The signature of ramified coverings*, Global Analysis (Papers in honor of K. Kodaira), Princeton (1969) 253–265.

[HJ] F. Hirzebruch and K. Jänich, *Involutions and singularities*, Algebraic Geometry, Bombay Coll. Oxford Univ. Press (1969) 219–240.

[HNK] F. Hirzebruch, W.D. Neumann and S.S. Koh, *Differential manifolds and quadratic forms*, Lec notes in pure and applied math vol. 4 Marcel Dekker, ISBN: 0-8247-1309-5.

[Hit] N.J. Hitchin, *On the curvature of the rational surfaces*, In Differential Geometry, Proc. Symp. Pure Math, vol. 27, AMS (1975) 65–80.

[Ho] K. Honda, *On the classification of tight contact structures I*, Geom. Topol. 4 (2000) 309–368.

[HL] C. Ho and T-J Li , *Luttinger surgery and Kodaira dimension*, Asian J. Math. 16(2) (2012) 299–318.

[HLe] M. Hutchings and Yi-Jen Lee, *Circle-valued Morse theory, Reidemeister Torsion, and Seiberg–Witten invariants of 3-manifolds*, Topology vol. 38, no. 4 (1999) 861–888.

[Iv] N.V. Ivanov, *Mapping class groups. Handbook of Geometric Topology*, 523–633, North-Holland, Amsterdam, (2002).

[I] Z. Iwase, *Dehn surgery along a torus T^2-knot II,* Jap. J. Math. 16(2) (1990) 171–196.

[Ka] A. Kas, *On the handlebody decomposition associated to a Lefschetz fibration*, Pac. J. Math. 89 (1980) 89–104.

[KT] P.K. Kim and J.L. Tollefson, *Splitting the PL involutions of nonprime 3-manifolds*, Mich. Math. J. 27(3) (1980) 259-274

[KLT] C. Kutluhan, Yi-Jen Lee, and C. Taubes, *HF=HM I: Heegaard Floer homology and Seiberg–Witten Floer homology.* arXiv:1007.1979 (2010)[math.GT].

[K1] R. Kirby, *A calculus of framed links in* S^3, Invent. Math. 45 (1978) 35–56.

[K2] R. Kirby, *Akbulut's corks and h-cobordisms of smooth, simply connected* 4-*manifolds*, Turk. J. Math. 20(1) (1996) 85–93.

[K3] R. Kirby, *The topology of 4-manifolds*, Lecture Notes in Math, 1374, Springer-Verlag (1989) ISBN 3-540-51148-2.

[KM] P. Kronheimer and T. Mrowka, *On genus of imbedded surfaces in the projective plane*, Math. Res. Lett.1 (1994) 797–808.

[KM2] P. Kronheimer and T. Mrowka, *Monopoles and three-manifolds*, Cambridge Univ. Press, New math. monographs 10 (2007) ISBN 978-0-521-88022-0.

[LP] F. Laudenbach and V. Poenaru, *A note on 4-dimensional handlebodies*, Bull. Math. Soc. France 100 (1972) 337–344.

[Lau] H.B. Laufer, *Normal two-dimensional singularities*, Ann. Math. series 71, ISBN: 9780691081007.

[La] B. Lawson, *The Theory of Gauge Fields in Four Dimensions*, CBMS Regional Conference Series in Math, no.58 (1985) ISBN-10: 0-8218-0708-0.

[LaM] B. Lawson and M. Michelson, *Spin Geometry*, Princeton University Press (1989).

[Le] Y. Lekili, *Wrinkled fibrations on near-symplectic manifolds*, (Appendix B by I. Baykur), Geom. Topol 13(1) (2009) 277–318.

[LM] Y. Lekili and M. Maydanskiy, *Floer theoretically essential tori in rational blowdowns*, arXiv:1202.5625v2 [math.SG].

[L] T. J. Li, *Symplectic 4-manifolds with Kodaira dimension zero*, J. Differ. Geom. 74 (2006) 321–352.

[LL] T.J. Li, and A. Liu, *Symplectic structure on ruled surfaces and a generalized adjunction formula*, Math. Res. Lett. 2(4) (1995) 453-471.

[LZ] T. J Li and W. Zhang, *Additivity and relative Kodaira dimensions* http://arxiv.org/abs/0904.4294v1.

[Li] W.B.R. Lickorish, *A representation of orientable combinatorial 3-manifolds*, Ann. Math. (2), 76 (1962) 531–540.

[Liv] C. Livingston, *Knot Theory*, The Carus Math. Monographs (MAA) ISBN 0-88385-027-3.

[LMa] P. Lisca and G. Matic, *Tight contact structures and Seiberg-Witten invariants*, Invent. Math.129 (1997) 509–525.

[Lit] R.A. Litherland, *Slicing doubles of knots in homology 3-spheres*, Invent. Math. 54 (1979) 69–74.

[LoP] A. Loi and R. Piergallini, *Compact Stein surfaces with boundary as branched covers of B4*, Invent. Math. 143 (2001) 325–348.

[Lu] K.M. Luttinger *Lagrangian tori in* $\mathbb{R}^4$, J. Diff. Geom. 42(2) (1995) 220–228.

[Ly] H. Lyon, *Torus knots in the complements of links and surfaces*, Mich. Math. J. 27 (1980) 39–46.

[Mar] M. Marcolli, *Seiberg-Witten gauge theory*, Texts and readings in math 17, Hindustan book agency, Text and reading math no. 17 (1999) ISBN-10: 8185931224.

[Ma] R. Matveyev, *A decomposition of smooth simply-connected h-cobordant 4-manifolds*, J. Differ. Geom. 44(3) (1996) 571–582.

[Mat] Y. Matsumoto, *An introduction to Morse theory*, Translation of math monographs v. 208 (2001) ISBN-10: 0821810227.

[Maz] B. Mazur, *On embeddings of spheres*, BAMS 65 (1959) 59–65.

[Maz1] B. Mazur, *A note on some contractible 4-manifolds*, Ann. of Math. vol. 73, no.1 (1961) 221–228.

[MeT] G. Meng and C. Taubes, *SW=Milnor Torsion*, Math. Res. Lett. 3 (1996) 661–674.

[MSa] D. McDuff and D. Salamon, *Introduction to Symplectic Topology*, Oxford Math Monographs, ISBN 019 850 451.

[Mc] R. McOwen, *Partial Differential Equations*, Prentice-Hall ISBN 0-13-121880-8

[M1] J. Milnor, *Morse Theory*, Princeton Univ. Press (AM-51) (1963) ISBN: 9780691080086.

[M2] J. Milnor, *On simply connected 4-manifolds*, Proc. Int. Symp. Algebraic Topology, Mexico (1958) 122–128.

[M3] J. Milnor, *Singular points of complex hypersurfaces*, Princeton Univ. Press (AM-61) (1969) ISBN: 9780691080659.

[M4] J. Milnor, *Lectures on the h-cobordism theorem*, Princeton Univ. Press, Math notes (1965) ISBN-10: 069107996X.

[M5] J.W. Milnor, *A duality theorem for Reidemeister torsion*, Ann. Math. (2) 76 (1962), 137-147.

[MH] J. Milnor and D. Husemoller, *Symmetric Bilinear Forms*, Springer-Verlag, Ergebnisse der Mathematik und ihrer Grenzgebiete (1973) ISBN-10: 038706009X.

[MS] J. Milnor and J.D. Stasheff, *Characteristic Classes*, Princeton Univ. Press (AM-76) (1974) ISBN: 9780691081229.

[Moo] J.D. Moore, *Lecture notes on Seiberg-Witten invariants*, Springer lecture notes in math 1629, ISBN 3 540-41221-2.

[Mo] J.M. Montesinos, *4-Manifolds, 3-fold covering spaces and ribbons*, TAMS, vol. 245, Nov (1978) 453–467.

[Mor] J.W. Morgan, *The Seiberg-Witten Equations and Applications to the Topology of Smooth Four-Manifolds*, Princeton Univ. Press (MN-44) (1996) ISBN: 9781400865161.

[MF] J. Morgan and R. Friedman, *Smooth Four-Manifolds and Complex Surfaces* Ergebnisse der Mathematik und ihrer Grenzgebiete, A Series of Modern Surveys in Mathematics, ISBN-13: 978-3540570585.

[MMS] J. Morgan, T. Mrowka and Z. Szabo, *Product formulas along T^3 for Seiberg–Witten invariants*, Math. Res. Lett. 4 (1997) 915–929.

[MST] J. Morgan, Z. Szabo and C.H. Taubes, *A product formula for the Seiberg–Witten invariants and the generalized Thom conjecture*, J. Differ. Geom. 44(4) (1996) 706-788.

[Mu] D. Mumford, *Algebraic Geometry 1: Complex projective varieties*, Springer (Classics in math), ISBN: 3-540-58657.

[NR] W.D. Neumann and F. Raymond, *Seifert manifolds, plumbing, μ-invariant and orientation reversing maps*, Algebraic and Geometric Topology, Lecture Notes in Math. 664, Springer, Berlin (1987) 163–196.

[Ni] L.I. Nicolaescu, *Notes on Seiberg–Witten theory*, Grad. Studies in Math. vol. 28 (2000), ISBN-13: 978-0821821459.

[Ni2] L.I. Nicolaescu, *Reidemeister torsion of* 3*-manifolds*, deGruyter studies in math. 30 (2003) ISBN 3-11-017383-2.

[OO] H. Ohta and K. Ono, *Simple singularities and topology of symplectically filling 4-manifolds*, Comm. Math. Helvitici 74 (1999) 575–590.

[OW] P. Orlik and P. Wagreich, *Isolated singularities of algebraic surfaces with $\mathbb{C}^*$ action*, Ann. Math. 93 (1971) 205–228.

[O] J. Osoinach Jr, *Manifolds obtained by surgery on an infinite number of knots in S^3*, Topology 45 (2006) 725–733.

[Oz1] B. Ozbağci, *Embedding fillings of contact 3-manifolds*, Expo. Math. 24 (2006) 161–183.

[Oz2] B. Ozbağci, *Lectures on the topology of symplectic fillings of contact 3-manifolds* (lecture notes).

[OSt] B. Ozbağci and A.I. Stipsicz, *Surgery on Contact 3-Manifolds and Stein Surfaces*, Springer ISBN 3-540-22944-2.

[OS] P. Ozsváth and Z. Szabó, *The symplectic Thom conjecture*, Ann. Math. 151 (2000) 93–124.

[OS2] P. Ozsváth and Z. Szabó, *Higher type adjunction inequalities*, Jour. of D.G. 55, (2000) no.3, 385–440.

[OS3] P. Ozsváth and Z. Szabó, *An introduction to Heegaard Floer homology,* Floer homology, gauge theory, and low-dimensional topology, 3-27, Clay Math. Proc. 5, AMS, Providence, RI (2006).

[OS4] P. Ozsváth and Z. Szabó, *Lectures on Heegaard Floer homology,* Floer homology, gauge theory, and low-dimensional topology, 29–70, Clay Math. Proc. 5, AMS, Providence, RI (2006).

[P] J. Park, *Seiberg-Witten invariants of generalized rational blow-downs*, B. Austral. Math. Soc. 56 (1997) 363–384.

[P2] J. Park, *Simply connected symplectic 4-manifolds with* $b_2^+ = 1$ *and* $c_1^2 = 2$, Invent. Math. 159(3) (2005) 657–667.

[Pa] B.D. Park, *Seiberg-Witten invariants and branched covers along tori,* PAMS 133(9) (2005) 2795–2803.

[Po] V. Poénaru, *On the geometry of differentiable manifolds*, Studies in modern topology, MAA studies in math v.5, (1968) 165–207.

[Roh] V.A. Rohlin, *New results in the theory of four-dimensional manifolds*, Doklady Acad. Nauk SSSR 84 (1952) 221–224.

[Ro] D. Rolfsen, *Knots and Links*, AMS Chelsea Publishing, AMS, ISBN 0-8218-3436-3.

[RW] Y. Ruan and S. Wang, *Seiberg-Witten invariants and double branched covers of 4-manifolds,* Commun. Anal. Geom. 8 (2000) 477–515.

[R] L. Rudolph, *An obstruction to sliceness via contact geometry and "classical" gauge theory,* Invent. Math.119, (1995) 155–163.

[Sal1] D. Salamon, *Removable singularities and vanishing theorem for Seiberg–Witten invariants*, Turk. J. Math. 20 (1996) 61–73.

[Sal2] D. Salamon, *Spin geometry and Seiberg–Witten invariants*, University of Warwick notes (1996).

[Sa] N. Saveliev, *Lectures on the Topology of 3-Manifolds: An Introduction to the Casson Invariant*, De Gruyter text book (1999) ISBN 3-11-016271-7.

[Sc] A. Scorpan, *The Wild World of 4-Manifolds*, AMS, ISBN-10: 0-8218-3749-4.

[Sh] P. Shanahan, *The Atiyah-Singer index theorem*, Lec. Notes Math. 638 Springer-Verlag, ISBN: 0-387-08660-9.

[S] S. Smale, *Generalized Poincare's conjecture in dimensions greater than four*, Ann. Math. (2) 74 (1961) 391–406.

[Sp] E. Spanier, *Algebraic Topology* , Springer, ISBN 0-387-90646-0.

[Str] R.J. Stern, *Will We Ever Classify Simply-Connected Smooth 4-Manifolds?*, CMI/AMS Book Series, Floer Homology, Gauge Theory and Low Dimensional Topology (2006), 225–240. MR2249255.

[St] A.I. Stipsicz, *Surgery diagrams and open book decompositions of contact 3-manifolds*, Acta Math. Hungar. 108(1–2) (2005).

[Tan] K-T. Tan, *Unique factorization in group rings*, Pub. De L'Institut Mathématique. Nouvelle Série, tome 21 (1977) 197–200.

[Ta] C. Taubes, *The Seiberg–Witten invariants and symplectic forms*, Math. Res. Lett. 1 (1994) 809–822.

[Ta2] C. Taubes, *Seiberg–Witten and Gromov Invariants for Symplectic 4-Manifolds*, First International Press Lecture Series, vol. 2, International Press, Somerville, MA (2000) MR 1798809 (2002) 53115.

[Ta3] C. Taubes, *The Seiberg–Witten invariants and 4-manifolds with essential tori*, Geom. Topol. 5 (2001) 441–519.

[Ta4] C. Taubes, *More constraints on symplectic forms from Seiberg–Witten invariants*, Math. Res. Lett. 2 (1995) 9–13.

[T] M. Teragaito, *A Seifert manifold with infinitely many knot-surgery decriptions*, IMRN, vol (2007) Article ID rnm028.

[To] I. Torisu, *Convex contact structures and fibered links in 3-manifolds*, Int. Math. Res. Notices 9 (2000) 441–454.

[Tra] B. Trace, *On attaching 3-handles to a 1-connected 4-manifold*, Pac. Jour. Math. 99 (1982) 175–181.

[Tr] A.G. Tristram, *Some cobordism invariants of links*, Proc. Camb. Phil. Soc. 66 (1969) 251–264.

[Tu] V.G. Turaev, *Reidemeister torsion in knot theory*, Russ. Math. Surv. 41(1) (1986) 119–182.

[W] A.H. Wallace, *Modifications and cobounding manifolds*, Canad. J. Math. 12 (1960) 503–528.

[Wa1] C.T.C Wall, *On simply connected 4-manifolds*, J. London Math. Soc. 39 (1964) 141–149.

[Wa2] C.T.C Wall, *Surgery on compact manifolds*, Academic Press, London (1970) MR0431216.

[Wg] S. Wang, *A vanishing theorem for Seiberg–Witten invariants*, Math. Res. Lett. 2 (1995) 305–310.

[Wh] J.H.C. Whitehead, *On simply connected 4-dimensionl polyhedra*, Comm. Math. Helvitici 22 (1949) 48–92.

[Wil] L. Williams, *On handlebody structures of rational balls*, arXiv:1406.1575v1.

[Wi] E. Witten, *Monopoles and four manifolds*, Math. Research Letters 1 (1994) 769–796.

[Y] K. Yasui, *Corks, exotic 4-manifolds and knot concordance*, arXiv:1505.02551v2.

[Z] C. Zeeman, *On the dunce hat*, Topology 2 (1963) 341–358.

Index

The manufacturer's authorised representative in the EU for product safety is Oxford University Press España S.A. of El Parque Empresarial San Fernando de Henares, Avenida de Castilla, 2 - 28830 Madrid (www.oup.es/en or product.safety@oup.com). OUP España S.A. also acts as importer into Spain of products made by the manufacturer.
Printed and bound by CPI Group (UK) Ltd, Croydon, CR0 4YY
14/09/2026
02216284-0006